OCEAN CIRCULATION PHYSICS

This is Volume 19 in
INTERNATIONAL GEOPHYSICS SERIES
A series of monographs and textbooks
Edited by WILLIAM L. DONN

The complete listing of books in this series is available from the Publisher upon request.

OCEAN CIRCULATION PHYSICS

Melvin E. Stern

Graduate School of Oceanography
University of Rhode Island
Kingston, Rhode Island

ACADEMIC PRESS

New York London Toronto Sydney San Francisco 1975

A Subsidiary of Harcourt Brace Jovanovich, Publishers

ACADEMIC PRESS, INC.
111 Fifth Avenue, New York, New York 10003

United Kingdom Edition published by
ACADEMIC PRESS, INC. (LONDON) LTD.
24/28 Oval Road, London NW1

Library of Congress Cataloging in Publication Data

Stern, Melvin E
Ocean circulation physics.

(International geophysics series, no.)
Includes bibliographies.
1. Oceanic mixing. 2. Hydrodynamics. I. Title.
II. Series.
GC299.S77 551.4'701 74-1645
ISBN 0-12-666750-0

PRINTED IN THE UNITED STATES OF AMERICA

83 84 85 9 8 7 6 5 4 3 2

Contents

Preface

This book might have been called "Fluid Mechanics for Oceanographers" since it has been given as a one-year graduate course. It might also have been called "Geophysical Fluid Dynamics," since the ideas pertaining to the ocean circulation have been isolated in highly simplified models. Indeed, there was a personal motivation in putting together all the "isolations and dissections."

These goals may seem somewhat conflicting, and therefore every effort has been made to make the material self-contained for a diligent nonspecialist. Thus the basic principles of fluid dynamics are reviewed in the context in which they occur, but some prior familiarity with the subject is really assumed.

The partitioning of the book also requires some commentary. Part One is confined mainly to the rotational dynamics of one and two homogeneous layers in the field of gravity, the focal point being the wind driven circulation theory (Chapter VII). These simple layer models also provide insights into the analogous inertial effects which occur in a continuously stratified fluid (Part Two, Chapter IX). Therefore, the emphasis in Part Two is on the thermodynamics and mixing effects. Although each section in Part Two begins on an intelligible level, the material becomes much more difficult to understand because of the appeal to similarity arguments. Chapter VI of Part One is also somewhat complex, and perhaps this chapter should be read just prior to Section 10.5 of Part Two.

The purpose of the book is to communicate ideas and concepts; thus the comparison of theory with observation is confined to broad features of the ocean and to order of magnitude calculations. Likewise, no attempt has been made to cite original sources or to make the references complete.

It is a pleasure to thank Mrs. Anne Barrington for typing several orders of manuscript, and the John Simon Guggenheim Foundation for a generous fellowship. This book is happily dedicated to the Geophysical Fluid Dynamics Summer Study Program.

PART ONE

CHAPTER I

Wave Generation

1.1 Review of Ideal Fluid Theory

The interaction of wind and wave at the surface is the basic mechanism for transferring energy and momentum to the ocean, and therefore we start this text on the general circulation with the theory of wind generated waves (Sections 1.3 and 1.6). The Euler equation (1.1.3) derived below provides the basis for the wave theory and for much of the discussion of rotating fluids in subsequent chapters.

If $\rho(x, y, z, t)$ denotes the density and $\mathbf{V}(x, y, z, t)$ the velocity at the point x, y, z in a cartesian inertial frame, then the continuity equation

$$(\partial\rho/\partial t) + \nabla \cdot (\rho\mathbf{V}) = 0 \tag{1.1.1}$$

is derived from the following examination of the mass budget in a fixed cubical control volume. Let the infinitesimal sides δx, δy, δz of the cube be parallel to the corresponding coordinate axis and consider a face of cross-sectional area $\delta y\,\delta z$ lying perpendicular to the x axis. Let u, v, w denote the corresponding

x, y, z components of $\mathbf{V}$. Then $\rho u\, \delta y\, \delta z$ is the flux of mass per unit time through the face, and

$$[\delta x\, \partial(\rho u)/\partial x]\, \delta y\, \delta z$$

is the difference in mass flux between two such parallel faces separated by a distance δx. The mass flux through the remaining two pairs of faces of the cube is computed in an analogous way. Therefore, the total mass per unit time leaving the entire surface of the cube is $\nabla \cdot (\rho \mathbf{V})\, \delta x\, \delta y\, \delta z$. Since the rate of decrease of mass inside the fixed volume is given by $-(\partial \rho/\partial t)\, \delta x\, \delta y\, \delta z$, the conservation principle requires $-(\partial \rho/\partial t)\, \delta x\, \delta y\, \delta z = \nabla \cdot (\rho \mathbf{V})\, \delta x\, \delta y\, \delta z$, and Eq. (1.1.1) follows.

The Lagrangian derivative $d\rho/dt$ is a measure of the change in density of a material parcel of fluid, whereas the Eulerian derivative $\partial \rho/\partial t$ gives the density change at a fixed point. The relationship between the two derivatives is obtained by noting that the $x(t)$, $y(t)$, $z(t)$ coordinates of a moving parcel change at the rates $dx/dt = u$, $dy/dt = v$, $dz/dt = w$. By expanding the Lagrangian derivative, we have

$$\frac{d\rho(t, x(t), y(t), z(t))}{dt} = \frac{\partial \rho}{\partial t} + u\frac{\partial \rho}{\partial x} + v\frac{\partial \rho}{\partial y} + w\frac{\partial \rho}{\partial z} = \frac{\partial \rho}{\partial t} + \mathbf{V} \cdot \nabla \rho$$

$$= \frac{\partial \rho}{\partial t} + \nabla \cdot (\rho \mathbf{V}) - \rho\, \nabla \cdot \mathbf{V}$$

and (1.1.1) can then be written as

$$\rho^{-1}\, d\rho/dt + \nabla \cdot \mathbf{V} = 0 \tag{1.1.2}$$

If $\mathbf{k}$ denotes a unit vector pointing vertically upward and g the acceleration of gravity, then the equation of motion for an ideal fluid is

$$\rho\, d\mathbf{V}/dt = -\nabla p - \rho g \mathbf{k} \tag{1.1.3}$$

where $p(x, y, z, t)$ is the pressure field. Equation (1.1.3) is derived by equating the instantaneous acceleration of the center of gravity of the aforementioned cube to the sum of the forces (per unit volume) acting on it. Thus, $-\rho g \mathbf{k}$ is the gravity force per unit volume acting on the cube of volume $\delta x\, \delta y\, \delta z$. Since the pressure force on a face of the cube perpendicular to the x axis is $p\, \delta y\, \delta z$, and since the pressure difference between two faces separated by a distance x has the magnitude

$$[\delta x\, \partial p/\partial x]\, \delta y\, \delta z$$

we see that $-(\partial p/\partial x)\, \delta x\, \delta y\, \delta z$ is the $+x$ component of the total pressure force acting on the cube. Analogous considerations for the y and z directions imply that the total vector pressure force per unit volume is $-\nabla p$. The mass times the acceleration of the cube is given by $\rho\, \delta x\, \delta y\, \delta z\, d\mathbf{V}/dt$, and by equating $\rho\, d\mathbf{V}/dt$ to the pressure plus gravity forces, we obtain (1.1.3). It should be noted that the viscous stresses, which act on the sides of the cube, have been neglected in these ideal fluid equations (see Chapter V for the viscous equations of motion).

For the case of a homogeneous incompressible† fluid (ρ = constant), the first term in (1.1.2) vanishes, the force of gravity $\rho g\mathbf{k} = \nabla(\rho g z)$ can be expressed as the gradient of a potential, and (1.1.3) then simplifies to

$$d\mathbf{V}/dt = -\nabla(p/\rho + gz) \qquad (1.1.4)$$
$$\nabla \cdot \mathbf{V} = 0$$

The acceleration term appearing in (1.1.4) is also a material derivative, and the relationship between the Lagrangian and Eulerian derivatives for the x component of acceleration is

$$\frac{du}{dt} = \frac{\partial u}{\partial t} + u\left(\frac{\partial u}{\partial x}\right) + v\left(\frac{\partial u}{\partial y}\right) + w\left(\frac{\partial u}{\partial z}\right) = \frac{\partial u}{\partial t} + \mathbf{V} \cdot \nabla u$$

Analogous relations hold for the other acceleration components, and therefore the vector quantity may be written as

$$\frac{d\mathbf{V}}{dt} = \frac{\partial \mathbf{V}}{\partial t} + u\left(\frac{\partial \mathbf{V}}{\partial x}\right) + v\left(\frac{\partial \mathbf{V}}{\partial y}\right) + w\left(\frac{\partial \mathbf{V}}{\partial z}\right) = \frac{\partial \mathbf{V}}{\partial t} + (\mathbf{V} \cdot \nabla)\mathbf{V}$$

where $\mathbf{V} \cdot \nabla$ is the symbolic representation of the operator

$$u\, \partial/\partial x + v\, \partial/\partial y + w\, \partial/\partial z$$

Thus we see that (1.1.4) is a nonlinear partial differential equation. Because of the difficulty in solving such equations, we attach special importance to certain integral invariants such as the conservation of energy, which is derived below.

Since $\mathbf{V} \cdot d\mathbf{V}/dt = \frac{1}{2}\, d/dt\, (\mathbf{V}^2)$, the scalar product of $\mathbf{V}$, and (1.1.4) gives

$$(d/dt)(\tfrac{1}{2}\mathbf{V}^2) = -\mathbf{V} \cdot \nabla((p/\rho) + gz)$$

and because of the constant density, we can also write this as

$$(d/dt)(\tfrac{1}{2}\rho\mathbf{V}^2 + \rho g z) = -\nabla \cdot (p\mathbf{V})$$

The material derivative of any scalar function s in an incompressible flow can always be written as $ds/dt = \partial s/\partial t + \mathbf{V} \cdot \nabla s = \partial s/\partial t + \nabla \cdot (s\mathbf{V})$, and therefore the previous equation becomes

$$(\partial/\partial t)(\tfrac{1}{2}\rho\mathbf{V}^2 + \rho g z) + \nabla \cdot [\mathbf{V}(\tfrac{1}{2}\rho\mathbf{V}^2 + \rho g z + p)] = 0$$

By using Gauss's theory in the integration of this equation over any closed volume of the fluid, we get

$$(\partial/\partial t) \oiiint dx\, dy\, dz\, (\tfrac{1}{2}\rho\mathbf{V}^2 + \rho z g) = -\oiint d\mathbf{A} \cdot \mathbf{V}(\tfrac{1}{2}\rho\mathbf{V}^2 + \rho g z) - \oiint d\mathbf{A} \cdot \mathbf{V}p \qquad (1.1.5)$$

where $d\mathbf{A}$ is an outward pointing element of area on the bounding surface. The

† The justification for neglecting $\rho^{-1}\, d\rho/dt$, compared to the terms $\partial u/\partial x$, $\partial v/\partial y$, $\partial w/\partial z$, depends on the fact that the typical values of u, v, w in the problems at hand are all much smaller than the speed of sound waves in the liquid. For further discussion of the influence of compressibility, see Section 9.1.

left hand side of (1.1.5) gives the rate of increase of the sum of the kinetic plus potential energy in the fixed volume, but the potential energy term vanishes in this case because the mass of the fluid inside the volume is invariant. The first term on the right hand side of (1.1.5) is the energy convected into the volume by the normal component of $\mathbf{V}$ on the surface $d\mathbf{A}$. The last, or "pressure-work," term in (1.1.5) is familiar from thermodynamics and may be illustrated further by the following example. When a cellophane bag containing water is squeezed on the outside, the liquid will move, and the source of energy can be attributed to this "pressure work." Likewise, the mechanism for the generation of surface waves in Section 1.2 is the work done by the atmospheric pressure on the free surface of the ocean. The right hand side of (1.1.5) will vanish, however, when the bounding surface is rigid ($\mathbf{V} \cdot d\mathbf{A} = 0$), and in this case we have

$$\frac{\partial}{\partial t} \oiiint dx\, dy\, dz\, \tfrac{1}{2}\rho \mathbf{V}^2 = 0$$

Thus, the total kinetic energy of a bounded homogeneous fluid is constant in time.

In contrast with the "global" integral discussed above, we now derive a "local" integral invariant called the *circulation theorem.* Although its content is equivalent to (1.1.4), the theorem eliminates the pressure term and thereby introduces the *vorticity* in a transparent way. Let $C(t)$ be an arbitrary closed curve, constructed at time t, and let $C(t + dt)$ denote the configuration when each element on the curve is allowed to move with the same velocity $\mathbf{V}(x, y, z, t)$ as the fluid. Let $\delta\mathbf{V}$ denote the instantaneous difference in velocity between two points (P and Q in Fig. 1.1) separated by a distance $\delta\boldsymbol{\lambda}$. P will move a distance $\mathbf{V}\, dt$ to P' in the interval dt, and Q will move a distance $(\mathbf{V} + \delta\mathbf{V})\, dt$ to point Q'.

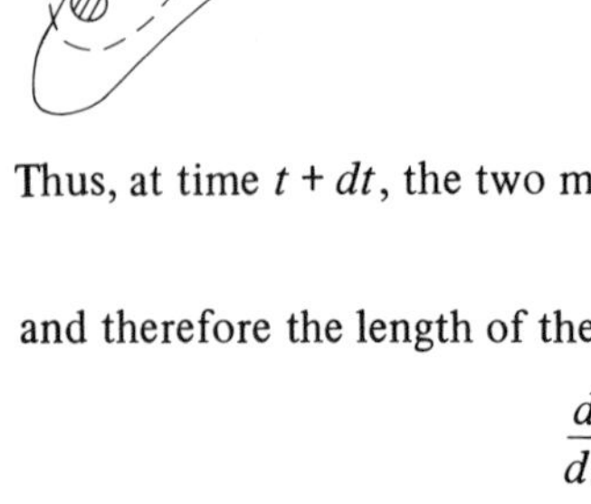

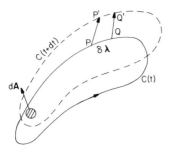

FIG. 1.1 The kinematics of the circulation theorem. $C(t)$ is a material curve which is displaced by the vector velocity V to $C(t + dt)$ in time dt. $\delta\boldsymbol{\lambda}$ is a typical element of arc and $d\mathbf{A}$ is a vector element of area on any surface which has $C(t)$ as its perimeter.

Thus, at time $t + dt$, the two material particles are separated by the distance

$$\delta\boldsymbol{\lambda}' = \delta\boldsymbol{\lambda} + (\delta\mathbf{V})\, dt$$

and therefore the length of the material arc changes at the rate

$$\frac{d}{dt}(\delta\boldsymbol{\lambda}) = \frac{\delta\boldsymbol{\lambda}' - \delta\boldsymbol{\lambda}}{dt} = \delta\mathbf{V}$$

If we now divide C into a large number of elementary arcs, the typical one being $\delta\boldsymbol{\lambda}$, then for each material arc we have

$$(d/dt)(\mathbf{V}\cdot\delta\boldsymbol{\lambda}) = (d\mathbf{V}/dt)\cdot\delta\boldsymbol{\lambda} + \mathbf{V}\cdot\delta\mathbf{V}$$

The sum (integral) of these gives the time rate of change of the closed line integral of $\mathbf{V}\cdot\delta\boldsymbol{\lambda}$, and thus we have

$$\frac{d}{dt}\oint_{C(t)} \mathbf{V}\cdot\delta\boldsymbol{\lambda} = \oint \frac{d\mathbf{V}}{dt}\cdot\delta\boldsymbol{\lambda} + \oint \mathbf{V}\cdot\delta\mathbf{V} = \oint_{C(t)} \frac{d\mathbf{V}}{dt}\cdot\delta\boldsymbol{\lambda} \tag{1.1.6}$$

where we have utilized the kinematical fact that the closed line integral of $\mathbf{V}\cdot\delta\mathbf{V} = \frac{1}{2}\,\delta(\mathbf{V}^2)$ vanishes. Equation (1.1.6) is a purely kinematic deduction, and the dynamical content of the *circulation theorem* emerges when we take the closed line integral of (1.1.4). The right hand side of the result will then vanish, because the closed line integral of the gradient of *any* scalar function vanishes, and we then have

$$\oint_{C(t)} \frac{d\mathbf{V}}{dt}\cdot\delta\boldsymbol{\lambda} = 0 \tag{1.1.7}$$

It then follows from (1.1.6) that

$$\frac{d}{dt}\oint_{C(t)} \mathbf{V}\cdot\delta\boldsymbol{\lambda} = 0 \tag{1.1.8}$$

and this states that the "circulation," or the closed line integral of $\mathbf{V}$ around any curve that moves with the fluid is independent of time.

The line integral of any vector field is related to the curl by Stokes's theorem, and therefore (1.1.8) can also be written as

$$\frac{d}{dt}\iint \boldsymbol{\zeta}\cdot d\mathbf{A} = 0 \tag{1.1.9}$$

where $d\mathbf{A}$ is a vector element of area (Fig. 1.1) on any surface having $C(t)$ as its perimeter, and where

$$\boldsymbol{\zeta} = \nabla \times \mathbf{V} = \begin{vmatrix} \mathbf{i} & \mathbf{j} & \mathbf{k} \\ \partial/\partial x & \partial/\partial y & \partial/\partial z \\ u & v & w \end{vmatrix} \tag{1.1.10}$$

$$\mathbf{j}\cdot\boldsymbol{\zeta} = \partial u/\partial z - \partial w/\partial x$$

gives the components of the vorticity. The following examples illustrate the usefulness of the circulation and vorticity theorems.

1.2 Review of Surface Gravity Waves

Suppose that a very deep layer of water is at rest at time $t < 0$, so that the circulation ($\int \mathbf{V} \cdot \delta\lambda$) vanishes identically. If variable atmospheric pressure forces are applied to the free surface at $t > 0$, then the circulation and the vorticity $\nabla \times \mathbf{V}$ must still vanish identically, according to (1.1.9). Thus, the velocity field induced by the surface pressures must be irrotational ($\nabla \times \mathbf{V} = 0$) as well as nondivergent ($\nabla \cdot \mathbf{V} = 0$). We confine attention to two-dimensional disturbances in a vertical plane, with u, w denoting the x, z components of $\mathbf{V}$. Equation (1.1.10) gives the magnitude of the vorticity, and therefore the cartesian equations to be satisfied are

$$\partial u/\partial z - \partial w/\partial x = 0, \qquad \partial u/\partial x + \partial w/\partial z = 0$$

or equivalently,

$$\partial^2 w/\partial x^2 + \partial^2 w/\partial z^2 = 0 \tag{1.2.1}$$

The vertical velocity $w(x, z, t)$ must satisfy $w(x, -\infty, t) = 0$ in *deep* water, and we now turn to the boundary condition at the free surface ($z = 0$).

In this section, the discussion is confined to the case of *free* waves, such as will occur if the above-mentioned atmospheric pressure forces are removed at some $t > 0$. We are then required to determine the subsequent evolution of the waves. Thus, if $z = \bar{\eta}(x, t)$ denotes the vertical height of the free surface above the $z = 0$ datum, then the material derivative of $\bar{\eta}$ gives the vertical velocity

$$w(x, \bar{\eta}, t) = (d/dt)\,\bar{\eta}(t, x) = \partial\bar{\eta}/\partial t + u(x, \bar{\eta}, t)\,\partial\bar{\eta}/\partial x \tag{1.2.2a}$$

of a parcel on the free surface. Moreover, the pressure must be uniform (zero) on the free surface, if we neglect the effect of surface tension (see Section 1.6).

Although the field equation (1.2.1) is linear, we note that the boundary condition (1.2.2a) is not. However, if $|\bar{\eta}|$ be sufficiently small, then the quadratic term $u\,\partial\bar{\eta}/\partial x$ will be much smaller than the linear term $\partial\bar{\eta}/\partial t$. If this linearizing approximation be introduced into (1.2.2a) together with

$$w(x, \bar{\eta}, t) = w(x, 0, t) + \bar{\eta}(\partial w/\partial z)(x, 0, t) + \ldots \simeq w(x, 0, t)$$

then the balance required by the linear terms is

$$w(x, 0, t) = \partial\bar{\eta}/\partial t \tag{1.2.2b}$$

The neglect of the nonlinear terms is valid when they are small compared to the linear terms, or

$$1 \gg \frac{|u\,\partial\bar{\eta}/\partial x|}{|\partial\bar{\eta}/\partial t|} \sim \frac{|u|}{c} \tag{1.2.2c}$$

where $c = |\partial\bar{\eta}/\partial x|^{-1}\,|\partial\bar{\eta}/\partial t|$. The numerator of (1.2.2c) is the particle speed and

c is the phase speed for a wave having the form

$$\bar{\eta} = \text{Re}\ \eta(t)e^{ikx} \tag{1.2.3}$$

The conventional symbol Re denotes the real part of an expression, $2\pi/k$ is the wavelength of $\bar{\eta}(x, t)$, and $\eta(t)$ is the complex amplitude (including the frequency). In the theory given below, all quadratic terms in wave amplitude will be systematically discarded, this procedure being valid when the particle velocity is small compared to the phase velocity (1.2.2c).

The velocity fields u, w corresponding to (1.2.3) are also proportional to $\exp ikx$, and substitution into (1.2.1) shows that w must also vary as $\exp \pm kz$. The appropriate sign $\exp +kz$ is determined by the fact that w must be bounded at $z = -\infty$, and the amplitude of w is then determined by using (1.2.2b). Accordingly, we find that the field equation, the continuity equation, and the kinematical boundary condition are satisfied by

$$w(x, z, t) = \text{Re}\,(\partial\eta/\partial t)\,e^{ikx}\,e^{kz}, \qquad u(x, z, t) = \text{Re}\,i(\partial\eta/\partial t)\,e^{ikx}\,e^{kz} \tag{1.2.4}$$

Let us now compute the pressure perturbation p associated with (1.2.4), and then relate the result to the dynamic boundary condition. The horizontal component of (1.1.4) is $\rho\, du/dt = -\partial p/\partial x$, and since the linearized value of the acceleration is

$$du/dt = \partial u/\partial t + u\,\partial u/\partial x + w\,\partial u/\partial z \simeq \partial u/\partial t$$

we obtain

$$\rho\,\partial u/\partial t = -\partial p/\partial x \tag{1.2.5a}$$

Likewise, the linearization of the vertical component of (1.1.4) gives

$$\rho\,\partial w/\partial t = -\partial p/\partial z - g\rho \tag{1.2.5b}$$

If $z = -\epsilon$ denotes some datum level located a small distance $\epsilon \sim |\bar{\eta}|$ beneath the troughs of the wave, then the vertical integral of (1.2.5b) from $z = -\epsilon$ to the free surface is

$$\int_{-\epsilon}^{\bar{\eta}} dz\,\frac{\partial w}{\partial t} = p(x, -\epsilon, t) - \rho g(\bar{\eta} + \epsilon)$$

and therefore

$$\frac{\partial p}{\partial x}(x, -\epsilon, t) = \rho g\,\frac{\partial \bar{\eta}}{\partial x} + \rho\,\frac{\partial}{\partial x}\int_{-\epsilon}^{\bar{\eta}} dz\,\frac{\partial w}{\partial t} \tag{1.2.6a}$$

$$\simeq \rho g\,\frac{\partial \bar{\eta}}{\partial x} \tag{1.2.6b}$$

since the last term in (1.2.6a) is quadratically small. When (1.2.5a) is evaluated at $z = -\epsilon \to 0$ and when (1.2.6b) is used, we get

$$(\partial/\partial t)u(x,0,t) = -g\,\partial\bar{\eta}/\partial x \tag{1.2.7}$$

When (1.2.3) and (1.2.4) are substituted in (1.2.7), the equation for the wave amplitude becomes

$$\partial^2\eta/\partial t^2 + gk\eta = 0 \tag{1.2.8}$$

and the solution

$$\eta = \exp \pm[i(gk)^{1/2}t]$$

corresponds to waves propagating in either (±) horizontal direction with frequency $(gk)^{1/2}$.

1.3 Generation of Gravity Waves by Atmospheric Pressure Fluctuations

Let us now compute the forced surface displacements $\bar{\eta}(x, t)$ produced by the *continuous* application of atmospheric pressure forces

$$P_a{}^k = \mathrm{Re}(-\rho_w/k)f_k(t)\exp ik(x - Ut) \tag{1.3.1a}$$

where $2\pi/k$ denotes the wavelength, U the propagation speed of the atmospheric disturbance, and $f_k(t)$ an amplitude function, as discussed below. The water density is ρ_w, and the $-\rho_w/k$ factor in (1.3.1a) is merely introduced for normalizing convenience at a later stage.

All the formulas up to and including (1.2.5b) apply to the present problem, since the water motions are always irrotational. The dynamical boundary condition (1.2.6b), however, must be modified so as to include the effect of (1.3.1a). Accordingly, we will now show that the horizontal pressure gradient $\partial p/\partial x$ just beneath the wave troughs (Fig. 1.2) is equal to the pressure gradient $\partial P_a^k/\partial x$ in the air plus the horizontal gradient in the weight of the intervening fluids, or

$$\partial p/\partial x = \partial P_a{}^k/\partial x + g(\rho_w - \rho_a)\,\partial\bar{\eta}/\partial x \tag{1.3.1b}$$

To show this, we first note that the pressure at point e (Fig. 1.2) exceeds that at point c by $\rho_w g\,d\bar{\eta}$, and the pressure at b exceeds that at point f by $\rho_a g\,d\bar{\eta}$. It is permissible to use these hydrostatic relations because the error [cf. (1.2.6a)] is quadratically small in the wave amplitude. When these two hydrostatic relations are subtracted, and when the result is divided by dx, we obtain (1.3.1b). Substitution of (1.3.1b) in the horizontal momentum equation (1.2.5a) then gives

$$-\rho_w(\partial u/\partial t)(x,0,t) = \partial P_a{}^k/\partial x + g(\rho_w - \rho_a)\,\partial\bar{\eta}/\partial x$$

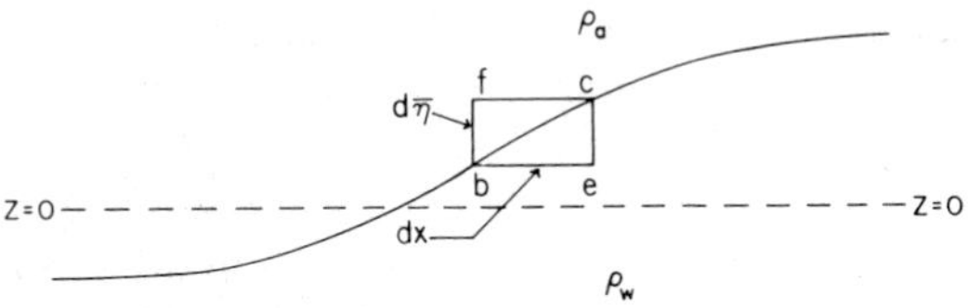

FIG. 1.2 Schematic diagram for relating horizontal pressure gradients on either side of air (subscript a)–water (subscript w) interface. The solid curve is the interface, $\bar{\eta}$ is the displacement of point b above the $z = 0$ datum, and $\bar{\eta} + d\bar{\eta}$ is the displacement of point c.

and by inserting (1.2.3), (1.2.4), and (1.3.1), we then have

$$d^2\eta/dt^2 + \omega^2\eta = f_k(t)e^{-ikUt} \tag{1.3.2}$$

where $\omega = [gk(1 - \rho_a/\rho_w)]^{1/2}$. It will now be shown that the general solution of the inhomogeneous wave equation (1.3.2) is

$$\eta(t) = \omega^{-1} \int_{-\infty}^{t} d\theta\, f_k(\theta)\, e^{-ikU\theta} \sin \omega(t-\theta) \tag{1.3.3}$$

The graph of $f_k(t)$ as a function of time can always be decomposed into the sum of a large number of discrete rectangular pulses, the typical one of which vanishes outside the time interval $\theta < t < \theta + d\theta$ and assumes the constant value $f_k(\theta)$ inside. If $\delta\eta(t)$ denotes the response produced by such a pulse, then we clearly have to take $\delta\eta = 0$ for $t < \theta$. Moreover, after the pulse ceases $(t > \theta + d\theta)$, $\delta\eta$ must behave like the free solution ($\sin \omega(t - \theta)$) of (1.3.2). It is also apparent that the amplitude of $\delta\eta$ must be proportional to the time $d\theta$ in which the pulse acts, and $\delta\eta$ must also be proportional to the value of the right hand side of (1.3.2) at the time $t = \theta$. These considerations suggest that the response to a single pulse should have the form

$$\delta\eta = \begin{cases} 0 & (t < \theta) \\ A\, d\theta\, f_k(\theta) e^{-ikU\theta} \sin \omega(t-\theta) & (t > \theta + d\theta) \end{cases}$$

where A is some numerical constant of proportionality. To obtain the total response $\eta(t)$ at time t, we must sum the values of $\delta\eta(t)$ with respect to all the pulses which act *prior* to t. Thus, the pulse acting at time $t - \theta$ produces the response given above, and by summation over all θ (with fixed t), we then obtain

$$\eta(t) = A \int_{-\infty}^{t} d\theta\, f_k(\theta) e^{-ikU\theta} \sin \omega(t-\theta)$$

By direct substitution in (1.3.2), the reader may now verify that this integral is a solution, provided $A = 1/\omega$ is chosen for the constant of proportionality. Thus, we have shown that the forced solution of (1.3.2) is given by (1.3.3), and the question now arises as to the amount of energy which is pumped into the water waves.

Let us form an "energy" equation from Eq. (1.3.2) by multiplying it with $d\eta^*/dt$, where the asterisk denotes a complex conjugate. When the following identities are used

$$\mathrm{Re}\,\eta\frac{d\eta^*}{dt} = \frac{1}{2}\frac{d}{dt}(\eta\eta^*), \qquad \mathrm{Re}\left(\frac{d\eta^*}{dt}\frac{d^2\eta}{dt^2}\right) = \frac{1}{2}\frac{d}{dt}\left(\frac{d\eta^*}{dt}\frac{d\eta}{dt}\right)$$

and when the wave "energy" is defined by

$$E_k = \frac{1}{2}\left(\frac{d\eta^*}{dt}\frac{d\eta}{dt} + \omega^2\eta\eta^*\right) \tag{1.3.4}$$

then the result can be written as

$$dE_k/dt = \mathrm{Re}\left[(d\eta^*/dt)f_k(t)e^{-ikUt}\right] \tag{1.3.5}$$

By using (1.3.3), we then obtain

$$dE_k/dt = \mathrm{Re}\int_{-\infty}^{t} d\theta\; f_k{}^*(\theta)f_k(t)e^{-ikU(t-\theta)}\cos\omega(t-\theta)$$

$$= \mathrm{Re}\int_{0}^{\infty} f_k{}^*(t-\xi')f_k(t)e^{-ikU\xi'}\cos\omega\xi'\,d\xi'$$

If the air pressure is applied at $t = 0$ so that $f^*(t - \xi') = 0$ for $\xi' > t$, then we have

$$dE_k/dt = \int_{0}^{t} f_k{}^*(t-\xi')f_k(t)e^{-ikU\xi'}\cos\omega\xi'\,d\xi' \tag{1.3.6}$$

The energy E_k (1.3.4) is positive definite, and in the free ($f_k = 0$) case, E_k is conserved. For simple harmonic forcing, with f_k = constant, the solution of (1.3.2) $\eta \propto \exp -ikUt$ is simple harmonic, except for the resonant case

$$kU = \omega \tag{1.3.7}$$

or equivalently,

$$k = g(1-\rho_a/\rho_w)U^{-2} \tag{1.3.8}$$

When resonance occurs, the water wave continually abstracts energy from the air, and E_k will increase as t^2.

How does the foregoing calculation bear on the problem of wave generation at the surface of the ocean? When the wind starts to blow over a calm surface, we are aware of an increase in the "gusts" (turbulent velocity fluctuations), and the concomitant atmospheric pressure fluctuations can thereby impart energy to the water, as discussed above. These atmospheric pressure fluctuations tend to be advected downstream with some speed U that depends on the mean wind in the atmospheric boundary layer. Equation (1.3.1a) represents the typical Fourier component k of the atmospheric pressure, and (1.3.6) gives the corresponding energy induced in the water. But we must now take into consideration the fact that the amplitude of each Fourier component of air pressure varies irregularly over large intervals of time. Thus, the temporal variation in $f_k(t)$ is just as unpredictable as the individual gusts and eddies in the air velocity, and such quantities are called random functions. Accordingly, the phase speed of the air wave will drift and thereby supply less energy than in the simple harmonic case.

1.4 Statistical Resonance Theory

In principle, one would determine $f_k(t)$ by measuring the pressure distribution on the free surface at any given time and by then taking a spatial Fourier transform. Suppose we then plot successive values of the complex vector $f_k(t)$ on an Argand diagram. Although this vector must vary smoothly over "short" intervals of time, it will vary very erratically over time intervals which are large compared to the time between individual atmospheric gusts. In particular, the phase angle of the vector will be randomly distributed about its mean angle [but the amplitude of $f_K(t)$ will be bounded in accord with the rms value of the turbulent fluctuations in the air].

In order to describe the average properties of such random processes, one must somehow construct an ensemble of "indistinguishable and independent realizations," the typical one of which responds as in Section 1.3. The arithmetical average of any dynamical quantity taken over a large number of realizations is called an *ensemble average* and denoted by the symbol $\langle\,\rangle$. When we perform such an average of (1.3.6), the mean wind speed U is absolutely identical in each member of the ensemble, and thus the average increase of energy in number k is

$$\langle dE_k/dt\rangle = \mathrm{Re}\int_0^t d\xi'\,\langle f_k^*(t-\xi')f_k(t)\rangle e^{-ikU\xi'}\cos\omega\xi' \qquad (1.4.1)$$

From (1.3.1a), we see that $\langle f_k^*(t)f_k(t)\rangle$ represents the mean square pressure *gradient* in the air, corresponding to wave number k. To obtain the water energy from (1.4.1), we must form a correlation between the air pressure waves at two times separated by an interval ξ, and thus we introduce the correlation coefficient

$$\gamma(\xi) \equiv \frac{\langle f_k^*(t-\xi) f_k(t)\rangle}{\langle f_k^*(t) f_k(t)\rangle} \tag{1.4.2}$$

having the property $\gamma(0) = 1$. We mentioned previously that the phase of f varies randomly in the complex plane over large time intervals, and thus we also have the property $\Upsilon(\infty) = 0$. Before demonstrating that γ is real for all ξ, we must mention an important assumption which is concealed in (1.4.2).

From observations of turbulent flow over rigid boundaries (pipes, channels, etc.), we know that a statistically steady or stationary state can be realized in which the average properties (stress, rms velocity, etc.) are independent of the *absolute* value of time. Likewise, if (1.4.2) be evaluated for turbulent flow over a *rigid* plate, then its value will depend only on the relevant *lag* time ξ and not on t. This property is valid in the free surface problem only to the extent that the atmospheric pressure fluctuations are independent of the waves they induce in the water, and this condition is fulfilled as long as the water waves have infinitesimal amplitude. In that case, the air flow "sees" an essentially flat surface, and thus the statistical properties of $f_k(t)$ can be determined independently of the motions induced in the water. However, if the forced response of the free surface becomes large, then the waves will present a "rough" boundary to the air flow and thereby modify $f_k(t)$. Thus, when the waves reach large amplitude, (1.4.2) will depend on t as well as ξ.

Let us now return to the case of infinitesimal amplitude water waves, so that we may assume that (1.4.2) is independent of absolute time. This assumed *stationarity* of the air turbulence also implies that there is no difference between *lag* and *lead* correlations, so that (1.4.2) must be symmetric in ξ, or

$$\gamma(\xi) = \gamma(-\xi) \tag{1.4.3}$$

Likewise, the complex conjugate of (1.4.2), or

$$\gamma^*(\xi) = \frac{\langle f_k^*(t) f_k(t-\xi)\rangle}{\langle f_k^*(t) f_k(t)\rangle}$$

must have the same value when we replace t by $t + \xi$, and $t - \xi$ by t (in the numerator). Upon doing this, we have

$$\gamma^*(\xi) = \frac{\langle f_k^*(t+\xi) f_k(t)\rangle}{\langle f_k^*(t) f_k(t)\rangle} = \gamma(-\xi) \tag{1.4.4}$$

where (1.4.2) has been used in obtaining the last line. From (1.4.3) and (1.4.4), it follows that

$$\gamma^*(\xi) = \gamma(\xi)$$

and therefore $\gamma(\xi)$ is real for all ξ. Consequently, when (1.4.2) is used in (1.4.1), we obtain

$$dE_k/dt = \langle f_k^*(t) f_k(t)\rangle \int_0^t d\xi\, \gamma(\xi) \cos kU\xi \cos \omega\xi$$

and for large t, this becomes

$$dE_k/dt = \langle f_k^*(t) f_k(t)\rangle \int_0^\infty d\xi\, \gamma(\xi) \cos kU\xi \cos \omega\xi \qquad (1.4.5)$$

From (1.4.5), we conclude that the asymptotic rate of increase of energy in each water wave is independent of time, but the growth rate varies with wave number. For example, when $kU = \omega$ [cf. (1.3.8)], the product of the two trigonometric terms in (1.4.5) is always positive, whereas the integrand oscillates in ξ when $kU \neq \omega$. Therefore, a maximum growth rate (1.4.5) will occur near $kU = \omega$ if $\langle f_k^* f_k\rangle$ varies slowly with k and if $\gamma(\xi)$ varies slowly with ξ. This means that the response of the water is dominated by that wavelength k^{-1}, whose free propagation speed is equal to the wind speed U. The reader is referred to Phillips (1966) for a more complete discussion of the stochastic resonance problem and for comparison with observations. In Section 1.6, we will discuss a different mechanism for generating surface gravity waves.

1.5 Stability of Parallel Shear Flow

In this section, we consider some of the properties of rotational ($\nabla \times \mathbf{V} \neq 0$) flow. The motion is assumed two-dimensional, with the vector velocity $\mathbf{V}$ having the components u, w in the x, z directions, and the vorticity vector $\nabla \times \mathbf{V}$ points in the y direction with magnitude $\zeta(t, x, z)$. The vorticity (1.1.10) and the continuity equations are given by

$$\zeta = \partial u/\partial z - \partial w/\partial x \qquad \text{and} \qquad \partial u/\partial x + \partial w/\partial z = 0 \qquad (1.5.1)$$

and we will show that ζ is conserved. By applying the circulation theorem to a two-dimensional curve $C(t)$ which encloses an infinitesimal area δA in the x, z plane, it follows from (1.1.9) that $\zeta\, \delta A$ is independent of time. Since the flow is two-dimensional, the conservation of mass requires that the area δA enclosed by the material curve $C(t)$ be independent of time. The invariance of δA and $\zeta\, \delta A$ then implies that $\zeta(t, x, y)$ be independent of t, and thus we have

$$0 = d\zeta/dt = \partial\zeta/\partial t + u\, \partial\zeta/\partial x + w\, \partial\zeta/\partial z \qquad (1.5.2)$$

The reader can also derive this analytically by taking the curl of the two-dimensional momentum equations.

The vorticity equation (1.5.2) and the continuity equation (1.5.1) are identically satisfied for a flow having $w = 0$ and $u = \bar{U}(z)$, and therefore any *laminar* shear flow is an inviscid equilibrium state. Let us now investigate the stability of the equilibrium by applying an arbitrary small perturbation $w = w'(x, z, 0)$, at time $t = 0$, and by then computing the subsequent $w'(x, z, t)$ from (1.5.2). If $\bar{U}(z)$ is stable, then we require that w' have infinitesimal amplitude at all t, but the basic state $\bar{U}(z)$ is unstable if we find *any* amplifying w'.

We proceed by tentatively assuming the basic state to be stable, so that w' and the perturbation horizontal velocity, $u' = u(x, z, t) - \bar{U}(z)$, are small for all t. Since the basic field is nondivergent, the perturbation velocities u' and w' must also be nondivergent, and this relation is expressed by (1.5.3) below. Since the total vorticity, $\zeta = \partial\bar{U}/\partial z + (\partial u'/\partial z - \partial w'/\partial x)$, is the sum of the basic vorticity $\partial\bar{U}/\partial z$ and the perturbation vorticity (1.5.5), the substitution of $w = w'$ and $u = \bar{U} + u'$ into (1.5.1) and (1.5.2) gives (1.5.4), and thus we have

$$\partial u'/\partial x + \partial w'/\partial z = 0 \tag{1.5.3}$$

$$0 = \partial\zeta'/\partial t + \bar{U}(z)\,\partial\zeta'/\partial x + w'\,\partial^2\bar{U}/\partial z^2 + [u'\,\partial\zeta'/\partial x + w'\,\partial\zeta'/\partial z] \tag{1.5.4}$$

$$\zeta' \equiv \partial u'/\partial z - \partial w'/\partial x \tag{1.5.5}$$

The bracketed term in (1.5.4) is now neglected because it is quadratic in the amplitude of the infinitesimal perturbation, and thus the linearized vorticity equation is

$$0 = \partial\zeta'/\partial t + \bar{U}(z)\,\partial\zeta'/\partial x + w'\,\partial^2\bar{U}/\partial z^2 \tag{1.5.6}$$

For any function ψ, the functions

$$w' = \partial\psi/\partial x \qquad \text{and} \qquad u' = -\partial\psi/\partial z \tag{1.5.7}$$

satisfy the continuity equation (1.5.3) identically. In terms of this *stream function*, (1.5.5) becomes $\zeta' = -\nabla^2\psi$, and (1.5.6) becomes

$$(\partial/\partial t + \bar{U}(z)\,\partial/\partial x)\,\nabla^2\psi - (\partial^2\bar{U}/\partial z^2)\,\partial\psi/\partial x = 0 \tag{1.5.8}$$

Our linearizing assumption will be justified and the basic flow will be stable if *all* of the initial value solutions of (1.5.8) are bounded functions of t. On the other hand, if some solutions should amplify, then the *linearized* theory (1.5.8) cannot be used to compute the large amplitudes at $t \to \infty$, but we can conclude that the basic state is unstable. What relation does such an initial value calculation have to the problem of the stability of real flows? Such flows are subjected to small perturbations arising from the "noise" background, and we should rightfully include such a small "noise force" into the field equations. But this procedure would be too grotesque, and we therefore introduce the random noise effect as an initial perturbation having arbitrary form. If any such perturbation should amplify in time, then we may confidently expect that the basic state will not be realized. The instability process will give rise to a new state, the prediction of which then requires nonlinear considerations (cf. Chapter X).

Before proceeding to a discussion of the solutions to the linearized vorticity equation (1.5.8), it is of interest to derive the corresponding momentum equations. When the linearized value of the vertical acceleration

$$dw/dt = \partial w'/\partial t + U(z)\,\partial w'/\partial x$$

and the linearized value of the horizontal acceleration

$$du/dt = \partial u'/\partial t + U\,\partial u'/\partial\, x + w'\,\partial \bar{U}/\partial z$$

are equated to the gradients of the pressure perturbation p', we obtain

$$\partial u'/\partial t + \bar{U}(z)\,\partial u'/\partial x + w'\,\partial \bar{U}/\partial z = -\,(\partial/\partial x)p'/\rho \tag{1.5.9a}$$

$$\partial w'/\partial t + U(z)\,\partial w'/\partial x = -\,(\partial/\partial z)p'/\rho \tag{1.5.9b}$$

as the linearized expression of the momentum equation (1.1.4). An important energy equation is now formed by multiplying (1.5.9a) with u', by multiplying (1.5.9b) with w', and by adding the results. Simplification by use of (1.5.3) then yields

$$\frac{\partial}{\partial t}\left[\frac{(u')^2+(w')^2}{2}\right] + u'w'\,\frac{\partial \bar{U}}{\partial z} = -\,\frac{\partial}{\partial x}\left(\frac{\bar{U}}{2}\,[(u')^2+(w')^2]\right) - \frac{\partial}{\partial x}\left(\frac{u'p'}{\rho}\right) - \frac{\partial}{\partial z}\left(\frac{w'p'}{\rho}\right) \tag{1.5.10}$$

Let us assume the perturbation fields to be cyclic (periodic) in x and use a bar to denote a horizontal average over an integral number of wavelengths, so that terms of the form $\overline{(\partial/\partial x)(\ )}$ will vanish. The boundaries at $z = 0$ and $z = H$ are assumed to be rigid ($w' = 0$), so that the vertical integral of the last term in (1.5.10) will also vanish, and thus we have the energy equation

$$\frac{\partial}{\partial t}\,\frac{1}{2}\int_0^H dz\,[\overline{(u')^2}+\overline{(w')^2}] = \int_0^H (\overline{-u'w'})\,\frac{\partial \bar{U}}{\partial z}\,dz \tag{1.5.11}$$

Thus, we see that the "perturbation kinetic energy," as represented by the integral on the left side of (1.5.11), can increase only if there is a correlation between the shear of the basic current and the *Reynold's stress* $\overline{u'w'}$ The latter represents an average vertical transport w' of horizontal momentum u', or a correlation between u' and w'. We will now show that a finite value of the correlation coefficient is difficult to maintain because of the distorting effect of $\partial U/\partial z$ on the perturbation, and it will appear that the basic shear is by no means a *sufficient* condition for a monotonic increase of the perturbation energy.

The point will be illustrated by examining the temporal evolution of a small perturbation in a (Couette) flow having uniform basic shear or $\partial^2 \bar{U}/\partial z^2 = 0$. In this case, the perturbation vorticity equation (1.5.6) reduces to

$$\left(\frac{\partial}{\partial t} + \bar{U}(z)\frac{\partial}{\partial x}\right)\zeta' = 0 \tag{1.5.12}$$

By direct substitution, one readily verifies that the general solution of (1.5.12) is

$$\zeta'(t, z, x) = Q(x - \bar{U}(z)t, z) \tag{1.5.13}$$

where $Q(x, z)$ is the given initial $(t = 0)$ value of ζ'. The interpretation of (1.5.13) is faciliated by making an analogy with the spread of a spot of dye in a strictly $(w' = 0)$ laminar Couette flow. If the dye is nondiffusive, and if $C(t, x, z)$ denotes the concentration, then C is conserved, or $0 = dC/dt = \partial C/\partial t + \bar{U}\,\partial C/\partial x$. This equation for C is identical in form to (1.5.12), and thus the distribution of ζ' can be inferred from the more familiar example of a spreading dye spot in a strictly laminar shear flow. Such a spot will move with the local $\bar{U}(z)$, and thereby be "drawn out" into a long thin line parallel to the flow. The curves of constant C will then become parallel to the shear flow as $t \to \infty$. Likewise, ζ' will become independent of x as $t \to \infty$. Since $\nabla^2\psi = -\zeta'$, the stream function will become independent of x, and therefore $w' = \partial\psi/\partial x \to 0$ as $t \to \infty$. We conclude that the Couette flow is stable to all forms of initial perturbation, and that some variation in the basic vorticity is a necessary condition for the instability of a shear flow.

In order to consider the instability for flows having $\partial^2 \bar{U}/\partial z^2 \neq 0$, we return to (1.5.8), and investigate the behavior of normal modes

$$\psi = \mathrm{Re}\,\bar{\psi}(z)e^{ik(x-ct)} \tag{1.5.14}$$

We are required to compute a (possibly complex) phase speed $c(k)$ as a function of the given wavelength, and the eigenfunction equation obtained by substituting (1.5.14) in (1.5.8) is

$$(\bar{U}(z)-c)((d^2\bar{\psi}/dz^2)-k^2\bar{\psi})-\bar{U}''(z)\bar{\psi}(z) = 0 \tag{1.5.15}$$

where $U'' = d^2\bar{U}/dz^2$. We assume horizontal walls at $z = 0$ and $z = H$, so that the boundary condition $0 = w = \partial\psi/\partial x$ reduces to

$$\bar{\psi}(0) = \bar{\psi}(H) = 0 \tag{1.5.16}$$

If one of the eigenvalues should be complex, or

$$c = c_r + ic_i, \qquad c_i > 0 \tag{1.5.17}$$

then the associated perturbation (1.5.14) will increase exponentially in time, and the basic flow will be unstable. Supposing this to be the case, we write (1.5.15) as

$$\frac{d^2\bar{\psi}}{dz^2} - k^2\bar{\psi} - \frac{\bar{U}''\bar{\psi}(z)}{(U - c_r - ic_i)} = 0 \tag{1.5.18}$$

This equation cannot be solved in general, but an important integral can be obtained by multiplying it with the complex conjugate eigenfunction $\bar{\psi}^*$. By utilizing (1.5.16) in a partial integration and by taking the imaginary part of the result, we then obtain

$$\operatorname{Im}\int_0^H dz\,\frac{\bar{\psi}^*\bar{\psi}\bar{U}''}{(U-c_r)-ic_i} = 0 \qquad \text{and} \qquad c_i\int_0^H dz\,\frac{\bar{\psi}^*\bar{\psi}\bar{U}''}{(U-c_r)^2+c_i^2} = 0 \tag{1.5.19}$$

For $c_i > 0$, this equation can only be satisfied if U'' changes sign or, equivalently, if $\bar{U}(z)$ has an inflection point (Rayleigh's theorem). On the other hand, if $\bar{U}''$ has only one sign, then c cannot be complex, no normal mode can amplify, and the basic flow is said to be stable. We shall return to the interpretation of this result in subsequent chapters.

1.6 Gravity Waves Generated by Shear Flow Instability

We will now carry through a stability calculation for the basic flow shown in Fig. 1.3, this being a highly idealized model of the wind distribution over the surface ($z = 0$) of the sea. Although this model does not do justice to the continuous curvature of a real wind profile (Miles, 1962), we can, nonetheless, obtain some insight into the wave generating mechanism.

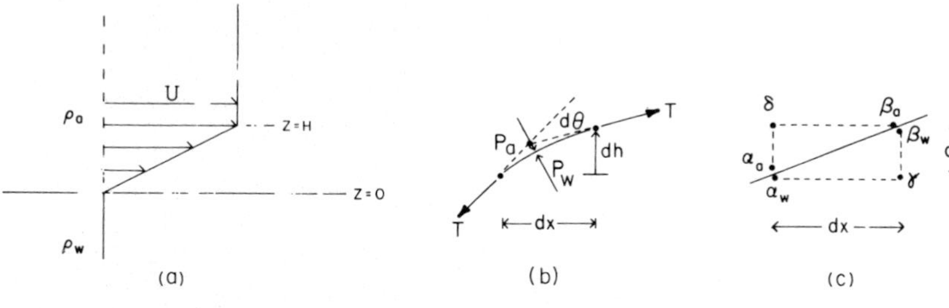

FIG. 1.3 Generation of surface gravity waves due to the instability of the air flow in a boundary layer. (a) Schematic diagram of a basic horizontal current $\bar{U}(z)$, with ρ_a denoting the air density and ρ_w the water density. (b) T is the surface tension acting on the ends of a surface film of width dx, p_w is the water pressure, and p_a is the air pressure. (c) α_a and β_a are two points in the air; α_w and β_w are the corresponding points on the opposite sides of the free surface. Points γ and δ complete the rectangle used to relate the pressures at the four points.

The water (density = ρ_w) is at rest in the undisturbed state of Fig. 1.3, the wind increases from $\bar{U}(0) = 0$ to $\bar{U}(H) = U$ in the air (density = ρ_a), and $\bar{U}(z) = U$ above $z = H$. Thus, the basic vorticity increases (discontinuously) from zero to U/H at $z = 0$, and decreases to zero again at $z = H$. Since $\partial^2 \bar{U}/\partial z^2$ vanishes in each of the three regions (these being separated by the two discontinuity surfaces), the perturbation vorticity equation (1.5.6) reduces to (1.5.12) for each region. Moreover, a particular solution of (1.5.12) is $\zeta' = 0$, and therefore we proceed to investigate the piecewise irrotational normal modes having horizontal wavelength $2\pi/k$. By properly connecting these solutions at $z = (0, H)$, we will then obtain an equation for the phase speed c as a function of k, U, H, g and the surface tension T. It will also be shown that some wavelength amplifies when U exceeds a certain critical value, this being equal to the minimum phase speed of *free* gravity–capillary waves. We will also be in a position to discuss another kind of instability (Kelvin–Helmholtz waves) which occurs when we let $H \to 0$ in Fig. 1.3 and which is important for the discussion in Chapter IX.

The form of the vertical velocity $w'(x, z, t)$ field and the perturbation height h of the free surface is again denoted by

$$w'(x, z, t) = \operatorname{Re} \hat{w}(z) e^{ik(x-ct)} \tag{1.6.1}$$

$$h(x, t) = \operatorname{Re} \hat{h} e^{ik(x-ct)} \tag{1.6.2}$$

where $\hat{h}$ is an amplitude constant. Since the perturbation is piecewise irrotational, (1.2.1) will be satisfied and $\hat{w}(z)$ will be a hyperbolic function, the precise form of which is given after the following consideration of the boundary conditions.

If F is a quantity that changes discontinuously at $z = 0$, or $z = H$, then the symbol $[F]_z$ will be used to denote the jump in F, or

$$F(z^+) - F(z^-) \equiv [F]_z \tag{1.6.3}$$

and, in particular, we have the jumps

$$[\rho]_0 = \rho_a - \rho_w; \qquad [d\bar{U}/dz]_0 = U/H; \qquad [d\bar{U}/dz]_H = -U/H$$

Because w' is the material derivative of the vertical height of a parcel and because $\bar{U}$ is continuous, it follows that w' must be continuous at the discontinuity surfaces. If the kinematical relation $w'(x, h, t) = dh/dt$ be linearized to

$$w'(x, 0, t) = \left(\frac{\partial}{\partial t} + \bar{U}(0) \frac{\partial}{\partial x}\right) h \tag{1.6.4}$$

and if (1.6.1) and (1.6.2) be substituted, then we have

$$ik\hat{h} = \frac{\hat{w}(0)}{\bar{U}(0) - c} \tag{1.6.5}$$

Although w' is continuous across $z = (0, H)$, $\partial w'/\partial z$ is not, and the proper connection condition must be determined from a momentum balance across the interface. The matching problem is related to the one that arose in Sections 1.2 and 1.3, wherein we used the continuity of pressure. Now, however, we also want to take into account the effect of surface tension T (dynes/cm), and therefore the pressure will not be continuous across the free surface. (Of course, there is no surface tension at $z = H$, so that pressure continuity is the relevant condition there.) Referring to Fig. 1.3b, we see that the difference $p_w - p_a$ in the pressure on the water and the air sides of the interface is proportional to the local curvature of the interface, or

$$p_w - p_a = -T\,\partial^2 h/\partial x^2 \tag{1.6.6}$$

for small perturbations. Now consider two sets of air–water points, such as α_w, α_a and β_w, β_a in Fig. 1.3c. If these pairs are separated horizontally by dx and vertically by dh, then the difference in the value of (1.6.6) at the two positions is

$$p(\alpha_w) - p(\alpha_a) - (p(\beta_w) - p(\beta_a)) = T\,\frac{\partial^3 h}{\partial x^3}\,dx \tag{1.6.7}$$

It is permissible, as shown in Section 1.2, to use the hydrostatic approximation to relate the pressures at points β_w and γ in Fig. 1.3c, and thus we have

$$p(\beta_w) = p(\gamma) - \rho_w g\,dh, \qquad p(\alpha_a) = p(\delta) + \rho_a g\,dh$$

By substituting these relations in (1.6.7), we get

$$p(\alpha_w) - p(\gamma) = p(\delta) - p(\beta_a) + dx\,T\frac{\partial^3 h}{\partial x^3} - (\rho_w - \rho_a)g\,dh$$

Since

$$p(\alpha_w) - p(\gamma) = -dx\,\partial p_w/\partial x \qquad \text{and} \qquad p(\delta) - p(\beta_a) = -dx\,\partial p_a/\partial x$$

the above equation becomes

$$[\partial p/\partial x]_0 = T\,\partial^3 h/\partial x^3 + g[\rho]_0\,\partial h/\partial x \tag{1.6.8}$$

where the notation of (1.6.3) is used.

By relating the horizontal pressure gradients to the horizontal accelerations of the air or water, and by subtracting the two momentum equations, we get

$$[\partial p/\partial x]_0 = [-\rho\,du/dt]_0$$

When $u = \bar{U}(z) + u'$ is substituted, and the right hand side of the above equation is linearized, we get

$$\left[\frac{\partial p}{\partial x}\right]_0 = -\left[\rho\left(\frac{\partial}{\partial t} + \bar{U}\,\frac{\partial}{\partial x}\right)u' + \rho w'\frac{\partial \bar{U}}{\partial z}\right]_0$$

$$= -\mathrm{Re}\left[\rho(\bar{U}-c)ik\hat{u} + \rho\hat{w}\,\frac{\partial \bar{U}}{\partial z}\right]_0 e^{ik(x-ct)}$$

$$= \mathrm{Re}\left[\rho(\bar{U}-c)\,\frac{\partial \hat{w}}{\partial z} - \rho\hat{w}\,\frac{\partial \bar{U}}{\partial z}\right]_0 e^{ik(x-ct)} \tag{1.6.9}$$

where the continuity equation has been used in obtaining the last line. We now combine (1.6.9), (1.6.8), and (1.6.5) to obtain

$$\left[\rho(\bar{U}-c)\,\frac{\partial \hat{w}}{\partial z} - \rho\hat{w}\,\frac{\partial \bar{U}}{\partial z}\right]_0 = \left(\frac{g[\rho]_0 - k^2 T}{\bar{U}(0)-c}\right)\hat{w}(0) \tag{1.6.10}$$

and this provides the relation for connecting the values of $\partial\hat{w}/\partial z$ on the two sides of the water interface.

The corresponding condition at $z = H$ can be most simply obtained by setting (1.6.9) equal to zero, and thus we have

$$[(\bar{U}-c)\,\partial\hat{w}/\partial z - \hat{w}\,\partial\bar{U}/\partial z]_H = 0 \tag{1.6.11}$$

Equations (1.6.10) and (1.6.11) and $[\hat{w}]_0 = 0 = [\hat{w}]_H$ provide the connection conditions for the piecewise irrotational perturbations.

Since $\nabla^2 w' = 0$ in each of these regions, the ordinary differential equation for the eigenfunctions (1.6.1) is $d^2\hat{w}/dz^2 - k^2\hat{w} = 0$. By convention we take k to be a positive number, and therefore the solutions satisfying

$$\hat{w}(\pm\infty) = 0 = [\hat{w}]_0 = [\hat{w}]_H$$

are

$$\hat{w}(z)/\hat{w}(0) = e^{kz}, \qquad z \leqslant 0 \tag{1.6.12}$$

$$\hat{w}(z)/\hat{w}(0) = \cosh kz + A\sinh kz, \qquad 0 \leqslant z \leqslant H \tag{1.6.13}$$

$$\hat{w}(z)/\hat{w}(0) = e^{-k(z-H)}(\cosh kH + A\sinh kH), \qquad z \geqslant H \tag{1.6.14}$$

where the constants of integration A and $\hat{w}(0)$ are determined as follows.

When Eqs. (1.6.13) and (1.6.14) are used in (1.6.11), we obtain (1.6.15). Likewise, (1.6.16) is obtained from (1.6.10) by using (1.6.12) and (1.6.13), and thus we have the two relations

$$A = -\frac{1 + \tanh kH - (U/kH(U-c))}{1 + \tanh kH - ((U\tanh kH)/kH(U-c))} \tag{1.6.15}$$

$$A = \frac{\rho_w}{\rho_a} - \frac{U}{kcH} - \frac{g(\rho_w - \rho_a) + Tk^2}{\rho_a kc^2} \tag{1.6.16}$$

Subtraction then yields the following cubic equation for the phase speed:

$$1-\frac{g(\rho_w-\rho_a)+Tk^2}{kc^2\rho_w}=\left(\frac{\rho_a}{\rho_w}\right)\frac{U}{kcH}$$

$$-\left(\frac{\rho_a}{\rho_w}\right)\frac{1+\tanh kH-(U/kH(U-c))}{1+\tanh kH-((U\tanh kH)/kH(U-c))} \quad (1.6.17)$$

Since the air–water density ratio ($\rho_a/\rho_w = 10^{-3} \ll 1$) is small, we will use this approximation in the following discussion of the c roots. In the limiting case of $\rho_a = 0$, (1.6.17) degenerates into a quadratic equation, whose roots $c = c_0$ correspond to the phase speed of *free* gravity–capillary waves. The positive root

$$c_0(k) = ((g + Tk^2/\rho_w)/k)^{1/2} \quad (1.6.18)$$

has a minimum at wave number

$$k = k_{0m} = (g\rho_w/T)^{1/2} \quad (1.6.19)$$

and the corresponding value of $c = c_{0m}$ is

$$c_{0m} = \sqrt{2}(gT/\rho_w)^{1/4} \quad (1.6.20)$$

The question now arises as to whether the c root of (1.6.17) can have a small imaginary part when ρ_a/ρ_w is small but finite. If k, $c_0(k)$, U, and H are such that the denominator of the last term in (1.6.17) does *not* vanish, then we can compute c by means of a simple power series having the form

$$c = c_0(k) + c_1'(k)(\rho_a/\rho_w) + c_2'(\rho_a/\rho_w)^2 + \ldots \quad (1.6.21)$$

In this case, the expansion coefficients (c_1', c_2', . . .) are determined by substituting (1.6.21) in (1.6.17) and by equating terms having like powers of ρ_a/ρ_w. But such a procedure is bound to yield *real* values of c_1', c_2', . . . and therefore the air–water interaction will only modify the *phase* speed of free waves.

The situation is entirely different, however, when the denominator in (1.6.17) vanishes, or

$$1 + \tanh kH - \frac{U \tanh kH}{kH(U - c_0(k))} = 0 \quad (1.6.22)$$

because the last term in (1.6.17) then becomes large, even for small ρ_a/ρ_w. Thus, the simple power series (1.6.21) does not converge, and we must seek a different asymptotic expansion in the vicinity of those wave numbers that satisfy (1.6.22). Accordingly, we try an expansion having the form

$$c = c_0(k) + c_1(k)(\rho_a/\rho_w)^{1/2} + \ldots \quad (1.6.23)$$

where $c_1(k)$, . . . are the undetermined coefficients of the expansion in $\frac{1}{2}$-integral powers of ρ_a/ρ_w. When (1.6.23) is substituted in (1.6.17), and when (1.6.18) is noted, we find that the largest nonvanishing terms are proportional to $(\rho_a/\rho_w)^{1/2}$. By collecting all such terms we obtain the quadratic equation

$$\left(\frac{2c_0^2}{c_0^3}\right)c_1 = \left(\frac{1}{c_1}\right)\frac{1+\tanh kH-(U/kH(U-c_0))}{(U\tanh kH/kH(U-c_0)^2)}$$

for evaluating c_1. Simplification of this with the aid of the subsidiary condition (1.6.22) then gives

$$\begin{aligned}\frac{2c_1^2}{U^2} &= \frac{c_0}{U}\left(\frac{U-c_0}{U}\right)\frac{1+\tanh kH-(U/kH(U-c_0))}{(U/kH(U-c_0))\tanh kH}\\ &= \frac{c_0}{U}\left(\frac{U-c_0}{U}\right)\frac{(\tanh kH)-1}{\tanh kH}\\ &= -\frac{c_0}{U}\,\frac{1-\tanh kH}{kH(1+\tanh kH)}\end{aligned} \tag{1.6.24}$$

Since c_0, U, and k have been assumed to be positive, we see that c_1 is pure imaginary, and the positive imaginary root of (1.6.24) corresponds to an amplifying surface gravity wave. Thus, the basic flow is unstable, *provided* (1.6.22) can be satisfied jointly with (1.6.18), and in order to investigate this consistency condition, we first write (1.6.22) as

$$\frac{U}{c_0(k)} = \left(1-\frac{\tanh kH}{kH(1+\tanh kH)}\right)^{-1} \tag{1.6.25}$$

Since the right hand side of (1.6.25) increases from unity (when $kH \gg 1$) to an infinite value (when $kH \to 0$), and since c_0 has a minimum value (1.6.20), it follows that (1.6.25) cannot be satisfied if $U < c_{0m}$. Thus, if $U < c_{0m}$, the simple series (1.6.21) will converge for all k, and consequently no gravity wave will be amplified. On the other hand, if U exceeds the minimum speed of free gravity–capillary waves, or

$$U \geqslant \sqrt{2}(Tg/\rho_w)^{1/4} \simeq 20 \text{ cm/sec} \tag{1.6.26}$$

then (1.6.25) will certainly be satisfied for some k if H is large, or

$$H \gg 1/k_{0m} \tag{1.6.27}$$

The latter condition merely implies that the vertical thickness of the atmospheric shear layer must be large compared to the wavelength (~centimeter) of gravity–capillary waves, and this restriction is trivial. We therefore conclude that if the wind speed assumes the critical value (1.6.26) then only the wave $k = k_{0m}$ will satisfy (1.6.25), and consequently capillary waves (k_{0m}) amplify at slightly greater wind speeds. The longer gravity waves can be amplified at wind speeds much greater than the critical value (1.6.26). The phase speed of all these amplifying waves is essentially equal to their corresponding value (1.6.18) in the absence of the wind.

The Kelvin–Helmholtz instability is a different species, and this mode will be isolated by considering the discontinuous profile obtained by letting

$$H \to 0 \tag{1.6.28}$$

and also by setting $T = 0$ in (1.6.17). Although the largest terms in (1.6.17) are of order $1/H$, the sum of all such terms vanishes, and thus the leading term in the expansion of (1.6.17) in powers of (1.6.28) is found to be

$$1 - \frac{g(\rho_w - \rho_a)}{\rho_w k c^2} = \frac{\rho_a}{\rho_w}\left[\frac{U-c}{c} - \frac{U}{c}\frac{U-c}{c} + \ldots + O(kH)\right]$$

or

$$c^2\left(1 + \frac{\rho_a}{\rho_w}\right) - 2Uc\left(\frac{\rho_a}{\rho_w}\right) + \frac{\rho_a}{\rho_w}U^2 - \frac{g(\rho_w - \rho_a)}{\rho_w k} = 0 \tag{1.6.29}$$

The roots of this quadratic equation are complex when

$$U^2 > g(\rho_w{}^2 - \rho_a{}^2)/k\rho_w\rho_a \tag{1.6.30}$$

and we conclude that short waves (large k) will always amplify when the basic velocity profile has a jump discontinuity. The reader can show that the phase speed of these waves is entirely different from the speed of free gravity waves.

References

Phillips, O.M. (1966). "The Dynamics of the Upper Ocean." Cambridge Univ. Press, London.

Miles, J. W. (1962). On the Generation of Surface Waves by Shear Flows, Part 4. *J. Fluid Mech.* **13**, 433–448.

CHAPTER II

Rotating Fluids

2.1 Review of Coriolis Force

If a fluid is in a state of solid body rotation about an axis fixed in space, and if $2\pi/\omega$ denotes the time for one complete revolution, then the absolute azimuthal speed of a parcel is $V_0 = \omega R$, where R is the perpendicular distance from the parcel to the axis. If $\mathbf{k}$ denotes a unit vector along the axis, and $\mathbf{r}_0$ the vector distance of the parcel from an origin 0 on the axis, then the vector velocity is $\mathbf{V}_0 = \boldsymbol{\omega} \times \mathbf{r}_0$, where $\boldsymbol{\omega} = \omega\mathbf{k}$ is the angular velocity vector. If $R\, d\theta\, dR$ denotes a polar element of area in a plane perpendicular to $\mathbf{k}$, then the "circulation" of V_0 about the perimeter of that area is readily computed to be $\partial/\partial R\ [\omega R(R\, d\theta)]$, and the circulation per unit area is therefore 2ω. From Stokes's theorem, we then obtain $\nabla \times \mathbf{V}_0 = 2\boldsymbol{\omega}$ as the vorticity of a fluid in solid body rotation.

By applying similar kinematic considerations to a triad $[\mathbf{i}, \mathbf{j}, \mathbf{k}]$ of cartesian unit vectors which rotate about the origin 0 with constant angular velocity $\boldsymbol{\omega}$, we find that the velocity of their endpoints is given by

$$d\mathbf{i}/dt = \boldsymbol{\omega} \times \mathbf{i}, \qquad d\mathbf{j}/dt = \boldsymbol{\omega} \times \mathbf{j}, \qquad d\mathbf{k}/dt = \boldsymbol{\omega} \times \mathbf{k}$$

Consequently, if $[x(t), y(t), z(t)]$ denote the $[\mathbf{i}, \mathbf{j}, \mathbf{k}]$ coordinates of a moving parcel, and if

$$\mathbf{V} = [u, v, w] = [dx/dt, dy/dt, dz/dt]$$

denotes the relative velocity, then the absolute velocity of the parcel is

$$\frac{d\mathbf{r}}{dt} = \mathbf{i}\frac{dx}{dt} + \mathbf{j}\frac{dy}{dt} + \mathbf{k}\frac{dz}{dt} + x\frac{d\mathbf{i}}{dt} + y\frac{d\mathbf{j}}{dt} + z\frac{d\mathbf{k}}{dt}$$

$$= \mathbf{i}u + \mathbf{j}v + \mathbf{k}w + (x\omega \times \mathbf{i} + y\omega \times \mathbf{j} + z\omega \times \mathbf{k})$$

$$= \mathbf{V} + \boldsymbol{\omega} \times \mathbf{r}$$

A second differentiation and the use of a similar procedure give the following expression for the absolute acceleration:

$$\frac{d^2\mathbf{r}}{dt^2} = \mathbf{i}\frac{du}{dt} + \mathbf{j}\frac{dv}{dt} + \mathbf{k}\frac{dw}{dt} + (u\boldsymbol{\omega} \times \mathbf{i} + v\boldsymbol{\omega} \times \mathbf{j} + w\boldsymbol{\omega} \times \mathbf{k}) + \boldsymbol{\omega} \times \frac{d\mathbf{r}}{dt}$$

$$= \frac{d\mathbf{V}}{dt} + (\boldsymbol{\omega} \times \mathbf{V}) + \boldsymbol{\omega} \times [\boldsymbol{\omega} \times \mathbf{r} + \mathbf{V}]$$

$$= \frac{d\mathbf{V}}{dt} + 2\boldsymbol{\omega} \times \mathbf{V} + \boldsymbol{\omega} \times [\boldsymbol{\omega} \times \mathbf{r}]$$

where $d\mathbf{V}/dt = \mathbf{i}u + \mathbf{j}v + \mathbf{k}w$ denotes the relative acceleration in the coordinate system that rotates with angular velocity $\boldsymbol{\omega}$. The Coriolis acceleration is the term $2\boldsymbol{\omega} \times \mathbf{V}$, and the centripetal acceleration is $\boldsymbol{\omega} \times (\boldsymbol{\omega} \times \mathbf{r})$. Since the centrifugal acceleration of magnitude $\omega^2 R$ is directed from the parcel toward the axis of rotation, we have $\boldsymbol{\omega} \times (\boldsymbol{\omega} \times \mathbf{r}) = -\nabla(\omega^2 R^2/2)$. Therefore, when $d^2\mathbf{r}/dt^2$ is set equal to the sum of the pressure gradient and gravity forces, the equations of motion in the rotating system can be written as

$$d\mathbf{V}/dt + 2\boldsymbol{\omega} \times \mathbf{V} = -\nabla\phi \qquad \text{and} \qquad \nabla \cdot \mathbf{V} = 0 \tag{2.1.1}$$

where

$$\phi = (p/\rho) - (\omega^2 R^2/2) + gz^*(x, y, z)$$

and where gz^* denotes the gravitational potential function in the x, y, z coordinates of the rotating system.

From (2.1.1), we see that a uniformly rotating ($\mathbf{V} = 0$) fluid must have a uniform value of ϕ, and since p is constant on the free surface, the equation for the shape of that surface is given by $gz^*(x, y, z) - \omega^2 R^2/2 = \text{constant}$. For the case of a laboratory fluid rotating about a vertical z axis, the potential function is merely $z^* = z$, and therefore the parabolic height of the free surface is given by $z = \omega^2 R^2/2g$ plus a constant.

The determination of the equilibrium shape of the free surface for the case of a self-gravitating spheroid like the earth is a more complicated problem, because the gravity potential gz^* varies with latitude as well as with the distance from the center of the earth. While referring the reader to a geodosy text for a proper treatment of the subject, we can make a rough estimate of the difference $\Delta R = R_e - R_p$ between the equatorial radius R_e and the polar radius of the surface of the ocean. In this estimate, we neglect the variation of gz^* with latitude, and thus the difference in gravitational potential at the surface of the polar and equatorial ocean is $gz_e{}^* - gz_p{}^* \sim g\,\Delta R$. The corresponding difference between the centrifugal potentials is $\omega^2 R_e^2$, and the constancy of ϕ then implies

$$\Delta R \sim \omega^2 R_e{}^2/2g \sim 20 \text{ km}$$

The order of magnitude of this variation in the height of the sea is much larger than any "oceanographic" effect, and thus we can consider the geoids (level surfaces) as given and fixed.

2.2 Rotational Rigidity and Inertia Waves

Before considering the effect of the Coriolis force on the kinematics and dynamics of rapidly rotating fluids, we take note of some "peculiar" properties of rapidly rotating solids. Thus, we recall that gyroscopes and bicycles are stabilized by the rotation of their parts and are thereby able to resist forces that would otherwise tilt them or cause them to fall. Also, a flexible body like a chain or a string is endowed with a kind of transverse rigidity when whirled about one of its endpoints. In order to spin faster, a rotating ice skater must do work as he brings his arms closer to his body. We shall now show that a similar kind of resistance to deformation can occur in a rotating fluid, in consequence of which material columns of fluid lying parallel to the axis of rotation tend to remain parallel when subjected to forces that would otherwise "bend" the column.

Let us first examine the behavior of a rotating fluid when disturbed by an infinitesimal velocity field $\mathbf{V}$. Thus, we linearize (2.1.1) by replacing $d\mathbf{V}/dt$ with $\partial\mathbf{V}/\partial t$. Let $\phi'(x, y, z, t)$ denote the perturbed value of ϕ in a cartesian system with the z axis pointing in the direction of $\boldsymbol{\omega}$, and let $f = 2|\boldsymbol{\omega}|$ denote the Coriolis parameter. The x, y, z components of relative velocity are again denoted by u, v, w, and the corresponding components of $2\boldsymbol{\omega} \times \mathbf{V}$ are then given by $-fv, fu, 0$. The linearized equations of motion are

$$\partial u/\partial t - fv = -\partial\phi'/\partial x \qquad (2.2.1)$$

$$\partial v/\partial t + fu = -\partial\phi'/\partial y \qquad (2.2.2)$$

$$\partial w/\partial t = -\partial\phi'/\partial z \qquad (2.2.3)$$

$$\partial u/\partial x + \partial v/\partial y + \partial w/\partial z = 0 \qquad (2.2.4)$$

The elimination of v in (2.2.1) and (2.2.2) leads to the first equation in (2.2.5), and the elimination of u in (2.2.1) and (2.2.2) leads to the second equation in (2.2.5), or

$$\left(\frac{\partial^2}{\partial t^2}+f^2\right)u=-\frac{\partial^2\phi'}{\partial x\,\partial t}-f\frac{\partial\phi'}{\partial y} \quad \text{and} \quad \left(\frac{\partial^2}{\partial t^2}+f^2\right)v=-\frac{\partial^2\phi'}{\partial y\,\partial t}+f\frac{\partial\phi'}{\partial x} \tag{2.2.5}$$

The horizontal velocity components are now eliminated from (2.2.5) by taking the x derivative of the first equation and the y derivative of the second equation, and then adding the results. Using (2.2.4) and simplifying, we have

$$\left(\frac{\partial^2}{\partial t^2}+f^2\right)\frac{\partial w}{\partial z}=\left(\frac{\partial^2}{\partial x^2}+\frac{\partial^2}{\partial y^2}\right)\frac{\partial\phi'}{\partial t} \tag{2.2.6}$$

When (2.2.3) is used to eliminate ϕ', the result is

$$\left(\frac{\partial^2}{\partial t^2}+f^2\right)\frac{\partial^2 w}{\partial z^2}+\left(\frac{\partial^2}{\partial x^2}+\frac{\partial^2}{\partial y^2}\right)\frac{\partial^2 w}{\partial t^2}=0 \tag{2.2.7}$$

In an unbounded rotating fluid, (2.2.7) has the plane wave solution

$$w=\exp i(kx+ly+mz+\Omega t)$$

where k, l, and m are arbitrary wave numbers, and the frequency obtained by substitution in (2.2.7) is

$$\Omega=f(1+k^2/m^2+l^2/m^2)^{-1/2} \tag{2.2.8}$$

The wave fronts will be parallel to the rotation axis and w will be independent of z when $m=0$. In this case, (2.2.8) reduces to $\Omega=0$, thereby implying a vanishing of the rotational restoring force for parcels displaced parallel to the axis. On the other hand, when $k=l=0$, the wave fronts are perpendicular to the axis, $\Omega=f$, and the restoring force attains its maximum value. In the latter case, (2.2.1) gives $fv=\partial u/\partial t$, so that u and v are 90° out of phase. The x and y displacements of a material parcel, obtained by integrating u and v, are also 90° out of phase, and therefore the trajectory of a material parcel is a (inertia) circle.

The restoring force discussed above has no counterpart in a nonrotating fluid. Thus, if an impulsive torque be applied to a rotating fluid, the parcels will, in general, oscillate back and forth with an amplitude that decreases as the Coriolis parameter f increases. We have seen, however, that the frequency of free plane waves varies according to the angle between the wave fronts and the axis, and therefore the restoring force on displaced parcels is anisotropic. This property is now illustrated further by examining the deformations in a thin dye streak which is introduced as a tracer. If the line of dye is *not* parallel to the rotation axis, then it will be greatly distorted by a velocity perturbation which *is* parallel to the axis, since no restoring force acts on such perturbations. A velocity disturbance of the latter type will *not,* however, change the linear shape of the dye if the latter is oriented parallel to the axis. The line of dye will only be stretched. Moreover, the same vertical line of dye will suffer only small changes in shape when the perturbation velocity has an oblique orientation to the rotation axis, because of the rotational restoring force that acts on such perturbations. By applying these qualitative considerations to a material column of fluid lying parallel to the axis of rotation, we can conclude that such a column will tend to remain nearly parallel to the axis when subject to torques that ordinarily would "bend" the column. This rotational constraint, sometimes called the Taylor–Proudman theorem, is illustrated again below, and then the more general dynamical consequences are examined.

Consider a thin layer of rotating fluid bounded by infinite horizontal surfaces at $z = 0$ and $z = H$. Let a rigid obstacle of infinitesimal height $\hat{h}$ be placed on the lower ($z = 0$) surface and then towed with uniform speed U in the $-x$ direction. For simplicity, we assume that the height of the obstacle varies sinusoidally in x with wavelength $2\pi/k$, so that its horizontal motion induces the vertical velocity

$$w(x, z = 0^+, t) = \hat{h}Uk \sin k(x + Ut)$$

just above ($z = 0^+$) the moving obstacle. This given field of vertical velocity and $w(x, H, t) = 0$ then provide the boundary conditions for the solution of (2.2.7). Accordingly, we find that the forced linear response is

$$w(x, z, t) = \hat{w}(z) \sin k(x + Ut)$$

where

$$\frac{d^2\hat{w}}{dz^2} + \frac{(k^2U^2)k^2}{f^2 - U^2k^2}\hat{w} = 0 \tag{2.2.9a}$$

and

$$\hat{w}(0) = \hat{h}Uk, \qquad w(H) = 0 \tag{2.2.9b}$$

Equation (2.2.9a) has exponential or sinusoidal solutions, depending on the sign of $f^2 - U^2k^2$. For slowly rotating ($f \to 0$) fluids, the disturbance $\hat{w}(z)$ decreases exponentially with z, and thus the horizontal velocity varies considerably with z. In the rapidly rotating case, however, the limiting form of (2.2.9a) is $d^2\hat{w}/dz^2 = 0$, and thus w varies linearly with z. Equation (2.2.6) then implies that ϕ' is independent of z, and (2.2.5) implies that u and v are also independent of z, or

$$0 = \partial\phi'/\partial z = \partial u/\partial z = \partial v/\partial z = \partial^2 w/\partial z^2 \tag{2.2.10}$$

The independence of the horizontal velocity u, v on z implies that a material column of fluid will remain vertical as it passes over (or around) a moving obstacle.

As mentioned above, the validity of (2.2.10) depends on the coefficient of w in (2.2.9a). The relevant nondimensional value of that coefficient is obtained by introducing z/H as the nondimensional vertical coordinate in (2.2.9a), and thus the coefficient is

$$(kH)^2/((f/Uk)^2-1) \tag{2.2.11}$$

Equations (2.2.10) are only valid when (2.2.11) is much less than unity, a condition which can arise if either the aspect ratio kH or the *Rossby number* Uk/f is small. The aspect ratio measures the relative value of the vertical depth and horizontal length scale. The Rossby number measures the relative value of inertial and Coriolis forces.

Since the validity condition (2.2.11) does not depend on the amplitude of the perturbation, we expect that the principle of vertical rigidity (2.2.10) may have a more general range of applicability than is indicated by the particular (linear) example examined above. Accordingly, we now consider a similar problem, but one in which the effect of nonlinearity is retained.

2.3 The Conservation of Potential Vorticity

Let us start with the same geometry as in Section 2.2, viz., a homogeneous layer of fluid between two horizontal boundaries. These boundaries may now, however, have finite amplitude irregularities, and we let $h(x, y, t) \leqslant H$ denote the separation distance measured parallel to the rotation vector $\frac{1}{2}f\mathbf{k}$. If L denotes a typical horizontal scale of either the variability of the boundaries or the field of motion, then we will assume

$$H/L \ll 1 \tag{2.3.1}$$

Since this aspect ratio is small, we expect the principle of vertical rigidity to hold, whereby thin vertical columns of fluid will remain vertical, although the height h changes as the column moves. For such motions, the relative vector velocity

$$\mathbf{V} = \mathbf{V}_2(x, y, t) + w(x, y, z, t)\mathbf{k}$$

has the property

$$0 = \partial \mathbf{V}_2/\partial z = \partial \phi/\partial z = \partial^2 w/\partial z^2 \tag{2.3.2}$$

where $\mathbf{V}_2$ is the horizontal component of $\mathbf{V}$ and w is the vertical component. In making this separation of $\mathbf{V}$, we also find it convenient to introduce

$$\nabla_2 \equiv \mathbf{i}\,\partial/\partial x + \mathbf{j}\,\partial/\partial y$$

as a symbol for the horizontal part of the differential operator ∇. In this notation, the continuity equation can be written as

$$\partial w/\partial z = -(\partial u/\partial x + \partial v/\partial y) = -\nabla_2 \cdot \mathbf{V}_2 \tag{2.3.3}$$

Since $\partial \mathbf{V}_2/\partial z = 0$, the horizontal component of acceleration simplifies to

$$d\mathbf{V}_2/dt = \partial \mathbf{V}_2/\partial t + (\mathbf{V}_2 \cdot \nabla_2)\mathbf{V}_2 + w\, \partial \mathbf{V}_2/\partial z = \partial \mathbf{V}_2/\partial t + (\mathbf{V}_2 \cdot \nabla_2)\mathbf{V}_2 \qquad (2.3.4a)$$

and the corresponding component of (2.1.1) then becomes

$$d\mathbf{V}_2/dt + f\mathbf{k} \times \mathbf{V}_2(x, y, t) = -\nabla_2 \phi(x, y, t) \qquad (2.3.4b)$$

where $f\mathbf{k} = 2\boldsymbol{\omega}$. We note that the vertical component of (2.1.1), or $dw/dt = -\partial\phi/\partial z$, is not exactly compatible with (2.3.2), except in the uninteresting case $w \equiv 0$. In the next two paragraphs, however, we shall justify the use of the "hydrostatic" approximation

$$\partial\phi/\partial z = 0 \qquad (2.3.4c)$$

in a class of problems where the vertical acceleration dw/dt is negligibly small, but the vertical velocity [in (2.3.3)] is not.

Let us therefore estimate the error in (2.3.4b) which arises because of the approximation (2.3.4c). The exact equation $\partial\phi/\partial z = -dw/dt$ can be integrated to obtain the magnitude of the vertical variation in ϕ, and thus.we have

$$\int_0^H dz\, \frac{\partial \phi}{\partial z} \sim H\left|\frac{dw}{dt}\right|$$

Associated with this nonhydrostatic ϕ component, we have a horizontal pressure gradient of order

$$\nabla_2(H\, dw/dt) \qquad (2.3.5)$$

The order of magnitude of the vertical velocity obtained from (2.3.3) is $w \sim H\, \nabla_2 \cdot \mathbf{V}_2$, and we also note that

$$\nabla_2 \cdot \mathbf{V}_2 \sim |\mathbf{V}_2|/L \qquad (2.3.6a)$$

is an upper bound for the order of magnitude of $\nabla_2 \, . \, \mathbf{V}_2$. Consequently, the nonhydrostatic pressure force (2.3.5) is less than or equal to

$$|\nabla_2 H\, dw/dt| \sim |\nabla_2 (H^2/L)\, d\mathbf{V}_2/dt| \sim H^2/L^2 |d\mathbf{V}_2/dt| \ll |d\mathbf{V}_2/dt| \qquad (2.3.6b)$$

and is therefore negligible compared to the left hand side of (2.3.4b). Thus, the horizontal accelerations produced by the vertically averaged component of pressure are much larger than the nonhydrostatic effect, and (2.3.4) is a consistent asymptotic set in the same sense as the familiar "shallow water equations" (cf. Chapter III) that apply to a nonrotating fluid with a free surface.

Equations (2.3.4) are also valid, however, even when $H/L \sim 1$ ("nonshallow") provided the Rossby number is small, or

$$|\mathbf{V}_2|/fL \ll 1 \qquad (2.3.7)$$

and this proposition is important for laboratory experiments, wherein it is inconvenient to simulate the small aspect ratio that characterizes the geophysical prototype. In order to prove this proposition, we note that (2.3.7) implies that the Coriolis force is much larger than the inertial force, or $d\mathbf{V}_2/dt \ll f\mathbf{k} \times \mathbf{V}_2$ in (2.3.4b). As a first approximation to the horizontal balance of forces, we then obtain the *geostrophic* equation

$$f\mathbf{k} \times \mathbf{V}_2 = -\nabla_2 \phi \tag{2.3.8}$$

or

$$fv = \partial\phi/\partial x \qquad \text{and} \qquad fu = -\partial\phi/\partial y$$

and when ϕ is eliminated by cross differentiation, we get $\partial u/\partial x + \partial v/\partial y = 0$. For this case, we see that (2.3.6a) is a gross overestimate of the value of $\nabla_2 \cdot \mathbf{V}_2$, and of the associated value of w. Although each of the two terms $\partial u/\partial x$ and $\partial v/\partial y$ in $\nabla_2 \cdot \mathbf{V}_2$ may be of order $\mathbf{V}_2 L^{-1}$, their sum is smaller by a factor of (2.3.7). In the present case then, the value of (2.3.5) is smaller than the estimate given in (2.3.6b) by a factor of (2.3.7). Thus, we see that even when $H/L \sim 1$, the effect of the vertical acceleration in (2.3.4b) is still negligible compared to the $d\mathbf{V}_2/dt$ term. It is therefore legitimate to use (2.3.4b) to investigate the small (but very important) nongeostrophic effects that arise in the *second* approximation to the equations of motion. We shall return to this point in subsequent chapters on geostrophic flow, and for the time being, we only want to emphasize the region of validity of the hydrostatic dynamics, as given by Eqs. (2.3.4).

Having justified the proposition that vertical columns remain vertical, we can now apply the circulation theorem (1.1.9) to the motion of a material cylinder of height $h(x, y, t)$ and of infinitesimal cross-sectional area $\delta A(x, y, t)$. Thus $\zeta\, \delta A$ is conserved in the motion, where

$$\zeta = 2\omega + \zeta_r$$

is the absolute vorticity and

$$\zeta_r = \mathbf{k} \cdot \nabla_2 \times \mathbf{V}_2 = \partial v/\partial x - \partial u/\partial y$$

is the relative vorticity. Since the volume, or $h\, \delta A$, of the material cylinder is also independent of time, we see that ζ/h is independent of time, or

$$0 = \frac{d}{dt}\left(\frac{\zeta_r + 2\omega}{h}\right) = \left(\frac{\partial}{\partial t} + u\frac{\partial}{\partial x} + v\frac{\partial}{\partial y}\right)\frac{\zeta_r + 2\omega}{h} \tag{2.3.9}$$

or

$$\frac{d\zeta_r}{dt} = \frac{f + \zeta_r}{h}\frac{dh}{dt} \tag{2.3.10}$$

The quantity dh/dt equals the difference of the vertical velocity w at the two endpoints of the column, and therefore $\partial w/\partial z = h^{-1}\, dh/dt$. From (2.3.3) we then get

$$h^{-1}\, dh/dt = \partial w/\partial z = -(\partial u/\partial x + \partial v/\partial y) \tag{2.3.11}$$

or, equivalently,

$$\partial h/\partial t + \partial(uh)/\partial x + \partial(vh)/\partial y = 0 \tag{2.3.12}$$

An alternate form of the vorticity equation (2.3.10) is

$$d\zeta_r/dt = (2\omega + \zeta_r)\, \partial w/\partial z \tag{2.3.13}$$

and an alternate (analytical) derivation of (2.3.13) can be obtained from (2.3.4b) by eliminating ϕ, and by using (2.3.3). The quantity $(\zeta_r + f)/h$ is called the potential vorticity, and the above equations express the conservation of potential vorticity.

2.4 Illustrative Examples

(a) Consider a long channel having a vertical height H and a width L in the y direction. A liquid flows horizontally with uniform speed U at the entrance of the channel. Suppose there is a contracting section at some downstream position, so that the vertical height of the channel decreases from H to a value of $H - M$ at the downstream end of the contraction. If the channel is not rotating, then we readily see that the current at the downstream end will increase to the constant value of $UH/(H - M)$.

If the channel is rotating with angular velocity $\frac{1}{2}f\mathbf{k}$ about the vertical axis, then the given upstream current U has zero relative vorticity, and therefore the vertical columns of fluid have a *uniform* potential vorticity equal to f/H. These columns will shrink to a height of $H - M$ after they pass through the contraction, and according to (2.3.9), the vorticity must decrease to some value ζ_r. Let $u(y)$ denote the laminar velocity profile far downstream, with the sidewalls of the channel being at $y = \pm L/2$. Thus, the downstream relative vorticity is $\zeta_r = -\partial u/\partial y$. Since the potential vorticity is not only conserved but also uniform, we have

$$\frac{f - \partial u/\partial y}{H - M} = \frac{f}{H} \quad \text{and} \quad \frac{\partial u}{\partial y} = \frac{fM}{H} \tag{2.4.1}$$

The solution of (2.4.1) satisfying the mass transport condition

$$(H - M) \int_{-L/2}^{+L/2} u\, dy = UHL$$

is

$$u(y) = y\, fM/H + UH/(H - M) \tag{2.4.2}$$

Therefore, the minimum downstream velocity will occur at $y = -L/2$, and this velocity will vanish when

$$(M/H)((H-M)/M) = 2U/fL \tag{2.4.3}$$

When the channel contraction M satisfies (2.4.3), the flow will stagnate at one wall, and the subsequent modification of the laminar regime is such that the theory given above will no longer apply.

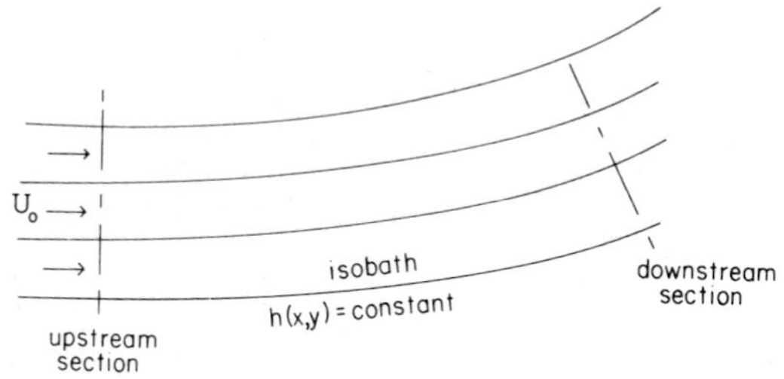

FIG. 2.1 Topographic steering of a current in a rapidly rotating fluid. We are given a uniform upstream current U_0 in the region where the isobaths are straight and parallel, and the flow at the downstream section is to be computed.

(b) We now consider the effect on a current produced by a transverse variation in the height h of the rigid bottom (Fig. 2.1). The solid curves of constant bottom elevation are called isobaths, and we assume the latter are straight and parallel at the upstream end of Fig. 2.1. In this region, the given horizontal velocity $U_0$of the fluid is constant and parallel to the isobaths, so that the relative vorticity vanishes. The problem is to determine the streamlines in the region where the isobaths curve. Because of inertia, we might expect that a vertical column of fluid will retain its linear path as it starts to cross the curved isobaths. But this would alter h in (2.3.9), thereby requiring a compensating change in ζ_r, and thus the horizontal streamlines must start to curve as soon as they reach the region in which the isobath curves. The relative vorticity will obviously depend on U_0/R, where R is the local radius of curvature of the isobath. In the limiting case $f \gg U_0R$ we can neglect ζ_r compared with $2\omega = f$ in (2.3.9). Consequently, the first approximation to the vorticity equation is $(d/dt)(f/h) = 0$, and therefore the vertical height of a column is conserved. Thus, the horizontal trajectory of a parcel is identical (to a first approximation) with the corresponding isobath. By proceeding to the second approximation (to the vorticity equation), we can estimate the extent to which a streamline will depart from the isobath on which it was initially located. As mentioned above, the order of magnitude of the induced vorticity is $\zeta_r \sim U_0/R$, and from (2.3.9), we can show that the compensating change δh in the thickness of the material column is of order

$$\delta h/h \sim U_0/fR$$

Such changes in the height of material columns are brought about by a crossing of the isobaths, and we conclude that the streamlines will follow the isobaths to a greater extent as the Rossby number U_0/fR decreases. In the ocean, the corresponding effects of bottom topography can have an important effect in steering currents (Warren, 1963).

2.5 The β Effect in a Rotating Spherical Annulus

We have shown that the relative velocity perpendicular to the axis of rotation tends to be independent of distance measured parallel to the axis, whereas the component of velocity parallel to the axis can vary linearly. As a first approximation to the vorticity equation, we have also shown that the thickness h of a column tends to remain constant, but we now examine the restoring force which arises when a column is forced to change its height. Suppose that the fluid located between concentric spherical boundaries (Fig. 2.2)

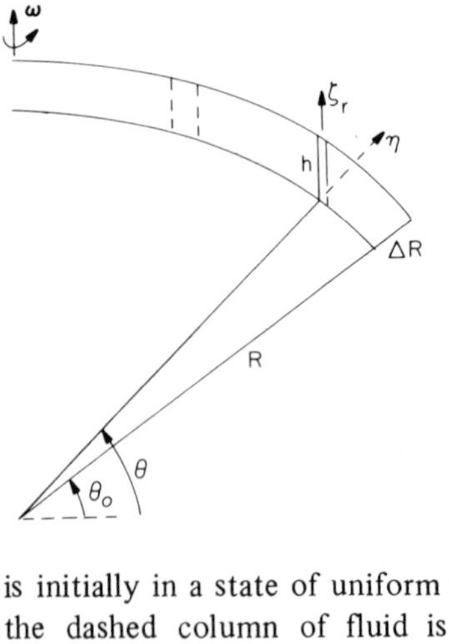

FIG. 2.2 A section of a thin spherical annulus. A column of fluid, indicated by the dashed lines, is displaced southward to latitude θ. The height h of the fluid is measured in a direction parallel to the rotation vector, ω. ζ_r is the relative vorticity in the same direction, whereas η is the component along the local normal. ΔR is the radial thickness of the annulus.

is initially in a state of uniform rotation ω about the North Pole, and suppose the dashed column of fluid is then displaced southward. If the gap width $\Delta R \ll R$ is small, then we see that the slopes of the two boundaries at the ends of the column are nearly equal.† Because of this boundary condition, the

† By means of a sketch, the reader can easily see that this geometrical approximation breaks down in a thin latitudinal zone on either side of the equator, and therefore the following theory does not apply in that region.

principle of vertical rigidity requires that all *three* components of relative velocity must be independent of distance measured parallel to ω. Since the radial velocity component vanishes on inner and outer spheres, it must therefore vanish inside the spherical annulus. The total relative velocity then consists of a component v directed toward higher latitudes θ and a component u directed eastward toward greater longitudes ϕ.

For such two-dimensional motions in a *thin* spherical gap, the continuity equation takes the form

$$\partial u/\partial\phi + \partial(v \cos\theta)/\partial\theta = 0 \tag{2.5.1}$$

as may be seen by computing the u and v fluxes through a surface area element of latitudinal length $R\,d\theta$ and longitudinal length $R\cos\theta\,d\phi$.

When the law of conservation of potential vorticity is applied to the column shown in Fig. 2.2, we see that $(\zeta_r + 2\omega)/h$ is conserved, where

$$\zeta_r = \eta/\sin\theta$$

is the component of relative vorticity along the polar axis, whereas η is the component along the local normal. From the geometry of Fig. 2.2, we also have

$$h = \Delta R/\sin\theta$$

and therefore the potential vorticity equals

$$\frac{2\omega + (\eta/\sin\theta)}{\Delta R/\sin\theta}$$

Thus the conservation of potential vorticity can be written as

$$(d/dt)(2\omega\sin\theta + \eta) = 0 \tag{2.5.2}$$

or

$$d\eta/dt + v\beta = 0 \tag{2.5.3}$$

where

$$\beta = (2\omega\cos\theta)/R \tag{2.5.4}$$

and $v = R\,d\theta/dt$ is the northward component of velocity. Thus, we see that a northward velocity of the column causes a compensating decrease in the radial component of relative vorticity. This represents the important dynamical effect of the curvature of the earth, and secondary kinematical effects associated with the spherical geometry are usually (Veronis, 1973) eliminated by means of the following "tangent plane" approximation.

If x measures distance eastward and if y measures distance northward, then the approximate continuity equation

$$\partial u/\partial x + \partial v/\partial y = 0 \tag{2.5.5}$$

follows from (2.5.1), provided that the typical horizontal scale L of the motion is so small compared to the radius R that we can treat the $\cos\theta$ term in (2.5.1) as constant. Thus we are now using a local Cartesian coordinate system in which the x, y plane is tangent to the sphere in the region of interest. The corresponding Cartesian approximation for the component of relative vorticity normal to the local x, y plane is

$$\eta = \partial v/\partial x - \partial u/\partial y \tag{2.5.6}$$

and the Cartesian approximation to (2.5.3) is

$$(\partial/\partial t + u\,\partial/\partial x + v\,\partial/\partial y)\eta + v\beta = 0 \tag{2.5.7}$$

Equations (2.5.5)–(2.5.7) are a complete set, and they can also be derived analytically from the momentum equations, when the appropriate kinematical approximations are introduced therein. Accordingly, we note that then the radial velocity vanishes, and consequently the Coriolis force has an easterly component $-2\omega\sin\theta v$ and a northerly component $2\omega\sin\theta\, u$. Therefore, the horizontal momentum equations are given by

$$\begin{aligned} \partial u/\partial t + u\,\partial u/\partial x + v\,\partial u/\partial y - (2\omega\sin\theta)v &= -\partial\Phi/\partial x \\ \partial v/\partial t + u\,\partial v/\partial x + v\,\partial v/\partial y + (2\omega\sin\theta)u &= -\partial\Phi/\partial y \end{aligned} \tag{2.5.8}$$

where $\partial\Phi/\partial x$ is the eastward component of the horizontal pressure gradient force. The equivalence of (2.5.8) with (2.5.7) may be established analytically by cross differentiation of (2.5.8) and by using (2.5.5).

In order to illustrate the utility of this β-plane approximation, we now consider the propagation of free waves in a zonal channel of uniform meridional width $L \ll R$ and centered at the mid latitude θ_0. Thus, the β parameter in (2.5.3) can be considered constant, with value $\beta = 2\omega\cos\theta_0/R$. For the case of infinitesimal perturbations from a state of solid body rotation, (2.5.3) can be linearized, and therefore we replace d/dt with $\partial/\partial t$. The use of (2.5.6) in (2.5.3) then gives

$$(\partial/\partial t)(\partial v/\partial x - \partial u/\partial y) + \beta v = 0 \tag{2.5.9}$$

According to (2.5.5), a stream function ψ can be introduced with

$$v = \partial\psi/\partial x \qquad \text{and} \qquad u = -\partial\psi/\partial y$$

so that (2.5.9) becomes

$$(\partial/\partial t)\,\nabla^2\psi + \beta\,\partial\psi/\partial x = 0 \tag{2.5.10}$$

Since $v(x, \pm L/2) = 0$ at the zonal walls, the boundary condition is $\psi(x, \pm L/2) = 0$, and therefore the first normal mode solution of (2.5.10) is

$$\psi = \cos(\pi y/L)\cos(kx - \hat{\omega}t)$$

where $2\pi/k$ is the wavelength in the zonal direction, and the frequency $\hat{\omega}$ relation obtained by substitution in (2.5.10) is

$$\hat{\omega} = -\beta k/(k^2 + \pi^2/L^2) \tag{2.5.11}$$

These *Rossby* waves clearly have negative (westward) phase velocity $\hat{\omega}/k$, but the group velocity

$$\partial\hat{\omega}/\partial k = -\beta \frac{\pi^2/L^2 - k^2}{(k^2 + \pi^2/L^2)^2}$$

is positive for large k. This means that waves with length shorter than $2\pi(\pi/L)^{-1}$ propagate energy to the east, even though phase is propagated to the west. For a given value of $\hat{\omega}$, (2.5.11) has two (or zero) k roots. Therefore, if disturbances of given frequency are generated at a point in the channel (by the application of external forces), then two different sets of waves will be generated, each of which corresponds to one of the k roots mentioned above. The short waves (large k) propagate energy to the east of the generating region, while the long waves propagate to the west.

2.6 Inertial Western Boundary Current

The problem sketched in Fig. 2.3 provides another illustration, and the ideas discussed will be useful in a subsequent discussion of the wind driven ocean

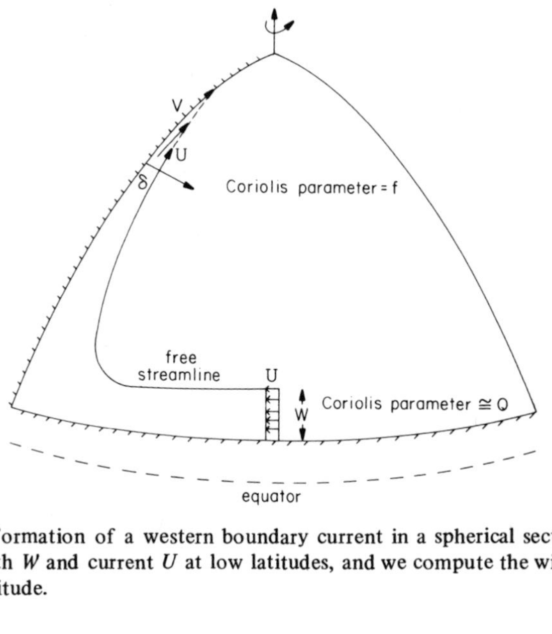

FIG. 2.3 Formation of a western boundary current in a spherical sector. The stream has a given width W and current U at low latitudes, and we compute the width δ of the jet at some high latitude.

circulation. Figure 2.3 is a top view of a bounded sector of a small gap spherical annulus. An isolated low latitude current of uniform westward speed U and width W is forced to move from a low latitude source (not shown) to a high latitude sink (also not shown), and we are required to trace the "free streamline" (vorticity discontinuity) that separates the moving stream from the adjacent water. In the latter region, the horizontal velocity $\mathbf{V}$ vanishes, and (2.1.1) implies that the dynamic pressure ϕ is constant.

By deriving a "Bernoulli equation" (below) we will show that $\mathbf{V}^2$ remains constant along the free streamline as the current moves northward. Taking the scalar product of (2.1.1) with $\mathbf{V}$, and noting $\mathbf{V} \cdot 2\boldsymbol{\Omega} \times \mathbf{V} = 0$, we get $d/dt(\mathbf{V}^2/2) = -\mathbf{V} \cdot \nabla\phi$. For steady state conditions, we have $\mathbf{V} \cdot \nabla\phi \equiv d\phi/dt$, and therefore

$$(d/dt)(\tfrac{1}{2}\mathbf{V}^2 + \phi) = 0 \tag{2.6.1}$$

Thus, the *Bernoulli* function $\frac{1}{2}\mathbf{V}^2 + \phi$ is conserved on any steady streamline and, in particular, along the free streamline. Since the pressure field must be continuous, and since ϕ is everywhere constant on the resting side of the free streamline, we conclude that $\mathbf{V}^2$ must be constant along the free streamline.

The northward deflection of the current (Fig. 2.3) by the western boundary produces an increase in the value of the Coriolis parameter $2\omega \sin\theta$, and therefore the relative vorticity of a parcel must also decrease, according to (2.5.2). Since the relative vorticity, as well as the Coriolis parameter at low latitudes, is small compared to the Coriolis parameter f at high latitudes, (2.5.2) implies that the relative vorticity of the current at the high latitude is essentially equal to $\eta = -f$. The only way in which such a large increase in relative vorticity can occur, with a given mass transport, is by means of a decrease in the width of the stream and a corresponding increase in the velocity along the coast. Thus, at high latitudes, the streamlines are essentially parallel to the western boundary. Let δ denote the width of the jet measured normal to the western boundary at the latitude associated with f, and let V denote the speed at the longitude of the western boundary. Since U is the speed on the free streamline, the essentially uniform vorticity across the stream is $\eta = (U - V)/\delta$, and therefore

$$(V - U)/\delta = f$$

The volume transport per unit depth at the high latitude is $\frac{1}{2}(V + U)\delta$, and the conservation of mass requires

$$\tfrac{1}{2}(V + U)\delta = UW \equiv \hat{T}$$

The simultaneous solution of the two equations gives

$$V = (2fUW)^{1/2}\,[1 + (U/2fW)]^{1/2}$$

The interesting case for discussion occurs when $U/2fW \ll 1$, in which case we have

$$V = (2f\hat{T})^{1/2} \tag{2.6.2}$$

and also

$$\delta = (2\hat{T}/f)^{1/2} \tag{2.6.3}$$

Thus, the width of the western boundary current (2.6.3) depends only on the total transport and the latitude.

2.7 Another Western Boundary Current

The following discussion of a laboratory experiment (Stommel *et al.*, 1958) will also prove useful in later (Chapter XII) work. In the pie-shaped basin of Fig. 2.4 the radius is a, the angular width is ϕ_0 radians, and the system rotates

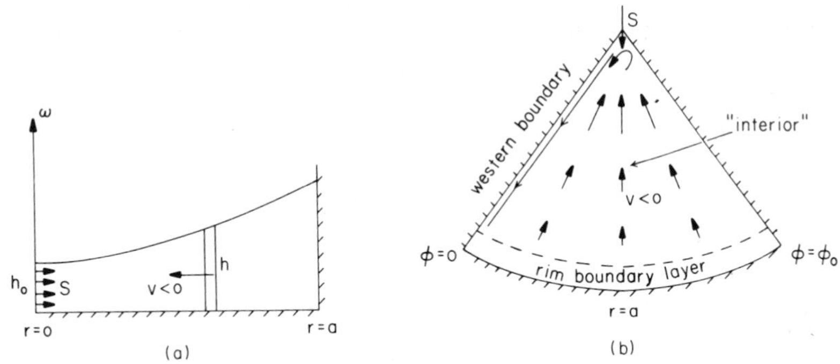

FIG. 2.4 Flow out of a source region S in a rotating pie-shaped basin. The water entering at the apex slowly raises the height h of the free surface, as shown by the vertical sectional view (a). The horizontal flow pattern is shown in the plan view (b).

about a vertical axis through the apex ($r = 0$) with angular velocity ω. S cm^3/sec of water are forced into the basin through a source region (such as a thin vertical porous pipe) inserted at $r = 0$. For small S, the vertical height $\bar{h}$ will depart only slightly from the equilibrium parabola

$$\bar{h} = h_0(t) + (\omega^2 r^2/2g) \tag{2.7.1}$$

where $h_0(t)$ is the instantaneous height at $r = 0$. When the uniform rate of rise

$\partial h_0/\partial t$ of the free surface is multiplied by the horizontal area $a^2\phi_0/2$, we obtain the rate of volume increase S, or

$$\partial h_0/\partial t = S/(a^2\phi_0/2) \tag{2.7.2}$$

The way in which the incoming water fills up the basin is, however, most unusual compared to that of the nonrotating case.

As the water emerges from the source region, the Coriolis force deflects the radial velocity into the azimuthal direction, and thus the inflow proceeds immediately to the "western" boundary, as shown in Fig. 2.4b. This relatively thin current then flows rimward along the radial wall. When the current reaches the rim ($r = a$), it is again deflected in the azimuthal direction. Only then does the fluid enter the "interior" region where it proceeds to fill up the main part of the basin. Why?

Let us first note that the law of conservation of potential vorticity (2.3.9) applies to any material column of fluid, whether it is bounded by rigid surfaces (Section 2.3) or by a free surface (Fig. 2.4). (See the following chapter for a systematic discussion of the dynamics of a fluid having a free surface.) Although this law may not apply to the fluid in the thin boundary layers (because of the importance of viscous forces), it does apply to the columns as soon as they enter the "interior region." Moreover, if S is small, then the relative vorticity ζ_r will also be small in the interior. In this region, then, the first approximation to (2.3.9) is $dh/dt = 0$. This implies that the vertical velocity vanishes in the interior although the basin is filling up! The paradox is removed when one considers that the rise in the local height of the free surface is produced by fluid columns leaving the rim ($r = a$) and conserving their height as they march inward toward relatively shallow water. If v denotes the radial component of velocity, then the linearization of $dh/dt = 0$ (for small S) yields

$$\partial\bar{h}/\partial t + v\,\partial\bar{h}/\partial r = 0 \tag{2.7.3}$$

where $\bar{h}(r, t)$ is given by (2.7.1). From (2.7.1)–(2.7.3), we then obtain

$$vr = -2gS/\phi_0\omega^2a^2 \tag{2.7.4}$$

This (2.7.4) radial flow is readily seen to be axisymmetric, irrotational ($\zeta_r = 0$), and nondivergent. Since the vertical component of velocity vanishes in the interior, the continuity equation then implies that the azimuthal velocity $u(r, \phi)$ must also be nondivergent, or $\partial u/\partial\phi = 0$. The solution of the latter equation that satisfies the "eastern" boundary condition, $u(r, \phi_0) = 0$, is $u(r, \phi) = 0$, and thus we conclude that the motion in the interior region is purely radial. The total amount of water transported radially inward by the interior solution (2.7.4) is

$$T(r) = -r\phi_0 v(r)\bar{h}(r) = \bar{h}(r)2gS/\omega^2a^2 = ((r^2/a^2) + (2gh_0/\omega^2a^2))S \tag{2.7.5}$$

Although the above solution is not valid near the apex ($r = 0$), we must now account for the transport, $T(0) = 2gh_0S/\omega^2a^2$, which is given by (2.7.5) as

$r \to 0$. Accordingly, we assume that this inward directed flux is added to the outward directed source flux S in an "apex boundary layer," and the sum $S + 2gh_0S/\omega^2a^2$ is then transported radially outward in the western boundary layer. Since this sum is precisely equal to the value of $T(a)$ in (2.7.5), we see that no water leaves the western boundary current until it reaches the rim. This conclusion agrees with the nondivergent velocities, $u(r, \phi) = 0$, which we found for the interior region of the basin.

The Coriolis force associated with the radial velocites in the interior must be balanced by an azimuthal pressure gradient $g/r\ \partial h'/\partial\phi$, where $h'(r, \phi)$ is the small departure of the height of the free surface from the parabola (2.7.1). Thus, the hydrostatic and geostrophic force balance requires

$$-2\omega v = (g/f)\ \partial h'/r\,\partial\phi$$

and we may solve for h' by substituting (2.7.4) in the above equation. Integrating in the ϕ direction, we find that

$$\Delta h' = 4S/\omega^2a^2 \qquad (2.7.6)$$

is the amount by which the water level on the eastern boundary exceeds that near the western boundary. Since this azimuthal height variation was neglected in (2.7.1), we see that the foregoing theory is only valid when (2.7.6) is small compared to (2.7.1), or when $4S/\omega^2a^2h_0 \ll 1$.

References and Supplementary Reading

Greenspan, H. P. (1968). "The Theory of Rotating Fluids." Cambridge Univ. Press, London and New York.

Stommel, H. M. (1965). "The Gulf Stream," 2nd ed. Univ. of California Press, Berkeley.

Stommel, H. M., Arons, A. B., and Faller, A. J. (1958). Some Examples of Stationary Planetary Flows. *Tellus* **10,** 179–187.

Veronis, G. (1973). Large Scale Ocean Circulation. *Adv. Appl. Mech.* **13,** 1–92.

Warren, B. A. (1963). Topographic Influences on the Gulf Stream. *Tellus* **15,** 167–183.

CHAPTER III

Density Currents

3.1 Review of Nonrotating Shallow Water Theory

Let $z = h(x, y, t)$ denote the height of the free surface of a liquid relative to the nonrotating bottom ($z = 0$), let $w^*(x, y, z, t)$ denote the vertical component of velocity, $\mathbf{V}_2^*$ the horizontal component, p^* the pressure field, and ∇_2 the horizontal part of the nabla operator. By using the upper boundary condition $p^*(x, y, h, t) = 0$ in the vertical component of (1.1.3), we obtain the integral

$$\rho^{-1} p^*(x, y, z, t) = g(h-z) + \int_z^h dz\, dw^*/dt \qquad (3.1.1)$$

The horizontal component of (1.1.3) is

$$d\mathbf{V}_2^*/dt = -\rho^{-1} \nabla_2 p^* \qquad (3.1.2)$$

and the continuity equation is

$$w^* = -\int_0^z dz\, \nabla_2 \cdot \mathbf{V}_2^* \qquad (3.1.3)$$

The right hand side of (3.1.2) is given by the hydrostatic approximation

$$\rho^{-1}\,\nabla_2 p^* = g\,\nabla_2 h(x, y, t) \tag{3.1.4a}$$

when the remaining term

$$\nabla_2 \int_z^h dz\, dw^*/dt \tag{3.1.4b}$$

obtained from the differentiation of (3.1.1) is small. If L denotes the horizontal length scale of $\mathbf{V}_2{}^*$, and if $h \sim H$, then the order of magnitude of (3.1.4b) is $H/L\, dw_*/dt$. Since (3.1.3) gives $w_* \sim (H/L)\mathbf{V}_2$, we see that (3.1.4b) is of order $H^2/L^2\; d\mathbf{V}_2{}^*/dt$. This term is negligible compared to the left hand side of (3.1.2) for a shallow layer, and therefore the approximate momentum equation is

$$d\mathbf{V}_2{}^*/dt = -g\,\nabla_2 h \tag{3.1.5}$$

Since the right hand side of (3.1.5) is independent of z, we will require $\mathbf{V}_2{}^*(x, y, t)$ to be independent of z, and therefore (3.1.3) is a linear function of z. At $z = h$, $w^* = dh/dt$ and (3.1.3) becomes

$$dh/dt = -h\,\nabla_2 \cdot \mathbf{V}_2{}^* \tag{3.1.6}$$

These equations provide a complete set for the determination of h and $\mathbf{V}_2$, and the reader is referred to Rouse (1961) and Benjamin (1962) for applications of this theory. The shallow water theory is generalized in the following section, and applications to rotating fluids follow.

3.2 The Hydrostatic Approximation in a Rotating Fluid

Suppose that the rotation vector, $\omega\mathbf{k} \equiv (f/2)\mathbf{k}$, and the gravity vector $-g\mathbf{k}$ are parallel. The equation for the horizontal component of relative velocity $\mathbf{V}_2(x, y, t)$ is then obtained by adding the Coriolis force $f\mathbf{k} \times \mathbf{V}_2$ to the left hand side of (3.1.2) and the centrifugal term $\nabla\omega^2 R^2/2$ to the right hand side, where $R(x, y)$ again (Section 2.1) denotes the distance of a point from the rotation axis. In the following paragraph, we show how the centrifugal term combines with the horizontal pressure gradient term $\nabla_2 p/\rho$ to give Eqs. (3.2.4) and (3.2.5).

In the absence of relative motion ($\mathbf{V}_2 = 0$), the pressure will be uniform on the parabolic level surface, $z = \omega^2 R^2/2g$, and we now examine the pressure difference between two nearby points (x, y, z) and $(x + \delta x, y, z + \delta z)$ on the same level surface when $\mathbf{V}_2 \neq 0$. Since the local slope of the parabola is $\partial(\omega^2 R^2/2g)/\partial x$, the vertical separation of the two points is

$$\delta z = \delta x(\partial/\partial x)(\omega^2 R^2/2g)$$

and the pressure difference is therefore

$$p(x+\delta x, y, z+\delta z)-p(x, y, z) = \frac{\partial p}{\partial x}\,\delta x + \frac{\partial p}{\partial z}\,\delta z = \delta x\left[\frac{\partial p}{\partial x} + \left(\frac{\partial p}{\partial z}\right)\frac{\partial}{\partial x}\,\frac{\omega^2 R^2}{2g}\right]$$

Further simplification can be obtained by using the hydrostatic approximation for $\partial p/\partial z$, and also by introducing the notation

$$\lim_{\delta x \to 0} \frac{p(x+\delta x, y, z+\delta z)-p(x, y, z)}{\delta x} = \left(\frac{\partial p}{\partial x}\right)_*$$

for the "horizontal pressure gradient on a constant level surface." We then have

$$\rho^{-1}(\partial p/\partial x)_* = \rho^{-1}\,\partial p/\partial x - (\partial/\partial x)(\omega^2 R^2/2)$$

By performing the same calculation for the y direction, we see that the relation between the pressure gradient on a level surface and the gradient on a constant z surface is

$$\rho^{-1}(\nabla_2 p)_* = \rho^{-1}\,\nabla_2 p - \nabla_2(\omega^2 R^2/2) \tag{3.2.1}$$

Since the sum of $d\mathbf{V}_2/dt$ and the Coriolis acceleration is equal to the negative value of the right hand side of (3.2.1), the horizontal momentum equation becomes

$$d\mathbf{V}_2/dt + f\mathbf{k} \times \mathbf{V}_2 = -\rho^{-1}(\nabla_2 p)_* \tag{3.2.2}$$

If $\eta(x, y, t)$ denotes the height of the free surface above a specified level surface, then $\rho g \eta$ gives the pressure variation thereon, and

$$(\nabla_2 p)_* = \rho g\,\nabla\eta(x, y, t)$$

is the pressure gradient on the constant level surface. Therefore, (3.2.2) becomes

$$(d\mathbf{V}/dt) + f\mathbf{k} \times \mathbf{V}(x, y, t) = -g\,\nabla\eta \tag{3.2.3}$$

and the subscript "2" has been discarded with the understanding that $\mathbf{V}(x, y, t)$ denotes the horizontal velocity in all that follows. The shallow water continuity equation [cf. (3.1.6)] is given by

$$dh/dt = -h\,\nabla \cdot \mathbf{V}$$

where h is the total vertical height of a fluid column, and dh/dt is the difference between the vertical velocity at the free and bottom surfaces.

If the rigid bottom is also a parabolic level surface, then $\nabla\eta = \nabla h$, and the two preceding equations then yield the complete set

$$(d\mathbf{V}/dt) + f\mathbf{k} \times \mathbf{V}(x, y, t) = -g\,\nabla h \tag{3.2.4}$$

$$(dh/dt) + h\,\nabla \cdot \mathbf{V} = 0 \tag{3.2.5}$$

3.3 Conservation Theorems

The law of conservation of potential vorticity (Section 2.3) can be immediately derived for the free surface case by noting that the volume $h\,\delta A$ of a material column is conserved, where δA is the cross-sectional area measured in a plane perpendicular to the axis of rotation. The circulation theorem also requires that $(f+\zeta)\,\delta A$ be conserved where

$$\zeta = \mathbf{k}\cdot\nabla\times\mathbf{V}(x, y, t)$$

is the vertical component of relative vorticity. Therefore, $(f+\zeta)/h$ is conserved, or

$$(d/dt)((f+\zeta)/h) = 0 \tag{3.3.1}$$

This can also be derived by taking the curl of (3.2.4), and by using (3.2.5) in the simplification.

Another useful (Bernoulli) invariant can be derived from (3.2.4) when the flowfield is steady, so that $dh/dt \equiv \mathbf{V}\cdot\nabla h$. Thus, we find that the scalar product of $\mathbf{V}$ with (3.2.4) gives

$$(d/dt)(\mathbf{V}^2/2 + gh) = 0 \qquad (3.3.2$$

Therefore, $\mathbf{V}^2/2 + gh$ is conserved on any streamline.

Although (3.3.2) does not apply to unsteady flows, we can derive a more general integral which expresses the conservation of kinetic plus potential energy. Thus, the scalar product of (3.2.4) with $h\mathbf{V}$ gives the first equation listed below while the second equation is obtained by multiplying (3.2.5) with h.

$$(h/2)\,d\mathbf{V}^2/dt = -(g/2)\mathbf{V}\cdot\nabla h^2$$

$$(g/2)\,dh^2/dt = -gh^2\,\nabla\cdot\mathbf{V} \equiv -\nabla\cdot(g\mathbf{V}h^2) + g\mathbf{V}\cdot\nabla h^2$$

The sum of these equations

$$\tfrac{1}{2}(d/dt)(h\mathbf{V}^2 + gh^2) - \tfrac{1}{2}\mathbf{V}^2\,dh/dt = -\nabla\cdot(g\mathbf{V}h^2) + (g/2)\mathbf{V}\cdot\nabla h^2 \tag{3.3.3}$$

can be further simplified by using (3.2.5) to express the second term in (3.3.3) as

$$-\tfrac{1}{2}\mathbf{V}^2\,dh/dt = \tfrac{1}{2}h\mathbf{V}^2\,\nabla\cdot\mathbf{V} = \tfrac{1}{2}\nabla\cdot(h\mathbf{V}^2\mathbf{V}) - \tfrac{1}{2}\mathbf{V}\cdot\nabla(h\mathbf{V}^2)$$

and so (3.3.3) becomes

$$\tfrac{1}{2}(d/dt)(h\mathbf{V}^2 + gh^2) - \tfrac{1}{2}\mathbf{V}\cdot\nabla(h\mathbf{V}^2 + gh^2) = -\nabla\cdot(\tfrac{1}{2}h\mathbf{V}^2\mathbf{V} + g\mathbf{V}h^2)$$

or

$$\tfrac{1}{2}(\partial/\partial t)(h\mathbf{V}^2 + gh^2) = -\nabla\cdot[\mathbf{V}(h\mathbf{V}^2/2 + gh^2)] \tag{3.3.4}$$

Here the first term is the rate of change of the vertically integrated kinetic energy, since the contribution of the vertical velocity is negligible in the shallow water theory. The potential energy of the vertical column is proportional to the vertical integrals of $gz\,dz$, or $gh^2/2$. Consequently, the left hand side of (3.3.4) gives the rate of change of total energy density. On the right hand side of (3.3.4), $\mathbf{V}(h\mathbf{V}^2/2)$ represents a flux of kinetic energy by the horizontal velocity, $\mathbf{V}(gh^2/2)$ is the flux of potential energy, and the remaining $\mathbf{V}(gh^2/2)$ represents the effect of "pressure work" in generating energy. If the fluid is bounded by vertical walls so that the normal component of $\mathbf{V}$ vanishes thereon, then the application of Gauss's theorem to (3.3.4) implies that the complete horizontal integral of $\frac{1}{2}(h\mathbf{V}^2 + gh^2)$ is independent of time.

If there are no bounding walls, and if we then have an isolated liquid "lens" whose free surface intersects the bottom on the curve $h(x, y, t) = 0$, then it is easily shown that the integration of (3.3.4) over the entire lens also yields the energy invariant

$$(\partial/\partial t) \oint\!\!\oint_{\text{lens}} dx\,dy\,(\tfrac{1}{2}h\mathbf{V}^2 + \tfrac{1}{2}gh^2) = 0 \tag{3.3.5}$$

because the integration can be commuted with $\partial/\partial t$ on the bounding perimeter $(h = 0)$ of the lens.

3.4 Inertia–Gravity and Kelvin Waves

As a first application of the foregoing theory, we consider the propagation of free surface waves in a rotating cylindrical annulus that is bounded by vertical walls and by a level (parabolic) bottom. Let L_0 denote the uniform radial gap width, and H the uniform vertical thickness of the undisturbed rotating equilibrium state. Let $h'(x, y, t) = h - H$ denote an infinitesimal height perturbation, and $\mathbf{V}(x, y, t)$ the associated horizontal velocity perturbation. Thus, we may linearize (3.2.4) and (3.2.5), and express the result as

$$(\partial\mathbf{V}/\partial t) + f\mathbf{k} \times \mathbf{V} = -g\,\nabla h' \qquad \text{and} \qquad (\partial h'/\partial t) + H\,\nabla \cdot \mathbf{V} = 0$$

The boundary conditions require that the normal component of $\mathbf{V}$ vanish on the annular walls.

It is not difficult to solve the foregoing equations in the cylindrical polar coordinate system which is appropriate to the annular geometry, but we prefer to use a "small gap approximation", which is sufficient for present purposes and indispensable in more complicated problems encountered later. According to this approximation, we can neglect the curvature of the annulus when L_0 is much smaller than the mean radius. Thus, if the cross channel direction is denoted by x, the azimuthal direction by y, and if u' and v' denote the x and y

components of $\mathbf{V}$, then the cartesian approximations to the equations given above are

$$\partial u'/\partial t - fv' = -g\,\partial h'/\partial x,\quad \partial v'/\partial t + fu' = -g\,\partial h'/\partial y,\quad \partial h'/\partial t + H(\partial u'/\partial x + \partial v'/\partial y) = 0 \tag{3.4.1}$$

The first equation in (3.4.2) is obtained by eliminating v' between the first two equations in (3.4.1), while the second equation in (3.4.2) is obtained by the elimination of u', and thus we have

$$\left(\frac{\partial^2}{\partial t^2} + f^2\right)u' = -g\frac{\partial^2 h'}{\partial x\,\partial t} - gf\frac{\partial h'}{\partial y} \quad \text{and} \quad \left(\frac{\partial^2}{\partial t^2} + f^2\right)v' = -g\frac{\partial^2 h'}{\partial y\,\partial t} + gf\frac{\partial h'}{\partial x} \tag{3.4.2}$$

By taking the x derivative of the first equation in (3.4.2), by taking the y derivative of the second equation, by adding the two results, and by using the last equation in (3.4.1), we get

$$-\frac{1}{H}\left(\frac{\partial^2}{\partial t^2} + f^2\right)\frac{\partial h'}{\partial t} = -g\,\nabla^2\frac{\partial h'}{\partial t} \tag{3.4.3}$$

At $x = (0, L_0)$, the boundary condition is $u' = 0$, and by using the first equation in (3.4.2), this condition is expressed as

$$f\,\partial h'/\partial y + \partial^2 h'/\partial x\,\partial t = 0 \qquad \text{on} \quad x = 0, \quad x = L_0 \tag{3.4.4}$$

Let us now consider the normal modes of (3.4.3) and (3.4.4). If $2\pi/k$ denotes the length of a wave in the y direction, and c the phase speed, then there are solutions of the form

$$h' = (Ae^{\gamma x} + Be^{-\gamma x})\sin k\,(y - ct) \tag{3.4.5}$$

where $A, B,$ and γ are constants which are determined as follows. The substitution of (3.4.5) in (3.4.3) yields

$$\gamma = +\left[\frac{f^2}{gH} + k^2\left(1 - \frac{c^2}{gH}\right)\right]^{1/2} \tag{3.4.6}$$

and the substitution of (3.4.5) into the two equations (3.4.4) gives

$$A(f - c\gamma) + B(f + c\gamma) = 0 \quad \text{and} \quad Ae^{\gamma L_0}(f - c\gamma) + Be^{-\gamma L_0}(f + c\gamma) = 0 \tag{3.4.7}$$

The eigenvalues c are obtained by setting the determinant of (3.4.7) equal to zero and by using (3.4.6) to eliminate γ.

We can readily see that $f - c\gamma = 0$ is one solution of the determinantal equation, and the use of (3.4.6) then gives

$$c = +(gH)^{1/2} \tag{3.4.8}$$

as the phase speed of this *Kelvin wave.* Since $f - c\gamma = 0$ for this mode, Eq. (3.4.7) implies $B = 0$, Eq. (3.4.5) becomes

$$h' = A\,[\exp xf(gH)^{-1/2}]\ \sin\{k(y - t(gH)^{1/2})\}$$

and substitution in the first of Eqs. (3.4.2) shows that u' vanishes identically. The $f + c\gamma = 0$ root of the determinantal equation yields a second Kelvin wave with $c = -(gH)^{1/2}$. Each of these two waves is trapped to, and decreases exponentially from, one of the two channel walls. Since (3.4.8) is independent of k, the group velocity equals the phase velocity of the Kelvin wave.

When $A \neq 0 \neq B$ in (3.4.7), the determinant will vanish only if $\exp \gamma L_0 = \exp -\gamma L_0$, or $\gamma L_0 = n\pi i$, where n is an integer. By substituting $\gamma = n\pi i L_0^{-1}$ in (3.4.6), the expression for the phase speed of these *inertia–gravity* waves is found to be

$$\frac{c^2}{gH} = 1 + \left[\frac{(f^2/gH) + (n\pi/L_0)^2}{k^2}\right] \tag{3.4.9}$$

Since γ is now imaginary, the eigenfunctions for the inertia–gravity waves vary sinusoidally across the annulus, in contrast with the exponential variation of the Kelvin modes. The inertia–gravity waves have larger phase speed (3.4.9) than the Kelvin waves and are also highly dispersive. Thus, we note that when $k \to 0$ in (3.4.9), the frequency kc approaches a constant, the group velocity vanishes, and therefore long inertia–gravity waves propagate energy very slowly. A comparison of (3.4.8) and (3.4.9) shows that the maximum group velocity occurs for the Kelvin wave, and the latter therefore determines the maximum speed with which a "signal" can be propagated down the rotating channel. The significance of the signal velocity becomes apparent if we consider an analogy with supersonic flow. When a projectile moves more rapidly than the speed of sound, a shock wave forms at the nose. Likewise, a hydraulic jump forms in a nonrotating open channel when the flow exceeds the maximum signal velocity for surface disturbances. The analagous effect in a rotating open channel is considered in the next section.

3.5 Flow through a Rotating Open Channel

The small gap annulus shown in the top view of Fig. 3.1 is the same as that used in Section 3.4, except that one of the side walls is curved so as to present a converging channel to a mean current which flows in the annulus. We are given the free surface height $h(x)$ at the upstream end as a function of the distance x from the straight wall. The corresponding value of the "geostrophic" current

$$v = (g/f)\,dh/dx \tag{3.5.1}$$

follows from (3.2.4), since the upstream flow is laminar and steady, or

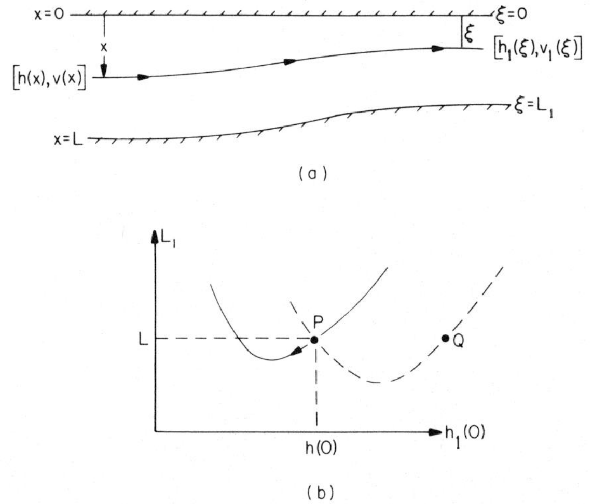

FIG. 3.1 (a) Top view of a flow $v(x)$ through a rotating open channel. We compute the profile $v_1(\xi)$ after the current passes through the converging section. (b) Graph of $h_1(0)$ as a function of L_1. The solid curve corresponds to subcritical upstream flow. See text.

$d\mathbf{V}/dt = 0$. The horizontal velocity vector $\mathbf{V}$ increases, however, in the converging section of the channel, and $d\mathbf{V}/dt \neq 0$ in this region. At a point far downstream from the converging section, the channel flow again approaches a laminar state ($d\mathbf{V}/dt = 0$), and we let $\xi = \xi(x)$ denote the perpendicular distance of a streamline from the straight wall. Let $h_1(\xi)$ denote the interfacial elevation associated with this streamline, and let $v_1(\xi)$ denote the downstream current. These fields are related by a geostrophic relation similar to (3.5.1), or

$$v_1(\xi) = (g/f)\, dh_1(\xi)/d\xi \tag{3.5.2}$$

Although the intervening fluid flow is not geostrophic, the endpoint states can be connected by the Bernoulli and mass conservation equations, as indicated below.

Consider the volume of fluid transported between the straight wall and the streamline shown in Fig. 3.1. At the downstream end, this transport equals the integral of $d\xi'\, v_1(\xi')h_1(\xi')$ from $\xi' = 0$ to $\xi' = \xi$, and by using (3.5.2), we obtain

$$2\int_0^{\xi} d\xi'\, v_1(\xi')h_1(\xi') = (g/f)[h_1^2(\xi) - h_1^2(0)] \tag{3.5.3}$$

The same computation performed at the upstream end yields a transport

$$2\int_0^{x} dx'\, v(x')h(x') = (g/f)[h^2(x) - h^2(0)] \tag{3.5.4}$$

and since the two transports (3.5.3) and (3.5.4) must be equal, we have

$$h_1(\xi) = [h^2(x)+\alpha]^{1/2} \tag{3.5.5a}$$

where

$$\alpha \equiv h_1{}^2(0)-h^2(0) \tag{3.5.5b}$$

A second relation between upstream and downstream parameters is provided by the Bernoulli equation (3.3.2), or $v_1{}^2(\xi) + 2gh_1(\xi) = v^2(x) + 2gh(x)$. By using (3.5.5a) to eliminate h_1, we then get

$$v_1 = [v^2(x)+2gh(x)-2g\{h^2(x)+\alpha\}^{1/2}]^{1/2} \tag{3.5.6}$$

When (3.5.5)–(3.5.6) are substituted in (3.5.2), the differential equation

$$\frac{f}{g}\,d\xi = \frac{d\{h^2(x)+\alpha\}^{1/2}}{[v^2+2gh(x)-2g\{h^2(x)+\alpha\}^{1/2}]^{1/2}} = \frac{(h^2+\alpha)^{-1/2}h(x)\,dh}{[v^2+2gh-2g\{h^2(x)+\alpha\}^{1/2}]^{1/2}} \tag{3.5.7}$$

is obtained, in which we are to solve for $\xi(x)$ when we are given $v(x)$ and $h(x)$. The constant of integration α can be determined by integrating (3.5.7) across the entire channel. Since ξ goes from 0 to L_1 as $h(x)$ goes from $h(0)$ to $h(L)$, the integration of (3.5.7) gives

$$\frac{fL_1}{g} = \int_{h(0)}^{h(L)} \frac{(h^2+\alpha)^{-1/2}h\,dh}{[v^2+2gh-2g\{h^2+\alpha\}^{1/2}]^{1/2}} \tag{3.5.8}$$

The functional relation between α [or $h_1(0)-h(0)$] and L_1-L is now discussed.

From (3.5.5b), we note that $\alpha = 0$ when $h_1(0) = h(0)$, and the integrand in (3.5.8) then reduces to dh/v or to $f\,dx/g$, if (3.5.1) is used. Thus, the right hand side of (3.5.8) reduces to fL/g, and consequently $L_1 = L$ when $\alpha = 0$. We now compute the value of α for small values of (L_1-L) by differentiating (3.5.8) with respect to α. When α is set equal to zero in the result, we have

$$\frac{f}{g}\left(\frac{\partial L_1}{\partial\alpha}\right)_{\alpha=0} = \frac{1}{2}\int_{h(0)}^{h(L)} dh\left[\frac{g}{hv^3}-\frac{1}{h^2v}\right]$$

We also note that the derivative of (3.5.5b) gives $d\alpha = 2h_1(0)\,dh_1(0)$, so that the above equation can be written as

$$\frac{f}{g^2h_1(0)}\frac{\partial L_1}{\partial h_1(0)} = \int_{h(0)}^{h(L)} \frac{dh}{vh}\left[\frac{1}{v^2}-\frac{1}{gh}\right] \tag{3.5.9}$$

For the case of subcritical flow, $v^2(x) < gh(x)$, Eq. (3.5.9) is positive, and the functional relationship between L_1 and $h_1(0)$ is indicated by the solid curve in Fig. 3.1b. The point P having coordinates $L_1 = L$, $h_1(0) = h(0)$ represents the upstream state, and for a converging channel $(L_1 - L < 0)$, the representative point will then move in the direction indicated by the arrow. Thus, we see that when $L_1 < L$, the downstream height $h_1(0)$ is less than the upstream height $h(0)$.

But for the case in which v is so large that (3.5.9) vanishes, or equivalently

$$\int_0^L \frac{dx}{h}\left[\frac{1}{v^2(x)} - \frac{1}{gh(x)}\right] = 0 \tag{3.5.10}$$

the representative point P for the upstream state is at the minimum of the curve (Fig. 3.1b), and therefore no solution exists when the channel width decreases downstream. The upstream flow is *critical* because a discontinuous change, or a hydraulic jump, will occur in this case. *Supercritical* flow occurs when $v^2(x)$ is sufficiently large compared with $h(x)$, so that the slope (3.5.9) of the curve at the representative point P is negative, as shown by the dashed curve in Fig. 3.1b. We expect that such a state will be unstable and will jump towards the state Q, in analogy with nonrotating supercritical flow (Rouse, 1961).

Whitehead *et al.* (1973) have discussed the application of these hydraulic concepts to experiments involving the free discharge of water through a rotating open channel. They also discuss the oceanographic problem posed by the flow through a strait or channel which connects two large basins containing water of slightly different densities. In the following section, we will generalize the shallow water theory, so as to describe the internal currents and waves that occur when two or more fluid layers of slightly different densities are in contact.

3.6 Internal Motions in Layers of Slightly Different Density

If there are three layers (Fig. 3.2) having densities $\rho_0 < \rho_1 < \rho_2$, then the equation of motion in each is given by (3.2.2), and the pressures are connected by the hydrostatic relation. Three level surfaces separated by uniform distances A and B are constructed in Fig. 3.2; p_0, p_1, and p_2 denote the pressures on the respective surfaces, $\eta(x, y, t)$ denotes the vertical distance from the upper interface to the mid-level surface, and $h(x, y, t)$ denotes the total vertical thickness of the intermediate layer. From the hydrostatic relations

$$p_2 - p_0 = \rho_0 g(A - \eta) + \rho_1 gh + \rho_2 g(B - (h - \eta))$$

$$p_1 = p_0 + g\rho_0(A - \eta) + g\rho_1\eta$$

we obtain the pressure gradient forces

$$(\nabla_2 p_2)_* - (\nabla_2 p_0)_* = g(\rho_1 - \rho_2)\,\nabla h + g(\rho_2 - \rho_0)\,\nabla\eta \tag{3.6.1}$$

$$(\nabla_2 p_1)_* = (\nabla_2 p_0)_* + g(\rho_1 - \rho_0)\,\nabla\eta \tag{3.6.2}$$

and we proceed to derive the equations of motion for the intermediate layer when its thickness h is much less than the mean thickness of the upper layer H_0 and the lower layer H_2.

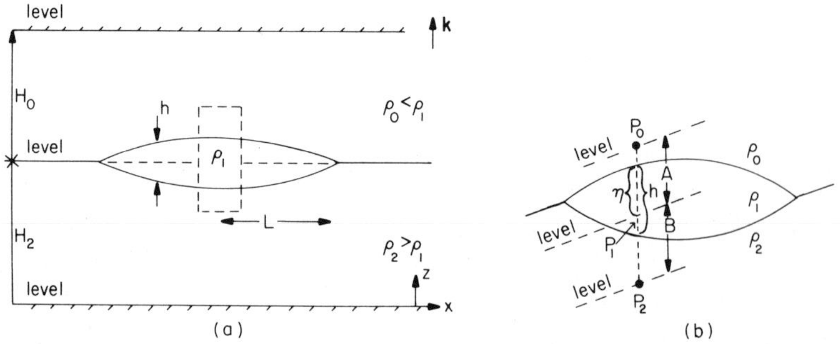

FIG. 3.2 (a) A vertical section of a collapsing lens of liquid of density ρ_1. The initial ($t = 0$) configuration is given by the dashed rectangle (with great vertical exaggeration), and there is no variation in the y direction. The width of the lens increases from $2L_0$ to $2L$ in time $t > 0$. (b) Enlargement of lens used to compute the hydrostatic pressures.

In this limiting case, the value of $\partial w/\partial z$ inside the thin layer will exceed the value outside by the factor $H_0/h \gg 1$, and the continuity equation then implies a correspondingly large value for the horizontal velocity $\mathbf{V}(x, y, t)$ *inside* the relatively thin layer. Likewise, the horizontal acceleration and the horizontal pressure gradients

$$(\nabla_2 p_0)_* \simeq 0 \simeq (\nabla_2 p_2)_* \tag{3.6.3}$$

outside the thin layer are small compared to $(\nabla_2 p_1)_*$. Therefore, (3.6.2) is approximated by

$$(\nabla_2 p_1)_* = g(\rho_1 - \rho_0)\,\nabla\eta$$

(3.6.1) is approximated by

$$0 = g(\rho_1 - \rho_2)\,\nabla h + g(\rho_2 - \rho_0)\,\nabla\eta$$

and thus we have

$$(1/\rho_1)(\nabla_2 p_1)_* = g^*\,\nabla h \tag{3.6.4}$$

where

$$g^* \equiv g((\rho_1 - \rho_0)/\rho_1)(\rho_2 - \rho_1)/(\rho_2 - \rho_0) \tag{3.6.5}$$

By setting the negative value of (3.6.4) equal to the sum of the relative and Coriolis accelerations, we obtain the equation of motion

$$(d\mathbf{V}/dt) + f\mathbf{k} \times \mathbf{V} = -g^* \nabla h \tag{3.6.6}$$

The shallow water continuity equation

$$h^{-1}\, dh/dt + \nabla \cdot \mathbf{V} = 0 \tag{3.6.7}$$

again follows from $\partial w/\partial z + \nabla_2 \cdot \mathbf{V} = 0$ because dh/dt is the difference in w between the two interfacial points which are separated by the vertical distance h.

The quantity g^* in (3.6.5) is called *reduced gravity* because (3.6.6) and (3.6.7) are identical to (3.2.4) and (3.2.5) when the value of g is reduced by the factor

$$((\rho_1 - \rho_0)/\rho_1)((\rho_2 - \rho_1)/(\rho_2 - \rho_0))$$

This factor is zero when ρ_1 equals either ρ_2 or ρ_0, and when ρ_1 lies midway between ρ_0, ρ_2, the factor attains the value of $\frac{1}{2}(\rho_1 - \rho_0)/\rho_1$. Thus, we see that $g^* \ll g$ when the density differences are small. Because of the correspondence between (3.6.6) and (3.2.4), the results obtained previously for the case of the single layer fluid may be carried over to the case of the internal density current.

Let us examine the problem of the collapsing density wedge, which is schematically indicated in Fig. 3.2a. Suppose the initial ($t = 0$) state to be one of rest, with the layer of density ρ_1 having a rectangular cross section of uniform height h_0 and width $2L_0 \gg h_0$. In the case of a nonrotating fluid ($f = 0$), this state is obviously not in equilibrium, and consequently the intermediate layer will spread laterally and collapse vertically, so that it assumes a lens-like shape (Fig. 3.2a) at $t > 0$. The energy equation (3.3.5) will be applied to this nonrotating collapse by replacing g with (3.6.5) and by assuming a two-dimensional motion in the vertical plane.

The kinetic energy vanishes in the initial state ($t = 0$) and the potential energy per unit distance normal to the page of Fig. 3.2 is $g^* h_0^2 L_0$, according to the integral in (3.3.5). In a nonrotating fluid, $h(x)$ will approach zero, and the width of the lens L will approach infinity, as $t \to \infty$. A characteristic collapse time T can be defined as that value of t for which the kinetic energy equals half (say) of the initial energy, or

$$\frac{1}{2} \int_{-L}^{+L} dx\, h\mathbf{V}^2 = \tfrac{1}{2} g^* h_0^2 L_0$$

As the lens collapses, its mass remains constant, and therefore its cross-sectional area has the constant value

$$\int_{-L}^{+L} h\, dx = 2h_0 L_0 \tag{3.6.8}$$

Since the speed of advance of the front of the lens is dL/dt, and since L increases by order L_0 after a time $t \sim T$, we have $dL/dt \sim L_0/T$. The horizontal velocities $\mathbf{V}$ inside the lens are also $|\mathbf{V}| \sim L_0/T$ at $t \sim T$, and therefore the energy equation given above yields the order of magnitude relation

$$(L_0/T)^2 \int_{-L}^{+L} h\,dx \sim g^* h_0^2 L_0$$

By using (3.6.8), the collapse time is then found to be

$$T \sim L_0/(g^* h_0)^{1/2} \tag{3.6.9}$$

The main oceanic thermocline has a typical vertical thickness of ~1 km, and the density changes by a fraction of order 10^{-3}. In a vertical distance of 10 m, the vertical variation of density is $\Delta\rho/\rho_0 = 10^{-5}$. Suppose, then, we use the foregoing model to estimate the collapse time of a lens with the same density factor, with $h_0 = 10^3$ cm, and with a horizontal dimension of (say) $L = 10^5$ cm. Thus, we have $g^* = 10^3 \times 10^{-5}$, and (3.6.9) gives $T \sim 1/\sqrt{3}$ days. Since this collapse time is comparable with the period of the earth, we conclude that the Coriolis force cannot be neglected in the investigation of the collapse of a "mesoscale water mass" having the dimensions cited above. We will now show that this inviscid wedge will not collapse completely in the case of a rotating ($f \neq 0$) fluid, and therefore the state of minimum potential energy will not be reached.

Let us look for an equilibrium solution in the two-dimensional model (Fig. 3.2) such that $d\mathbf{V}/dt = 0$, as $t \to \infty$, and in which the remaining terms of (3.6.6) are finite. The velocity $\mathbf{V}$ is directed normal to the page of Fig. 3.2, with the magnitude $v(x)$ then being determined by the geostrophic force balance

$$v(x) = (g^*/f)\,dh/dx \tag{3.6.10}$$

The cross-sectional area of the final $h(x)$ is related to the uniform thickness h_0 of the initial state by (3.6.8). Furthermore, the final value of the potential vorticity

$$(f + dv/dx)/h \tag{3.6.11}$$

of each column must equal the initial value for the same column. Since the initial vertical thickness of the lens is uniform h_0, and since the initial velocities vanish, the initial value of potential vorticity inside the lens is uniformly equal to f/h_0 and therefore (3.6.11) is uniformly equal to f/h_0. When (3.6.10) is used, we then have

$$f + (g^*/f)\,d^2h/dx^2 = fh/h_0 \quad \text{or} \quad [(d^2/dx^2) - (1/\lambda^2)](h - h_0)/h_0 = 0 \tag{3.6.12}$$

where

$$\lambda = (g^* h_0/f^2)^{1/2} \tag{3.6.13}$$

is the "Rossby radius of deformation" (Rossby, 1938). If the origin ($x = 0$) is taken on the axis of symmetry, then the solution of (3.6.12) which vanishes at the front ($x = \pm L$) of the lens is

$$\frac{h}{h_0} = 1 - \frac{\cosh x/\lambda}{\cosh L/\lambda} \tag{3.6.14}$$

By substituting (3.6.14) into (3.6.8), we then get

$$(L/\lambda) - \tanh(L/\lambda) = L_0/\lambda \tag{3.6.15}$$

and the value of the equilibrium width L is obtained from this and the given L_0, λ. The case of rapid rotation is defined by $\lambda \ll L_0$, and the value of L obtained from (3.6.15) is then only slightly greater than L_0. This means that the initially imbalanced state only spreads a small distance before it comes to geostrophic equilibrium.† This simple calculation indicates the extent to which the constraint of rotation inhibits the spreading of density currents on level surfaces.

Since the geostrophic equlibrium solution obtained above may be unstable (Chapter IV), it is instructive to give another argument which shows that a complete "collapse" is not possible for an *ideal* rotating fluid. Suppose the contrary to be true, so that $h \to 0$ for each fluid column. The conservation of potential vorticity then implies that the absolute vorticity $f + \zeta$ also tends to zero, and therefore the relative vorticity ζ tends to $-f$ *everywhere*. A complete collapse of the lens also implies that the width is $L \to \infty$. Therefore, the relative velocities would have to be large to order $fL \to \infty$. But this contradicts the conservation of energy, since the final kinetic energy can be no larger than the initial potential energy, or the average final speed can be no larger than

$$[g^* h_0^2 L_0 / \int h \, dx]^{1/2} = [g^* h_0^2 L_0 / h_0 L_0]^{1/2} = (g^* h_0)^{1/2}$$

Thus, we see that the assumption of complete collapse is inconsistent, and therefore the water mass can only reach the state of minimum potential energy through the action of dissipative effects (Chapter V).

3.7 Shallow Water Theory on a Rotating Sphere

The first consideration in the development of any shallow water theory is the determination of the "dynamical vertical," this being the direction z along which the horizontal velocity $\mathbf{V}$ is uniform. In the case of the bounded spherical

† It is interesting to compute the total energy in the final state, to show that this is *less* than the energy in the initial state, and to discuss the reasons why this situation is *not* inconsistent with the calculation given above.

annulus (Section 2.5), the force of gravity is not relevant, and the axis of rotation then determines the "dynamical vertical" (with the possible exception of low latitude motions). Another limiting case is provided by a liquid having a free surface and resting on the spherical surface of a *nonrotating* planet. Here the direction of the plumb line will obviously determine the dynamical vertical, for those motions whose horizontal scale R is large compared to the mean layer thickness H. But in the more general case, we have to consider the "competition" between the rotation rate Ω and the local force of gravity $g\mathbf{k}$. In all previous considerations of this chapter, Ω is parallel to $g\mathbf{k}$, and thus the two forces "cooperate" in defining the dynamical vertical. We now want to consider the case where $g\mathbf{k}$ varies with latitude θ, the vector $\mathbf{k}$ being taken normal to the local level surface.

It is plausible that the dynamical vertical will be determined by $g\mathbf{k}$ when the centrifugal acceleration is relatively small, and therefore our starting assumption is $\partial\mathbf{V}/\partial z = 0$, where z measures the normal distance from the bottom ($z = 0$) level surface. If $z = h$ denotes the height of the free surface, then the local horizontal mass transport is given by $h\mathbf{V}$, the free surface vertical velocity is $w^* = dh/dt$, and the continuity relationship connecting the two quantities [cf (2.5.1)] is given by

$$\partial(hu)/R\,\partial\phi) + (\partial(hv \cos\theta)/R\,\partial\theta) + (\partial h/\partial t)\cos\theta = 0 \qquad (3.7.1)$$

where R is the mean radius of the thin layer, v is the northward velocity, and u is the component in the direction of decreasing longitude.

The appropriate form of the vorticity equation is again derived by considering a material column having an infinitesimal horizontal cross-sectional area δA. The absolute circulation around the perimeter of δA is $(f + \zeta)\,\delta A$, where

$$f = 2\Omega \cdot \mathbf{k} = 2\Omega \sin\theta$$

now denotes the projection of the earth's vorticity on the local normal $\mathbf{k}$, and ζ denotes the normal component of relative vorticity. Since the volume $h\,\delta A$ of the column, as well as $(f + \zeta)\,\delta A$, is independent of time, we see that $(f + \zeta)/h$ is independent of time, or

$$(d/dt)(f+\zeta)/h = 0 \qquad (3.7.2a)$$

$$(d/dt)(f+\zeta) = (f+\zeta)/h\; dh/dt \qquad (3.7.2b)$$

$$(d/dt)(f+\zeta) = (f+\zeta)\;\partial w/\partial z \qquad (3.7.3a)$$

$$d\zeta/dt + v\beta = (f+\zeta)\;\partial w/\partial z \qquad (3.7.3b)$$

where $df/dt = v\beta$, and β is the northward gradient of the Coriolis parameter.

For future reference, we list here a special case of (3.7.3b) that is applicable when the velocity and vorticity are so small that (3.7.3b) can be linearized. If, in addition, the flow is steady, then the approximation to (3.7.3b) is

$$v\beta = f\,\partial w/\partial z \tag{3.7.4}$$

We now show that the corresponding momentum equation for the horizontal velocity on the sphere is

$$(d\mathbf{V}/dt) + 2\Omega \sin\theta \mathbf{k} \times \mathbf{V} = -g\,\nabla h \tag{3.7.5}$$

Thus we must justify the neglect of the vertical component of the Coriolis force

$$2\mathbf{\Omega} \times (\mathbf{V} + w\mathbf{k}) \tag{3.7.6}$$

in the computation of the hydrostatic pressure gradient

$$\nabla_2(p/\rho) = g\,\nabla h(x, y, t) \tag{3.7.7}$$

Moreover the horizontal component of (3.7.6) equals the sum of

$$(2\Omega \sin\theta)\mathbf{k} \times \mathbf{V} \tag{3.7.8}$$

and

$$(2\Omega \cos\theta) w\mathbf{i} \tag{3.7.9}$$

where **i** denotes an eastward pointing unit vector. But the neglect of (3.7.9) in (3.7.5) is generally† admissible for motions having large lateral scale L because $w \ll |\mathbf{V}|$. For such motions, the Coriolis force and pressure gradient forces are of the same order

$$\Omega|\mathbf{V}| \sim g|\nabla h|$$

and therefore the vertical component of (3.7.6) is

$$|\mathbf{k} \cdot 2\Omega \times \mathbf{V}| \sim |g\,\nabla h|$$

This vertical component of the Coriolis force tends to contribute a surface pressure of order

$$\int dz\, \mathbf{k} \cdot 2\Omega \times \mathbf{V} \sim |gh\,\nabla h|$$

and the horizontal gradient of this is of order $L^{-1}|gh\,\nabla h|$. The latter is small compared to (3.7.7) by order $h/L \ll 1$, and thus we conclude that (3.7.5) is the asymptotic shallow water equation for a sphere.

† When $\theta = 0$, (3.7.9) clearly cannot be neglected compared to (3.7.8), but it may still be negligible compared to the pressure gradient force, as in the example of Section 8.4. In Section 9.9, we show that there is a certain class of inertial oscillations for which (3.7.9) cannot be neglected.

References

Benjamin, T. B. (1968). Gravity Currents and Related Phenomena. *J. Fluid Mech.* **31**, part 2, 209–248.

Rossby, C. G. (1938). On the Mutual Adjustment of Pressure and Velocity Distributions in Certain Simple Current Systems, II. *J. Mar. Res.* **1**, 239–263.

Rouse, H. (1961). "Fluid Mechanics for Engineers." Dover, New York.

Whitehead, J. A., Leetmaa, A., and Knox, R. A. (1974). Rotating Hydraulics of Strait and Sill Flows. *Geophys. Fluid Dynamics* (in press).

CHAPTER IV

Quasi-Geostrophic Motion

4.1 The Transition from Shallow Water to Quasi-Geostrophic Theory

We encountered several examples in the preceding chapter where the local acceleration vanishes ($d\mathbf{V}/dt = 0$) and in which the remaining terms in (3.2.4) reduce to

$$f\mathbf{k} \times \mathbf{V} = -g\,\nabla h \tag{4.1.1}$$

The three vectors appearing in this geostrophic equation form an orthogonal triad, and by means of a sketch, we can easily obtain the explicit relation

$$\mathbf{V} = (g/f)\mathbf{k} \times \nabla h \tag{4.1.2}$$

for the relative velocity.

In those problems where $d\mathbf{V}/dt$ is small but finite, or

$$R = \frac{|d\mathbf{V}/dt|}{|f\mathbf{k} \times \mathbf{V}|} \ll 1 \tag{4.1.3}$$

the geostrophic equation gives a first approximation to the relation between the velocity and pressure fields, but (4.1.1) obviously provides no information about the space–time variation in either field. The dynamics must be obtained from the vorticity equation, the relevant asymptotic form of which is called the "quasi-geostrophic" equation (Section 4.2). We remark that, although each step in the hierarchy of approximations (Euler equations, shallow water equations, quasi-geostrophic equations) removes ("filters out") a certain class of phenomena, the advantage of such a procedure is that it allows us to isolate effects having different space–time scales. The quasi-geostrophic formalism is discussed in the following section, but first we consider a specific example of the way in which such a flow can be produced.

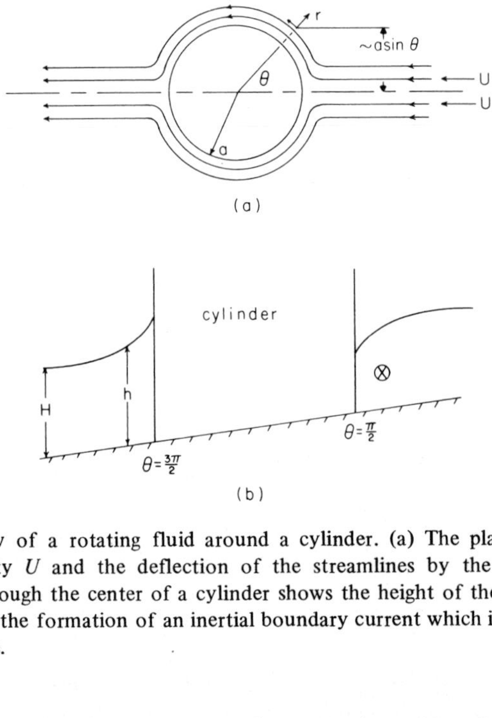

FIG. 4.1 Flow of a rotating fluid around a cylinder. (a) The plan view shows the undisturbed velocity U and the deflection of the streamlines by the cylinder. (b) The vertical section through the center of a cylinder shows the height of the free surface. The problem illustrates the formation of an inertial boundary current which is approximately in geostrophic balance.

The rigid bottom surface shown in the vertical section of Fig. 4.1b rotates with angular velocity $(f + \Delta f)/2$ about a distant vertical axis, and is part of the parabolic level surface in the $(f + \Delta f)/2$ system. The circular cylinder of radius a lies flush on the bottom, and its vertical axis rotates with angular velocity $f/2$ about the distant axis of the basin. The very small difference $\Delta f/2$ then gives rise to a relative velocity $\mathbf{V}$ in a coordinate system fixed to the cylinder. The shallow water equations are given by (3.2.3) [and not (3.2.4)], where η is the free surface height relative to a level surface in the $f/2$ system, h the free surface height relative to the bottom, and $\nabla h - \nabla\eta$ gives the slope of the bottom relative to the level surface. It is convenient to denote the latter vector difference by $f\mathbf{k} \times \mathbf{U}$, or

$$\nabla\eta - \nabla h = -f\mathbf{k} \times \mathbf{U}$$

where $\mathbf{U}$ is directed in the azimuth since $\nabla\eta - \nabla h$ is directed radially. By substituting this relation in (3.2.3), we obtain

$$(d\mathbf{V}/dt) + f\mathbf{k} \times \mathbf{V} = -g\,\nabla h + f\mathbf{k} \times \mathbf{U} \tag{4.1.4}$$

The physical significance of $|\mathbf{U}| = U$ appears when we note that $d\mathbf{V}/dt = 0$ and $h = H$ in the region far upstream from the cylinder. In this region, (4.1.4) reduces to $\mathbf{V} = \mathbf{U}$, and therefore U is the relative speed between the undisturbed fluid and the cylinder. We now turn to the problem of computing the flow in the vicinity of the cylinder.

In order to determine h and $\mathbf{V}$, one more equation besides (4.1.4) is necessary. Although the continuity equation (3.2.5) can be used for this purpose, it is more convenient to take the potential vorticity equation instead, since the latter can be readily integrated as follows. Since a is small compared to the basin dimension, the undisturbed flow has no relative vorticity, as well as a uniform thickness $h = H$. Thus, the potential vorticity $(f + \zeta)/h$ has a uniform value of f/H at points which are far upstream from the cylinder. The conservation law then implies $(f + \zeta)/h = f/H$ at all points, and thus the relative vorticity is $\zeta = f(h - H)/h$, or

$$\mathbf{k} \cdot \nabla \times \mathbf{V} = fh'/H \qquad \text{where} \qquad h' = h - H \tag{4.1.5}$$

We now seek the joint solution of (4.1.4) and (4.1.5) under steady state conditions, or $\partial\mathbf{V}/\partial t = 0$.

The relevant nondimensional numbers for this flow problem are

$$F \equiv fa/(gH)^{1/2} \qquad \text{and} \qquad \mu \equiv U/(gH)^{1/2} \tag{4.1.6}$$

For small U, $d\mathbf{V}/dt = \mathbf{V} \cdot \nabla\mathbf{V}$ is of order U^2 and is small compared to $f\mathbf{k} \times \mathbf{V}$ in (4.1.4). Thus, we shall consider first the limiting case $\mu \to 0$, in which the $d\mathbf{V}/dt$ term can be neglected, and (4.1.4) then becomes

$$f\mathbf{k} \times \mathbf{V} = -g\,\nabla h + f\mathbf{k} \times \mathbf{U} \qquad \text{or} \qquad \mathbf{V} = (g/f)\mathbf{k} \times \nabla h' + \mathbf{U} \tag{4.1.7}$$

As will be seen, the range of validity of this geostrophic approximation (4.1.7) depends on the value of both numbers in (4.1.6).

When (4.1.7) is substituted in (4.1.5), the result becomes

$$\nabla^2 h' - (h'/\lambda^2) = 0 \tag{4.1.8}$$

where $\lambda \equiv (gH)^{1/2}/f$ is the "radius of deformation" (Section 3.6). By using polar coordinates with the origin ($r = 0$) placed at the center of the cylinder, we then obtain

$$(r^{-1}(\partial/\partial r)r(\partial/\partial r) + r^{-2}(\partial^2/\partial\theta^2) - \lambda^{-2})h'(r, \theta) = 0 \qquad (4.1.9)$$

The boundary condition at $r = a$ requires the vanishing of the normal component of **V**, and therefore (4.1.7) implies

$$0 = -(g/fa)(\partial h'(a, \theta)/\partial\theta) - U\cos\theta \qquad (4.1.10)$$

while at large distances from the center of the cylinder we have $h'(\infty, \theta) = 0$. By separation of variables, the solution of (4.1.9) satisfying (4.1.10) is

$$h'(r, \theta)/H = -(Uaf/gH)(\sin\theta)\psi(r) \qquad (4.1.11a)$$

where†

$$(r^{-1}(d/dr)r(d/dr) - r^{-2} - \lambda^{-2})\psi(r) = 0, \qquad \psi(a) = 1, \quad \psi(\infty) = 0 \qquad (4.1.11b)$$

The solution of Bessel's equation (4.1.11b) depends on the ratio of a/λ, and we proceed immediately to the interesting case for which

$$a/\lambda = (f^2a^2/gH)^{1/2} = F \gg 1 \qquad (4.1.12)$$

In this case $r^{-2} \ll \lambda^{-2}$, $r^{-1}\, d\psi/dr \ll d^2\psi/dr^2$, and (4.1.11b) simplifies to

$$((d^2/dr^2) - \lambda^{-2})\psi = 0 \qquad (4.1.13a)$$

The solution of this equation satisfying the boundary condition listed in (4.1.11b) is

$$\psi(r) = e^{(a-r)/\lambda} \qquad (4.1.13b)$$

This solution for ψ and the corresponding value of (4.1.11a) become vanishingly small when the radial distance from the cylinder exceeds the radius of deformation λ, and therefore the perturbations h' and **V** are confined to a relatively thin ($\lambda \ll a$) *inertial boundary layer*, outside of which the streamlines (Fig. 4.1) are parallel. Since (4.1.11a) has a minimum at $\theta = \pi/2$, $r = a$, the value of the minimum layer thickness is given by

$$\min(h')/H = -Uaf/gH = -\mu F \qquad (4.1.14)$$

For $F \gg 1$, the perturbation (4.1.14) becomes large when $\mu \sim 1/F$. The

† Equation (4.1.9) also has a solution which is independent of θ and which leads to a nonvanishing mean value of h'. But this solution is rejected because no mass is "radiated" to $r = \infty$ during the transient stage in which the cylinder is accelerated from relative rest. Thus, the final state is established by Kelvin waves (Section 3.4) redistributing the mass around the cylinder, and (4.1.11a) is the only solution of (4.1.9) satisfying the condition that the mean value of $h'(r, \theta)$ vanishes. Moreover, this is the only solution which conserves the circulation around the material curve on the $r = a$ boundary.

relative velocities in the inertial boundary layer also will be large ($\sim(gH)^{1/2}$), the theory given above will cease to be valid, and therefore we now consider the nonlinear modifications. At any value of θ, the streamlines in the thin boundary layer are nearly straight and parallel to the local tangent to the cylinder, and therefore the acceleration term is of order $|d\mathbf{V}/dt| \sim |\mathbf{V}^2/a|$. The ratio of this term to the local Coriolis force $|f\mathbf{V}|$ is then of order $|\mathbf{V}/af|$. The maximum possible $\mathbf{V}$ occurs when the minimum value of h goes to zero, and from either the preceding considerations or from the Bernoulli equation, we estimate that $|\mathbf{V}| \sim (gH_m)^{1/2} \gg U$ when $F \gg 1$. Therefore, $|\mathbf{V}/af| \sim 1/F \ll 1$, and this means that the inertial terms are still negligible compared to either $g\, \partial h'/\partial r$ or the *radial* component of the Coriolis force in the boundary layer. We can therefore compute the volume transport in the boundary layer by multiplying the azimuthal velocity $f/g\, \partial h/\partial r$ with $h(r, \theta)$ and by integrating the result from $r = a$ to that value of r for which $h = H$. The transport is therefore equal to

$$(g/2f)(H^2 - h^2(a, \theta)) \tag{4.1.15}$$

By referring to Fig. 4.1a, we see that this transport is also equal to the amount of water flowing in the streamline tube which has the width $a \sin\theta$ at the upstream end. Since the vertical depth of this tube is H, and since the horizontal velocity is U, we obtain the mass transport budget.

$$(g/2f)(H^2 - h^2(a, \theta)) = UHa \sin\theta \tag{4.1.16}$$

For small values of $h - H = h'$, this is equivalent to (4.1.11a) at $r = a$, but (4.1.16) can also be applied to the large $\mathbf{V}$ case. Accordingly, we see that the minimum h occurs at $\theta = \pi/2$, and therefore h will vanish when $gH/2f = Ua$, or when

$$\mu = (2F)^{-1} \tag{4.1.17}$$

This gives the critical condition for the separation of the flow from the cylinder. The main points which we wish to emphasize, however, are that the flow is quasi-geostrophic even in the nonlinear case (4.1.16) and that the entire disturbance is confined to an inertial boundary layer (Fig. 4.1a).

4.2 Quasi-Geostrophic Vorticity Equation

When a parcel passes through a succession of states for which the pressure gradient force is nearly equal to the Coriolis force we have a "quasi-geostrophic" flow, and for the one layer model (3.2.4), the approximation is

$$\mathbf{V} \simeq (g/f)\mathbf{k} \times \nabla h \tag{4.2.1}$$

If L denotes the typical lateral scale of $\mathbf{V}(x, y, t)$, so that $|\mathbf{V} \cdot \nabla \mathbf{V}| \sim |\mathbf{V}^2| L^{-1}$, then an obvious necessary condition for quasi geostrophy is

$$R = |\mathbf{V}|/fL \ll 1 \tag{4.2.2}$$

but this condition is not sufficient because $\partial \mathbf{V}/\partial t$ might be comparable to $f\mathbf{k} \times \mathbf{V}$, as is the case for large-scale inertia–gravity waves (3.4.1). Consequently, we must also specify that the initial ($t = 0$) state satisfies (4.2.1). The small $\partial \mathbf{V}/\partial t$ will then depend on the small $\mathbf{V} \cdot \nabla \mathbf{V}$, and the temporal evolution of the geostrophic field is computed from the asymptotic vorticity equation derived below.

Equation (4.2.2) requires that the relative vorticity $\zeta = \mathbf{k} \cdot \nabla \times \mathbf{V}$ be small compared to f by a factor of R, and we also note that the lateral variations in layer thickness, obtained from the integration of (4.2.1) are

$$h - H_m \sim (fL/g)|\mathbf{V}| \qquad \text{or} \qquad (h - H_m)/H_m \sim RF^2 \tag{4.2.3}$$

where

$$F \equiv (f^2 L^2 / g H_m)^{1/2} \tag{4.2.4}$$

We now assume F to be of order unity or less (i.e. L is *not* large compared to the radius of deformation), so that the variation in layer thickness (4.2.3) is small to the same order as ζ/f.

The conservation of potential vorticity can be rewritten as

$$0 = \frac{d}{dt}\left(\frac{f + \zeta}{h}\right) = \frac{f}{H_m}\left(\frac{\partial}{\partial t} + \mathbf{V} \cdot \nabla\right) \frac{1 + \zeta/f}{1 + (h - H_m)/H_m} \tag{4.2.5}$$

and by expanding in powers of R we have

$$0 = \left(\frac{\partial}{\partial t} + \mathbf{V} \cdot \nabla\right)\left[1 + \frac{\zeta}{f} - \frac{h - H_m}{H_m} + \ldots + O(R^2)\right] \tag{4.2.6}$$

or

$$0 = \left(\frac{\partial}{\partial t} + \mathbf{V} \cdot \nabla\right)\left(\frac{\zeta}{f} - \frac{h}{H_m}\right) + O(R^3) \tag{4.2.7}$$

The leading (largest) terms in (4.2.6) or (4.2.7) are of order R^2 because the (nondimensional) magnitude of the operators $\partial/\partial t$ and $\mathbf{V} \cdot \nabla$ is small to order R. The neglected terms in (4.2.6), such as $(\partial/\partial t + \mathbf{V} \cdot \nabla)(h - H_m/H_m)^2$, are small to order R^3, when $R \ll 1$ and $F \sim 1$. Thus, the *fractional* errors in the asymptotic vorticity equation

$$\left(\frac{\partial}{\partial t} + \mathbf{V} \cdot \nabla\right)\left(\frac{\zeta}{f} - \frac{h}{H_m}\right) = 0 \tag{4.2.8}$$

and in the asymptotic momentum equation (4.2.1) are both of order R.

Therefore, if (4.2.1) be substituted for $\mathbf{V}$ wherever the latter appears in (4.2.8), then the resulting differential equation for h is also asymptotic when $R \ll 1$. Thus, we have shown that it is permissible to evaluate the velocity and the vorticity in (4.2.8) by means of the geostrophic approximation. Consequently, if $u = -gf^{-1}\ \partial h/\partial y$ and $v = g/f\ \partial h/\partial x$ denote the Cartesian components of (4.2.1), and

$$\mathbf{k} \cdot \nabla \times \mathbf{V} = \partial v/\partial x - \partial u/\partial y = (g/f)\, \nabla^2 h \tag{4.2.9}$$

denotes the geostrophic approximation to the vorticity, then substitution in (4.2.8) yields

$$\left(\frac{\partial}{\partial t} + \frac{g}{f}\mathbf{k} \times \nabla h \cdot \nabla\right)\left(\frac{g}{f^2}\nabla^2 h - \frac{h}{H_m}\right) = 0$$

or

$$\left(\frac{\partial}{\partial t} - \frac{g}{f}\frac{\partial h}{\partial y}\cdot\frac{\partial}{\partial x} + \frac{g}{f}\frac{\partial h}{\partial x}\frac{\partial}{\partial y}\right)\left(\frac{g}{f^2}\nabla^2 h - \frac{h}{H_m}\right) = 0 \tag{4.2.10}$$

This quasi-geostrophic vorticity equation can be used to compute $h(x, y, t)$ from any initial distribution $h(x, y\ 0)$, and the foregoing procedure will now be generalized to the case of multiple density layers (cf. Section 3.6). Let the subscript j denote a typical layer of density ρ_j and mean thickness H_j. Let $h_j(x, y, t)$ denote local thickness, $\mathbf{V}_j$ the horizontal velocity, ζ_j the relative vorticity, and $p_j(x, y, t)$ the pressure variation on a level surface in the jth layer. When the potential vorticity equation for each layer is expanded in the small Rossby number, the leading term is

$$(\partial/\partial t + \mathbf{V}_j \cdot \nabla)(\zeta_j - (h_j f/H_j) = 0 \tag{4.2.11}$$

where

$$\mathbf{V}_j = \rho_j^{-1} f^{-1} \mathbf{k} \times \nabla p_j \qquad \text{and} \qquad \zeta_j = \rho_j^{-1} f^{-1}\ \nabla^2 p_j$$

The layers are coupled by the hydrostatic relation connecting the p_j, h_j, and ρ_j in successive layers, as illustrated in Sections 3.6 and 4.4. In the very special case of the reduced gravity model considered in Section 3.6, we have shown that the equations for the thin layer are identical to those for a single layer of density ρ, provided g is replaced by g^* in (3.6.5). Therefore, the quasi-geostrophic equation for the model considered in Section 3.6 is given by (4.2.10), or

$$(\partial/\partial t + \mathbf{V} \cdot \nabla)(\zeta/f - h/H_m) = 0 \tag{4.2.12}$$

$$\mathbf{V} = (g^*/f)\mathbf{k} \times \nabla h \qquad \text{and} \qquad \zeta = (g^*/f)\,\nabla^2 h$$

where H_m is the mean thickness of a thin layer bounded by two deeper layers, and the fluctuations $h - H_m$ are required to be small in the sense of (4.2.3). In the following sections, we give some applications of the quasi-geostrophic vorticity equation.

4.3 Barotropic Instability

Figure 4.2 shows a vertical section through a small gap annulus, in which a geostrophic equilibrium current flows in the y direction. This current is confined to the layer of density ρ_1, whereas the underlying layer of density ρ_2 is at rest in the rotating system and has a depth which is much larger than h. In the region above the free surface, we have $\rho_0 = 0$, and thus the model is a special case of Section 3.6, for which the value of reduced gravity (3.6.5) becomes

$$g^* = g(\rho_2 - \rho_1)/\rho_1 \tag{4.3.1}$$

Let $\bar{h}(x)$ denote the undisturbed value of h, $\bar{V}(x)$ the corresponding geostrophic velocity, and $\bar{\zeta}(x)$ the undisturbed vorticity. These quantities are related by the last two equations in (4.2.12), or

$$\bar{V}(x) = (g^*/f)\,d\bar{h}/dx \qquad \text{and} \qquad \bar{\zeta}(x) = (g^*/f)\,d^2\bar{h}/dx^2 \tag{4.3.2}$$

If $\bar{V}(x)$ be sketched for the $\bar{h}$ shown in Fig. 4.2, then it will be seen to have the form of a jet which is directed into the plane of the page. The width of this jet (or the separation of the walls of the annulus) is denoted by L, and thus we can specify a Rossby number $R = |\bar{V}|/fL \ll 1$ in terms of the maximum velocity $|\bar{V}|$.

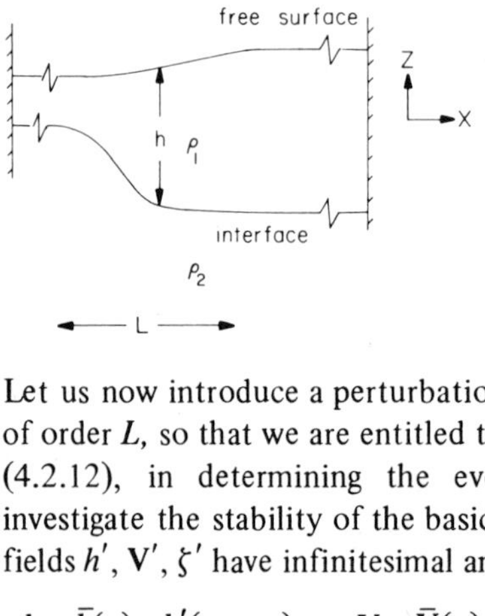

FIG. 4.2 A vertical section of the undisturbed geostrophic state used in the barotopic stability calculation. The upper layer, of density ρ_1, has a thickness h which is much less than that of the lower layer, of density ρ_2.

Let us now introduce a perturbation whose typical horizontal length scale is also of order L, so that we are entitled to use the quasi-geostrophic vorticity equation (4.2.12), in determining the evolution of the perturbation. In order to investigate the stability of the basic flow $\bar{V}$, we can assume that the perturbation fields h', $\mathbf{V}'$, ζ' have infinitesimal amplitude. By substituting the total fields

$$h = \bar{h}(x) + h'(x, y, t), \qquad \mathbf{V} = \bar{\mathbf{V}}(x) + \mathbf{V}'(x, y, t), \qquad \zeta = \bar{\zeta}(x) + \zeta'(x, y, t) \tag{4.3.3}$$

in (4.2.12), and by linearizing the result, we obtain

$$\left(\frac{\partial}{\partial t} + \bar{V}(x)\frac{\partial}{\partial y}\right)\left(\zeta' - \frac{fh'}{H_m}\right) + u'\frac{\partial}{\partial x}\left(\bar{\zeta} - \frac{f\bar{h}}{H_m}\right) = 0 \tag{4.3.4}$$

where

$$u' = -(g^*/f)\,\partial h'/\partial y \tag{4.3.5}$$

denotes the x component of $\mathbf{V}'$. The geostrophic perturbation vorticity is

$$\zeta' = \partial v'/\partial x - \partial u'/\partial y = (g^*/f)\,\nabla^2 h'$$

and thus (4.3.4) becomes

$$\left(\frac{\partial}{\partial t} + \bar{V}\frac{\partial}{\partial y}\right)\left[\nabla^2 h' - \frac{f^2 h'}{g^* H_m}\right] - \frac{\partial h'}{\partial y}\frac{\partial}{\partial x}\left[\bar{\zeta} - \frac{f\bar{h}}{H_m}\right] = 0 \tag{4.3.6}$$

Since the coefficients $\bar{h}$, $\bar{V}$, $\bar{\zeta}$ depend only on x, we may look for normal modes of the form

$$h' = \operatorname{Re} \hat{h}(x) e^{il(y-ct)} \tag{4.3.7}$$

where $2\pi/l$ denotes the length of the wave in the downstream direction, and where

$$c = c_r + ic_i$$

is the phase speed expressed in terms of its real and imaginary parts. The substitution of (4.3.7) in (4.3.6) yields the eigenfunction equation

$$\left(\frac{d^2}{dx^2} - l^2 - \frac{f^2}{g^* H_m}\right)\hat{h} - \hat{h}\,\frac{(d/dx)(\bar{\zeta} - f\bar{h}/H_m)}{\bar{V}(x) - c_r - ic_i} = 0 \tag{4.3.8}$$

The fluid is bounded by vertical walls upon which (4.3.5) must vanish, and (4.3.7) then implies

$$h = 0, \qquad \text{on the } x \text{ boundaries} \tag{4.3.9}$$

The eigenvalue problem (4.3.8) and (4.3.9) is quite similar to that which occurs in the previous study (Section 1.5) of the stability of two-dimensional shear flow, and again we shall use the Rayleigh integral technique to obtain a necessary condition for instability. Thus we take the product of (4.3.8) with the complex conjugate eigenfunction $\hat{h}^*$ and integrate the result between the x boundaries. Using (4.3.9) in a partial integration, and taking the imaginary part of the result, we then obtain

$$c_i \int dx \left[\frac{\hat{h}\hat{h}^*}{(\bar{V} - c_r)^2 + c_i^2}\right]\frac{d}{dx}\left(\bar{\zeta} - \frac{f\bar{h}}{H_m}\right) = 0$$

Therefore, a necessary condition for an amplifying ($c_i > 0$) wave is that the gradient of the basic potential vorticity, or

$$(d/dx)(\bar{\zeta} - (f\bar{h}/H_m)) = d^2\bar{V}/dx^2 - f^2\bar{V}/g^* H_m \tag{4.3.10}$$

must assume *both* positive and negative values within the cross stream interval of the jet. On the other hand, if (4.3.10) has only one sign, then a wave with finite c_i is not possible, and the basic state is said to be stable with respect to geostrophic modes.

Let us examine the foregoing criterion in the context of an unbounded free jet with $\bar{V}(x) > 0$ and $\bar{V}(\pm\infty) \to 0$. Suppose this state to be unstable, so that (4.3.10) will vanish at some finite x. What can we say about the stability of the class of similar basic flows with different values of H_m? Suppose we take a second example, having the same $\bar{V}(x)$ and the same $d\bar{h}/dx$ as the first flow, but with a smaller mean depth H_m. Since the value of (4.3.10) is reduced by decreasing H_m, we see that (4.3.10) will have a negative value everywhere if H_m is made sufficiently small. In this case, the second flow will certainly be stable. Thus, we conclude that, for sufficiently large values of the nondimensional parameter f^2L^2/g^*H_m, the jet is stable. On the other hand, when H_m is large, the Coriolis f term in (4.3.8) vanishes, and the eigenfunction equation for h is then formally identical to Eq. (1.5.18) for the stream function in the nonrotating two-dimensional shear flow problem. In the latter case, we know (Lin, 1955) that an unbounded symmetrical jet is definitely unstable, and therefore the analogous case of interest here also will have amplifying solutions. The conclusion is that the jet is definitely unstable for large H_m and definitely stable for small H_m. Therefore, a critical value of $F = f^2L^2/g^*H_m$ must exist (the numerical value of which depends on the particular shape of the jet), below which the jet is unstable.

The sloping interfaces in Fig. 4.2 imply that the basic state contains potential energy, whereas the nonrotating shear flows in Chapter I contain only kinetic energy. Does an amplifying quasi-geostrophic perturbation lower the center of gravity of the system, or does it reduce the kinetic energy of the basic flow? A simple insight into this fundamental question can be obtained by considering a special case in which the undisturbed current is uniform, or $d\bar{h}/dx$ is constant. The value of (4.3.10) for this model is constant, and therefore the flow is stable. Since this stable state contains available potential energy but no lateral shear, we may infer (Stern, 1961) that the available kinetic energy of the undisturbed flow will be the basic factor in those flows which do exhibit instability according to (4.3.10).

The model with constant $d\bar{h}/dx$ is reconsidered in the next section, but we will no longer assume that the bottom layer is infinitely deep. We will show that instability can then occur, and the source of instability will be the potential energy associated with the sloping interfaces. The latter instability is called "baroclinic," whereas the instability discussed in this section is called "barotropic." In Section 7.5, we will refer to observations of large scale oceanic eddies. These may very well be due to some kind of instability, and it then becomes important to determine the relative importance of the two mechanisms alluded to above.

4.4 Baroclinic Instability

In the undisturbed geostrophic equilibrium state shown in the vertical section of Fig. 4.3, the mean vertical thickness $H - H_m$ of the lower layer is comparable with the mean thickness H_m of the upper fluid, and the interface between the layers varies linearly from $x = 0$ to $x = L$. The radial width L of the annular channel is again assumed to be much less than the mean radius, so that the "small gap approximation" applies.

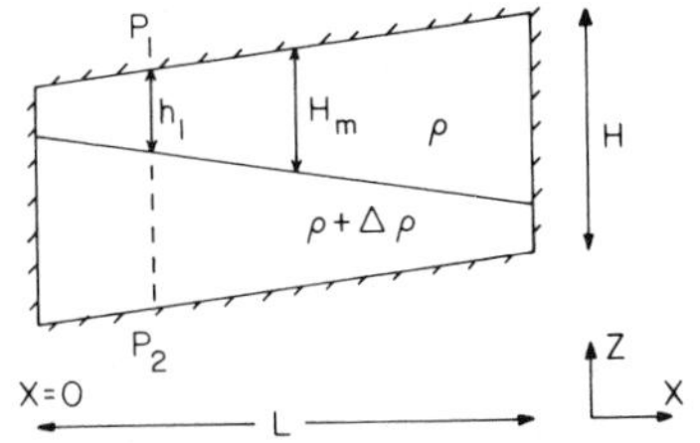

FIG. 4.3 Baroclinic instability in a small gap annulus. The vertical section corresponds to a geostrophic equilibrium state in which there is a constant difference in velocity between the upper ρ and lower $\rho + \Delta\rho$ layers of fluid. The slight slope of the rigid boundaries corresponds to that of the level surfaces in the rotating system.

Equations (4.2.11) can be used to compute the evolution of large scale (low Rossby number) quasi-geostrophic perturbations. The pressure $p_1(x, y, t)$ for the upper layer ($j = 1$) can be taken along the upper rigid surface, and $p_2(x, y, t)$ denotes the bottom pressure. The local thickness of the lower layer $h_2(x, y, t)$ equals $H - h_1$, where $h_1(x, y, t)$ is the thickness of the upper layer. If we let $\rho_1 = \rho$ and $\rho_2 = \rho + \Delta\rho$, then the hydrostatic relation is

$$p_2 = p_1 + \rho g h_1 + (\rho + \Delta\rho) g (H - h_1)$$

and the derivatives on the level surfaces are related by

$$\nabla p_2(x, y, t) = \nabla p_1 - g\,\Delta\rho\,\nabla h_1 \tag{4.4.1}$$

The geostrophic velocity in the upper layer is $\mathbf{V}_1 = \rho^{-1} f^{-1} \mathbf{k} \times \nabla p_1$, and in the lower layer $\mathbf{V}_2 = (\rho + \Delta\rho)^{-1} f \mathbf{k} \times \nabla p_2$. Since the vertical variation of density in the ocean is small, we confine attention to the case $\Delta\rho/\rho \ll 1$, and therefore the foregoing relations may be approximated by

$$\mathbf{V} = (1/\rho f)\mathbf{k} \times \nabla p_1 \tag{4.4.2}$$

$$\mathbf{V} = (1/\rho f)\mathbf{k} \times \nabla p_2 \tag{4.4.3}$$

The corresponding values of the relative vorticity are

$$\zeta_1 = (1/\rho f)\nabla^2 p_1 \qquad \text{and} \qquad \zeta_2 = (1/\rho f)\nabla^2 p_2 \tag{4.4.4}$$

Although $\Delta\rho$ has been neglected (compared to ρ) in the horizontal momentum relations (4.4.2), we obviously do not want to discard $\Delta\rho$ in (4.4.1), wherein it is multiplied by the large gravity factor. This selective procedure, called the Boussinesq approximation, recognizes the dynamically important effects of the small density variations on the vertical distribution of pressure. The significance of the buoyancy term becomes apparent when Eqs. (4.4.2) and (4.4.3) are subtracted. Using (4.4.1), we then obtain *Margules*' equation:

$$\mathbf{V}_1-\mathbf{V}_2=(g^*/f)\mathbf{k}\times\nabla h_1 \qquad \text{where} \qquad g^*=g\,\Delta\rho/\rho \tag{4.4.5}$$

In the undisturbed state (Fig. 4.3), $\nabla h=\nabla\bar{h}$ is constant, and therefore (4.4.5) implies that the difference U between the undisturbed geostrophic currents in the two layers is constant. If we take the basic current of the lower layer to be zero, then U equals the basic current of the upper layer. The relation between U and the interfacial slope, as obtained from (4.4.5), is

$$d\bar{h}/dx=fU/g^* \tag{4.4.6}$$

In the following investigation of the stability of the equilibrium state, we denote the infinitesimal perturbation fields by primed quantities. For example, $u_1'(x,y,t)$ denotes the cross stream (x) geostrophic velocity in the upper layer, and $\zeta_2'(x,y,t)$ the perturbation vorticity in the lower layer. The undisturbed thickness of the lower layer is $H-\bar{h}(x)$, and the perturbation height is $h_2'=-h_1'$. Therefore, the linearization of (4.2.11) yields the following equations for the upper and lower layers:

$$\left(\frac{\partial}{\partial t}+U\frac{\partial}{\partial y}\right)\left(\zeta_1'-\frac{fh'}{H_m}\right)-\frac{u_1'f}{H_m}\frac{d\bar{h}_1}{dx}=0 \tag{4.4.7}$$

$$\frac{\partial}{\partial t}\left(\zeta_2'+\frac{fh_1'}{H-H_m}\right)+\frac{u_2'f}{H-H_m}\frac{d\bar{h}_1}{dx}=0 \tag{4.4.8}$$

From (4.4.1), we see that

$$h_1'=(p_1'-p_2')/g^*\rho$$

and when this equation, together with the quasi-geostrophic relations

$$u_1'=-\frac{1}{\rho f}\frac{\partial p_1'}{\partial y}, \qquad u_2=-\frac{1}{\rho f}\frac{\partial p_2'}{\partial y}, \qquad \zeta_1'=\frac{1}{\rho f}\nabla^2 p_1', \quad \zeta_2'=\frac{1}{\rho f}\nabla^2 p_2' \tag{4.4.9}$$

are substituted in (4.4.7) and (4.4.8), we obtain

$$\left(\frac{\partial}{\partial t}+U\frac{\partial}{\partial y}\right)\left[\nabla^2 p_1'-\left(\frac{f^2}{g^*H_m}\right)(p_1'-p_2')\right]+\frac{f^2U}{g^*H_m}\frac{\partial p_1'}{\partial y}=0$$

$$\frac{\partial}{\partial t}\left[\nabla^2 p_2'+\frac{f^2}{g^*(H-H_m)}(p_1'-p_2')\right]-\frac{f^2U}{g^*H_m}\frac{\partial p_2'}{\partial y}=0 \tag{4.4.10}$$

where (4.4.6) has also been used. At the vertical walls ($x = 0, L$), we have $u_1' = 0 = u_2'$, and (4.4.9) implies $\partial p_1'/\partial y = 0 = \partial p_2'/\partial y$.

The normal mode solutions of (4.4.10) that satisfy the above boundary conditions are

$$p_1' = A_1 \sin(n\pi x/L)e^{ik(y-cUt)} \quad \text{and} \quad p_2' = A_2 \sin(n\pi x/L)e^{ik(y-cUt)} \tag{4.4.11}$$

where A_1 and A_2 are arbitrary amplitude constants, and $2\pi/k$ is the length of a wave propagating down the channel with phase speed cU. When (4.4.11) is substituted in (4.4.10), we get

$$(1-c)[-\lambda^2 A_1 + \alpha_1(A_2 - A_1)] + \alpha_1 A_1 = 0$$

$$-c[-\lambda^2 A_2 - \alpha_2(A_2 - A_1)] - \alpha_2 A_2 = 0$$

where

$$\lambda^2 = (n^2\pi^2/L^2) + k^2, \qquad \alpha_1 = f^2/g^*H_m, \qquad \alpha_2 = f^2/g^*(H - H_m)$$

The elimination of A_1 and A_2 then gives the following quadratic equation for c:

$$(1 + \lambda^2/\alpha_1 - 1/(1-c))(1 + \lambda^2/\alpha_2 - 1/c) = 1$$

This quadratic equation for c has imaginary roots when

$$4\alpha_1\alpha_2 \geqslant \lambda^4 \quad \text{or} \quad 4(f^2L^2/g^*H_m)(f^2L^2/g^*(H - H_m)) \geqslant (n^2\pi^2 + k^2L^2)^2 \tag{4.4.12}$$

The smallest possible value of the right hand side of (4.4.12) is π^4, which occurs for $n = 1$, $k = 0$ (long waves). Therefore, the c roots are imaginary and the baroclinic flow is unstable if

$$(fL/(g^*H_m)^{1/2})(fL/(g^*(H - H_m))^{1/2}) \geqslant \pi^2/2 \tag{4.4.13}$$

At the end of Section 4.3, we pointed out that, if the bottom layer is very deep ($H = \infty$), then a basic flow with $d\bar{h}/dx$ constant is stable. This conclusion is in agreement with that obtained from (4.4.13) when H is large. For finite H, however, Eq. (4.4.13) predicts that the baroclinic flow is unstable when f is sufficiently large. The disturbance which amplifies first, as the critical number (4.4.13) is approached, is the long geostrophic wave $n = 1$, $k = 0$. The reader is referred to the abundant laboratory and meteorological literature (Hide, 1969; Phillips, 1963) for further discussion of baroclinic instability, and in Chapter XII we mention its relevance in the context of the ocean thermocline.

4.5 Energetics of Geostrophic Flow

We have previously mentioned that the basic states of Figs. 4.2 and 4.3 contain available potential energy (APE, hereafter), due to the slope of the interface relative to the level surface, and available kinetic energy (AKE, hereafter), due to the lateral shear of the horizontal velocity in a layer. (The infinite vertical shear between layers is of no dynamical significance in the quasi-geostrophic theory for the following reason. By referring to the Kelvin–Helmholtz waves mentioned in Section 1.6, it will be seen that an infinite shear layer will generate small scale instability waves, but such scales have already been eliminated with the introduction of the hydrostatic approximation. The shallow water equations and the layer models are then used to obtain an insight into larger scale processes occurring in the continuously stratified thermocline. When these equations and these models are used, the vertical shear introduces no instability, and consequently when shear instabilities do occur in the quasi-geostrophic theory, they are due to the continuous *lateral* variation of the currents.) Although the basic instability mechanism in Section 4.3 is due to AKE, and although APE is fundamental in Section 4.4, the stability criterion (in both cases) depends on the ratio of the horizontal width L to the radius of deformation. In view of the importance of this ratio, another interpretation of its meaning is now given.

Potential energy is computed by integrating the product of local density with $gz\,dz$, and APE is computed by subtracting the potential energy which would occur in Fig. 4.3 when the interface is forced into a level position. By means of this straightforward calculation, we find that the available potential energy of the basic state shown in Fig. 4.3 is

$$\text{APE} = \tfrac{1}{2} \iint dx\, dy\, g\, \Delta\rho (h_1(x, y, t) - H_m)^2 \tag{4.5.1}$$

where $h - H_m = xfU(g\,\Delta\rho/\rho)^{-1}$ and where x measures distance from the midpoint ($x = 0$) of the channel. The integration is over the entire fluid, but if we consider a unit distance in the y direction, then

$$\text{APE} = \frac{g\,\Delta\rho}{2} \left(\frac{fU}{g(\Delta\rho/\rho)}\right)^2 \int_{-L/2}^{+L/2} x^2\, dx \tag{4.5.2}$$

On the other hand, the AKE per unit distance is

$$\text{AKE} = \rho \int_{-L/2}^{+L/2} \left(\frac{U^2 h_1}{2}\right) dx = \frac{\rho U^2 H_m}{2} \int_{-L/2}^{+L/2} \left(1 + \frac{xfU}{g(\Delta\rho/\rho)H_m}\right) dx \tag{4.5.3}$$

and the ratio of (4.5.2) and (4.5.3) is

$$\frac{\text{APE}}{\text{AKE}} = \frac{f^2L^2}{3g(\Delta\rho/\rho)H_m} \left[1 + \left(\frac{f^2L^2}{g(\Delta\rho/\rho)H_m}\right)\left(\frac{U}{fL}\right)\right]^{-1} \simeq \frac{f^2L^2}{3g(\Delta\rho/\rho)H_m}$$

provided the Rossby number U/fL is sufficiently small. Thus, we see that in a geostrophic flow, the ratio APE/AKE increases as the ratio of L to the radius of deformation $(g(\Delta\rho/\rho)H_m)^{1/2}f^{-1}$ increases. The latter has a typical magnitude of 10^2 km for the oceanic thermocline, whereas the typical lateral dimension of a basin is $L \sim 10^4$ km or larger. The relatively large amount of available potential energy in the ocean is a point to which we will return in subsequent discussions.

The last three sections may be summarized by noting that geostrophic currents having a large lateral width L tend to become unstable by the baroclinic mechanism of Section 4.4, whereas the barotropic mechanism of Section 4.3 becomes effective with thin jets. In both cases smaller scales of motion are produced by the instability. (See the remark made at the end of Section 3.6.)

4.6 Planetary Waves

The Rossby wave discussed in Section 2.5 is another example of a quasi-geostrophic motion, and we note that the maximum frequency obtained from (2.5.11) is $\hat{\omega} = -\beta L/2\pi$. Thus we see that the ratio of the inertial force $\partial\mathbf{V}/\partial t$ to the Coriolis force is indeed small to order $\beta L/f$, and this observation provides the point of departure for developing the planetary wave theory in models where the force of gravity becomes dynamically important.

Consider first the case (Section 3.7) of a homogeneous fluid having a free surface, and let H denote the uniform undisturbed thickness measured normal to the level surface. For an infinitesimal height perturbation h' the linearized vorticity equation (3.7.3b) becomes

$$\partial\zeta/\partial t + v\beta = H^{-1}f\,\partial h'/\partial t = H^{-1}(f_0 + \beta y)\,\partial h'/\partial t$$

where $f_0 = f - \beta y$ is the Coriolis parameter at the center of the zonal channel (Section 2.5). Since $\beta y/f_0 \sim L/R \ll 1$, where R now denotes the earth's radius, the previous equation may be approximated by

$$\partial/\partial t(\zeta - f_0 h'/H) + v\beta = 0 \qquad (4.6.1)$$

The first approximation to the momentum equation (in ascending powers of β) is given by $f_0\mathbf{k} \times \mathbf{V} = -g\,\nabla h'$ since the $\beta y\mathbf{k} \times \mathbf{V}$ term is obviously small, and we anticipate on the basis of the previous result (2.5.11) that $\partial\mathbf{V}/\partial t$ will also be small to the first order in $\beta L/f_0$. To this same order we may use the Cartesian β-plane approximation (Section 2.5), so that the components of the geostrophic velocity and the vorticity, respectively, are given by

$$f_0 v = g\,\partial h'/\partial x, \qquad f_0 u = -g\,\partial h'/\partial y, \qquad \zeta = (g/f_0)\,\nabla^2 h' \qquad (4.6.2)$$

Since the fractional error in these is of order $\beta L/f_0$, the substitution of (4.6.2) in (4.6.1) gives

$$\partial/\partial t(\nabla^2 h' - f_0{}^2 h'/gH) + \beta\partial h'/\partial x = 0 \qquad (4.6.3)$$

correct to order $(\beta L/f_0)^2$.

The boundary conditions at the zonal walls $(y = \pm L/2)$ are $0 = v = (g/f)\,\partial h'/\partial x$, and therefore the first normal mode of (4.6.3) is

$$h' = \cos \pi y/L \cos(kx - \hat{\omega} t), \qquad \hat{\omega} = \beta k/(k^2 + \pi^2/L^2 + f^2/gH) \tag{4.6.4}$$

Thus we see that when $f^2L^2/gH \sim 1$, the frequency of the planetary wave is reduced by the gravity effect.

The next step is to consider a two-layer model in which the thickness H_m of the light fluid is much less than the thickness $H - H_m$ of the heavy fluid. In this case we already know that interfacial perturbations will induce negligible pressures in the lower layer compared to those in the upper layer, and consequently (4.6.3) is applicable provided we replace H by H_m and g by g^*. The corresponding transformation of (4.6.4) then gives the frequency of the *baroclinic* Rossby mode when $H_m/H \ll 1$.

For finite layer depths the *barotropic* Rossby mode occurs when the interface is undisturbed, and when the horizontal velocity is continuous across that interface. Since the vertical velocity vanishes for this mode, the equation for the stream function is given by (2.5.10) and the frequency by (2.5.11).

It only remains to examine the effect of a finite H_m/H on the baroclinic mode. The reader can readily do this by inserting the appropriate β terms in (4.4.10), by deleting the irrelevant mean flow U, and by then computing the normal modes of the constant coefficient differential equations. In addition, see Fig. 23 in Stommel (1957).

References

Hide, R. (1969). Some Laboratory Experiments on Free Thermal Convection in a Rotating Fluid Subject to a Horizontal Temperature Gradient and Their Relation to the Theory of the Global Atmospheric Circulation. *In* "The Global Circulation of the Atmosphere" (G. A. Corby, ed.). Royal Meteorological Soc., London.

Lin, C. C. (1955). "The Theory of Hydrodynamic Stability." Cambridge Univ. Press, London and New York.

Phillips, N. (1963). Geostrophic Motion. *Rev. Geophys.* **1**, No. 2, 123.

Stern, M. E. (1961). The Stability of Thermocline Jets. *Tellus* **13**, No. 4, 503.

Stommel, H. (1957). A Survey of Ocean Current Theory. *Deep Sea Res.* **4**, 149.

CHAPTER V

Laminar Viscous Flow

5.1 Review of Navier–Stokes Equations

The foregoing chapters have been concerned mainly with the inertial properties of a fluid, and we now begin to consider the ways in which energy is supplied and dissipated. After reviewing the derivation of the viscous equations of motion for a fluid with constant density, we will discuss an important type of laminar flow which occurs above a rotating boundary. The extension in Section 5.3 provides a background for the treatment of the turbulent Ekman layer which is given in Chapter VII.

The reader will recall that the Euler equations (1.1.3) were derived by examining the forces which act on a cubical element $\delta x\ \delta y\ \delta z$ of fluid, and by taking account of those surface forces (pressure) which only act normal to the face of the cube. In a real fluid there are also tangential stresses, and their relationship to the macroscopic velocity field is determined by generalizing Newton's law of friction. This experimental law states that the viscous stress in water is a linear function of the shear and depends on nothing else. Thus, if we have an infinite water layer of thickness d and apply tractive forces on the parallel plates which bound the layer, then the force per unit area of the plate is proportional to the shear U/d, where U is the steady velocity difference between the plates. The experimental law holds only when d is large compared to the molecular dimensions, and such a restriction is fundamental in all of continuum mechanics. It is also a fact that Newton's law of friction breaks down when d is so large that the laminar flow becomes unstable, but this is not a restriction because we use the Navier–Stokes equations (5.1.2a) to compute the stress when the flow is not laminar.

On each of the six faces of the aforementioned cube, there are three orthogonal force components whose magnitudes depend on a scalar pressure p and on

$$\frac{\partial u}{\partial x}, \frac{\partial u}{\partial y}, \frac{\partial u}{\partial z}, \frac{\partial v}{\partial x}, \frac{\partial v}{\partial y}, \frac{\partial v}{\partial z}, \frac{\partial w}{\partial x}, \frac{\partial w}{\partial y}, \frac{\partial w}{\partial z} \tag{5.1.1}$$

where u, v, w are the x, y, z components of absolute velocity $\mathbf{V}$. Also, the relation of the surface forces to (5.1.1) must be linear, according to the elementary law mentioned above. Since the forces exerted upon opposite faces of the infinitesimal cube are oppositely directed, the resultant force on $\delta x\ \delta y\ \delta z$ must therefore be some linear combination of the *second* derivatives of u, v, w. However, the number of admissible combinations of second derivatives is restricted because the total viscous force must be a vector quantity, whose representation is independent of the Cartesian coordinate system used. The only second derivatives of $\mathbf{V}$ or of its components that form such a vector field are $\nabla^2\mathbf{V}$, $\nabla(\nabla\cdot\mathbf{V})$, and $\nabla\times\nabla\times\mathbf{V}$. The last vector, however, equals the difference between the first two, according to a vector identity, and thus there are only two independent vector fields: $\nabla^2\mathbf{V}$ and $\nabla\ \nabla\cdot\mathbf{V}$. Furthermore, in the case of an incompressible fluid, we have $\nabla\cdot\mathbf{V}=0$, and therefore $\nabla^2\mathbf{V}$ remains as the only field on which the viscous force can depend. According to this symmetry argument, the only generalized expression for the resultant surface force per unit volume that is consistent with the known empirical facts about water is $-\nabla p+\mu\nabla^2\mathbf{V}$, where μ is a constant depending on the physical properties. By setting this force equal to the difference between the acceleration and gravity force per unit volume, one obtains the Navier–Stokes equations

$$d\mathbf{V}/dt=-\nabla p_* + \nu\nabla^2\mathbf{V} \qquad \text{and} \qquad \nabla\cdot\mathbf{V}=0 \tag{5.1.2a}$$

where ρp_* denotes the sum of pressure and gravity potential, and where $\nu=\mu/\rho$ is the "kinematic" viscosity. For water, the value is $\nu \simeq 10^{-2}\ \text{cm}^2/\text{sec}$.

The validity of (5.1.2a) has been tested by determination of the drag on a small sphere (Stokes), by the determination of the torque on two differentially rotating cylinders (Couette), and by the determination of the point of instability of the fluid motion in the annulus between Couette's cylinders (Taylor). Although a direct quantitative test in a turbulent regime is virtually impossible, the fact that different flows with the same Reynolds number have similar properties leaves little doubt about the validity of (5.1.2a). Additional support is provided by the following energy equation formed from (5.1.2a).

Since the scalar product of the viscous force with $\mathbf{V} = (u, v, w)$ can be written as

$$-\nu \mathbf{V} \cdot \nabla^2 \mathbf{V} = -\nu(u\, \nabla^2 u + v\, \nabla^2 v + w\, \nabla^2 w)$$
$$= -\nu\, \nabla \cdot (u\, \nabla u + v\, \nabla v + w\, \nabla w) + \nu[(\nabla u)^2 + (\nabla v)^2 + (\nabla w)^2] \quad (5.1.2b)$$

then the scalar product of $\mathbf{V}$ with (5.1.2a) can be written as

$$\tfrac{1}{2}(\partial/\partial t)(\mathbf{V})^2 = -\epsilon - \nabla \cdot [p_* \mathbf{V} + \tfrac{1}{2}\mathbf{V}(\mathbf{V})^2 + \mathbf{S}] \quad (5.1.2c)$$

where the dissipation function ($\epsilon > 0$) equals the last term in (5.1.2b), and $\mathbf{S} = -\nu(u\, \nabla u + v\, \nabla v + w\, \nabla w)$. If (5.1.2c) be integrated over any volume of the fluid, the result can be expressed as follows. The total kinetic energy in the volume decreases at a rate given by the (positive definite) integral of ϵ and increases because of the flux of energy across the area which bounds the volume. The latter term equals the surface integral of the normal component of $p_* \mathbf{V} + \tfrac{1}{2}\mathbf{V}(\mathbf{V})^2 + \mathbf{S}$ at the boundary. The first of these three terms represents "pressure work" (cf. Chapter I), the second represents the net flux of kinetic energy across the bounding surface, and the third ($\mathbf{S}$) represents the work done by the viscous stress of the fluid outside the bounding surface. If the bounding surface is rigid and at rest ($\mathbf{V} = 0$), then $\mathbf{S}, p_* \mathbf{V}$, and $\tfrac{1}{2}\mathbf{V}(\mathbf{V})^2$ vanish at the boundary. In that case, the volume integral of (5.1.2c) implies that the rate of decrease of kinetic energy equals the volume integral of the dissipation ϵ, and consequently the kinetic energy inside a closed cavity decreases monotonically with time, in agreement with the second law of thermodynamics.

When the pressure is eliminated from (5.1.2a), we obtain the vorticity equation $\nabla \times d\mathbf{V}/dt = \nu\, \nabla^2(\nabla \times \mathbf{V})$. If the velocity field $\mathbf{V}$ is independent of y, so that $\nabla \times \mathbf{V}$ is directed along the y axis with magnitude

$$\zeta(x, z, t) = \partial u/\partial z - \partial w/\partial x \quad (5.1.3)$$

then the continuity equation is

$$\partial u/\partial x + \partial w/\partial z = 0 \quad (5.1.4)$$

and we shall show that the vorticity equation reduces to

$$d\zeta/dt = \nu\, \nabla^2 \zeta \quad (5.1.5a)$$

The right hand side of the vorticity equation $\nabla \times d\mathbf{V}/dt = \nu\, \nabla^2\, \nabla \times \mathbf{V}$ has a y component which is obviously identical to the right hand side of (5.1.5a), and for the case of a two-dimensional flow, the y component of $\nabla \times d\mathbf{V}/dt$ can be written as

$$\frac{\partial}{\partial z}\frac{du}{dt} - \frac{\partial}{\partial x}\frac{dw}{dt} = \frac{d}{dt}\left(\frac{\partial u}{\partial z}\right) + \frac{\partial \mathbf{V}}{\partial z}\cdot \nabla u - \frac{d}{dt}\left(\frac{\partial w}{\partial x}\right) - \frac{\partial \mathbf{V}}{\partial x}\cdot \nabla w$$

$$= \frac{d\zeta}{dt} + \left[\frac{\partial u}{\partial z}\frac{\partial u}{\partial x} + \frac{\partial w}{\partial z}\frac{\partial u}{\partial z}\right] - \left[\frac{\partial u}{\partial x}\frac{\partial w}{\partial x} + \frac{\partial w}{\partial x}\frac{\partial w}{\partial z}\right]$$

$$= \frac{d\zeta}{dt} + \frac{\partial u}{\partial x}\left(\frac{\partial u}{\partial z} - \frac{\partial w}{\partial x}\right) + \frac{\partial w}{\partial z}\left(\frac{\partial u}{\partial z} - \frac{\partial w}{\partial x}\right)$$

$$= \frac{d\zeta}{dt} + \left(\frac{\partial u}{\partial x} + \frac{\partial w}{\partial z}\right)\left(\frac{\partial u}{\partial z} - \frac{\partial w}{\partial x}\right) \tag{5.1.5b}$$

$$= \frac{d\zeta}{dt} \tag{5.1.5c}$$

where (5.1.4) has been used between (5.1.5b) and (5.1.5c). Thus, we have proved (5.1.5a).

If a fluid rotates with a solid body velocity $\mathbf{V}_0 = \boldsymbol{\omega} \times \mathbf{r}$ about an axis having unit vector $\mathbf{k}$, then it is intuitively obvious that the viscous force vanishes, or $\nabla^2 \mathbf{V}_0 = 0$. Therefore, if $\mathbf{V}$ now denotes an arbitrary field of *relative* velocity in a system that rotates about an axis with angular velocity $\boldsymbol{\omega} = \mathbf{k}f/2$, then $\nu\, \nabla^2 \mathbf{V}$ is the viscous force. When this is added to the rotating equations of motion (2.1.1), we have the following generalization of the Navier–Stokes equations for the rotating homogeneous fluid:

$$d\mathbf{V}/dt + f\mathbf{k} \times \mathbf{V} = -\nabla\phi + \nu\, \nabla^2 \mathbf{V} \qquad \text{and} \qquad \nabla \cdot \mathbf{V} = 0 \tag{5.1.6}$$

5.2 Laminar Ekman Flow

An equilibrium with $d\mathbf{V}/dt = 0$ is called laminar, and we now discuss such a flow for the case of a rotating viscous fluid. Suppose that the fluid is semi-infinite and bounded below by a rigid surface ($z = 0$) which is perpendicular to the rotation vector $\mathbf{k}f/2$. A relative motion is maintained by a uniform pressure gradient G acting in the x direction, or

$$\partial\phi/\partial x = G, \qquad \partial\phi/\partial y = 0, \qquad \partial\phi/\partial z = 0$$

and we now look for a steady solution of (5.1.6) in which $\mathbf{V}(z)$ is horizontal. If such a flow exists, it will automatically satisfy the continuity equation, as well as the laminarity condition ($d\mathbf{V}/dt = 0$), and consequently (5.1.6) reduces to $f\mathbf{k} \times \mathbf{V} = -\nabla\phi + \nu\, \nabla^2 \mathbf{V}$. The x, y components of $\mathbf{V}$ are denoted by $u(z)$ and $v(z)$, and therefore the Cartesian components of the momentum equation are

$$-fv = -G + \nu\, \partial^2 u/\partial z^2 \tag{5.2.1a}$$
$$fu = \nu\, \partial^2 v/\partial z^2 \tag{5.2.1b}$$

Although a fourth-order equation for either u or v can be obtained by elimination, it is much simpler to deal with a second-order equation for the complex quantity $u(z) + iv(z)$ which is obtained as follows. By multiplying (5.2.1b) with i and by adding the result to (5.2.1a), we get

$$(\nu(\partial^2/\partial z^2) - if)(u + iv) = G$$

If A and B denote constants of integration, then the general solution of the above equation is

$$u + iv = -(G/if) + A \exp -(i/2)^{1/2}(z/h_E) + B \exp +(i/2)^{1/2}(z/h_E) \tag{5.2.2}$$

where

$$h_E \equiv (1/\sqrt{2})(\nu/f)^{1/2} \qquad \text{and} \qquad i^{1/2} \equiv (1 + i)/\sqrt{2} \tag{5.2.3}$$

The solution which increases exponentially as $z \to \infty$ is now discarded by setting $B = 0$. The boundary condition at $z = 0$ is one of no slip ($u = 0 = v$), and therefore $A = G/if$. Thus, the appropriate solution is

$$u + iv = -(G/if)\quad [1 - \exp -(i/2)^{1/2}(z/h_E)] \tag{5.2.4}$$

The quantity h_E (5.2.3) is called the thickness of the *Ekman boundary layer.* For $z \gg h_E$, the real part of the solution vanishes, whereas the imaginary part of (5.2.4) approaches the constant velocity

$$v_0 = G/f \tag{5.2.5}$$

Thus, the viscous forces vanish above the Ekman boundary layer, and (5.2.5) merely expresses the balance between the pressure gradient and geostrophic velocity v_0 in that "interior" region. The viscous forces in the boundary layer cause a net transport *down* the pressure gradient, as given by the real part of the vertical integral of (5.2.4), or

$$T = \int_0^\infty u\, dz = \text{Re}\, \frac{G}{if} \int_0^\infty dz \exp -\left(\frac{i}{2}\right)^{1/2} \frac{z}{h_E} = \text{Re}\, \frac{G}{if}\left(\frac{2h_E}{1+i}\right) = -\frac{Gh_E}{f} \tag{5.2.6}$$

This transport is positive when $\partial\phi/\partial x < 0$. By eliminating G between (5.2.5) and (5.2.6), we also have the following relation between the boundary transport and the overlying geostrophic velocity

$$v_0 = -T/h_E \tag{5.2.7}$$

The negative sign means that the boundary transport is directed toward the left hand of a "northern hemisphere" observer who stands with his back to the geostrophic current. A less geopolitically oriented description of the foregoing relation is

$$\mathbf{T} = h_E \mathbf{k} \times \mathbf{V}_0 \tag{5.2.8}$$

where $\mathbf{V}_0$ is the geostrophic velocity vector, $\mathbf{T}$ the Ekman transport vector, and $\mathbf{k}$ the unit vector pointing along the axis of rotation.

5.3 Parametric Ekman Theory

The experimental realization of the flow discussed in Section 5.2 requires an approximation to the condition of horizontal uniformity. Furthermore, (5.2.4) applies to a vertically unbounded fluid, and consequently the mean depth H of the fluid used in the laboratory must be large compared to the thickness h_E of the Ekman boundary layer. The latter condition is readily attained in Faller's experiment (Greenspan, 1968), which is schematically shown in the vertical section of Fig. 5.1a. At the bottom of the outer and inner walls of this small gap

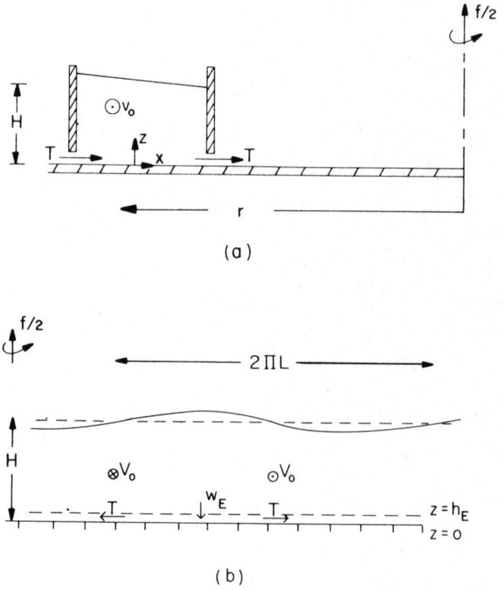

FIG. 5.1 (a) A laboratory realization of laminar Ekman flow in a small gap annulus which is filled with water to a mean height H. Water is pumped into the annulus through the clearance at the bottom of the outer wall and removed at the inner wall. The entire motion is axisymmetric, and v_0 denotes the geostrophic azimuthal velocity. (b) Schematic diagram for computing the viscous decay of a quasi-geostrophic current $\mathbf{V}_0$. The height h of the free surface has a mean value H and $h_E = (\nu/2f)^{1/2}$. The Ekman transport is denoted by T, and w_E denotes the Ekman suction velocity. The x wavelength is $2\pi L \gg H$.

annulus, there is a clearance space through which Q cm^3/sec of water is pumped radially inward along the lower boundary ($z = 0$). The entire flow is axisymmetric, and therefore the transport per unit distance in the azimuthal direction is $T = Q/2\pi r$, where r denotes the radius. In this geometry, the radial direction corresponds to the x direction in Section 5.2. If the width of the annulus is a small fraction of the mean radius, then the variations in r are small, and (5.2.7) then gives the azimuthally directed geostrophic velocity above the boundary layer, i.e. $v_0 = Q/(2\pi r h_E)$. Naturally, the theory of Section 5.2 does not apply near the vertical walls of the channel or in the vicinity of the clearances.

We now consider some of the corrections which are necessary to bring the theory of Section 5.2 into closer correspondence with the experimental conditions. The most important of these is the radial variation of the flow inside the annulus. It will be shown below, however, that if Q is sufficiently small and if r only changes by a factor of 2 (or so), then the geostrophic velocity at any radius is still given by the above relation, or

$$v_0(r) = Q/(2\pi r h_E) \tag{5.3.1}$$

In contrast to the ideal case of Section 5.2, the nonlinear term $\mathbf{V} \cdot \nabla \mathbf{V}$ does not vanish identically in the case of the circular geometry, but the ratio of that term to the Coriolis force is small to order $v/fr \ll 1$, when Q is sufficiently small. The omission of the nonlinear term from Eqs. (5.2.1) is thereby justified for the case of a finite circular geometry. The viscous force $\nu \nabla^2 \mathbf{V}$ contains the term $\nu\, \partial^2 \mathbf{V}/\partial z^2$, which is already included in (5.2.1), but now there will also be terms which are due to the radial variation of $\mathbf{V}$. The latter terms, however, are small compared to the value of $\nu\, \partial^2 \mathbf{V}/\partial z^2$ in the boundary layer by order ${h_E}^2/r^2 \ll 1$, and we will subsequently find that the viscous force still vanishes identically in the "interior" ($z \gg h_E$). Thus, the relevant viscous force for the circular geometry is also properly represented in (5.2.1), where u, v, G are now slowly varying functions of r, this radial dependency being determined by the requirements of the continuity equation. Since the vertical component of velocity is assumed to be zero, and since the flow is uniform in the azimuthal direction, the continuity equation for the radial component $u(r, z)$ becomes

$$(\partial/\partial r)[r u(r, z)] = 0 \tag{5.3.2}$$

Therefore, u must vary inversely with r, G must vary inversely with r, and v must also vary inversely with r. Consequently, the geostrophic vorticity $\partial v_0/\partial r + v_0/r$ vanishes, $\nabla^2 \mathbf{V}_0$ vanishes, and therefore the viscous force above the boundary layer vanishes, as previously asserted. Since (5.2.7) holds at each r, and since $2\pi r T$ is the net radial volume transport Q, we then obtain the relation (5.3.1).

It has been shown that the Ekman velocities in the circular geometry vary inversely with radius, and at any r its vertical structure is parametrically given by the primitive model of Section 5.2. We now extend this result to consider the coupling between the viscous boundary layer and the overlying quasi-geostrophic flow, for a more general horizontal variation. In order to fix ideas, consider the particular problem indicated in Fig. 5.1b. Here we have a single layer of water with a free surface located at height $h = H + h'$ above a level surface ($z = 0$). Suppose that the velocity $\mathbf{V}_0$ underneath the interface is geostrophically balanced at some initial ($t = 0$) time, or

$$f\mathbf{k} \times \mathbf{V}_0 = -g\,\nabla h \tag{5.3.3}$$

If viscous effects were entirely absent, then the state with $\partial \mathbf{V}_0/\partial z = 0$ would remain in geostrophic equilibrium. But in a real fluid, the total velocity $\mathbf{V}$ must vanish at $z = 0$, and therefore an Ekman boundary layer will develop between $z = 0$ and $z \sim h_E$. We assume that the lateral scale L is so large, and the geostrophic vorticity

$$\zeta_0 = \mathbf{k} \cdot \nabla \times \mathbf{V}_0 = (g/f)\,\nabla^2 h \tag{5.3.4}$$

so small, that the slowly varying velocity $\mathbf{V}_0(x, y, t)$ above the viscous boundary layer is quasi-geostrophic (Chapter IV) with negligible viscous force ($\nu\,\nabla^2\,\mathbf{V}_0$). We will now show that the important dynamical effects of viscosity on $\mathbf{V}_0$ arise indirectly, by means of the velocities emerging from the Ekman layer.

Since the fields are assumed to be slowly varying in x, y, t, the local force balance in the boundary layer is still given by Eqs. (5.2.1) where G now denotes the magnitude of the local pressure gradient.† Thus the relation between the Ekman transport down the pressure gradient and the local geostrophic velocity is still given by (5.2.8), or

$$\mathbf{T}(x, y, t) = h_E\mathbf{k} \times \mathbf{V}_0(x, y, t) \tag{5.3.5}$$

These vectors are sketched in Fig. 5.1b for the case in which $h(x, t)$ is independent of y. It is readily seen that we now have a novel situation in which the boundary layer transport is divergent ($\nabla \cdot \mathbf{T} \neq 0$). The magnitude of the horizontal divergence is determined from (5.3.5), or

$$\nabla \cdot \mathbf{T} = h_E(\partial u_0/\partial y - \partial v_0/\partial x) = -h_E\zeta_0 \tag{5.3.6}$$

In order to satisfy the continuity equation in the presence of this horizontally divergent boundary layer flow, we now require a field of vertical velocity w

† In Eqs. (5.2.1), the rotation vector and the normal to the boundary are exactly parallel, but in Fig. 5.1b, the rigid boundary is a level surface having a small horizontal slope. As in Section 3.7, the component of rotation normal to the level surface is denoted by $f/2$, and the vertical component of the Coriolis force is neglected. See also Section 9.9.

which increases from $w(x, y, 0, t) = 0$ to some finite value $w(x, y, z_E, t)$ at the "top" of the boundary layer, where

$$h_E < z_E \ll h \tag{5.3.7}$$

The precise location of the "top" is unimportant, as long as it satisfies (5.3.7). This $w(x, y, z_E, t) \equiv w_E$ is called the Ekman suction velocity, and its value is obtained by integrating the continuity equation, $\partial w/\partial z = -(\partial u/\partial x + \partial v/\partial y)$, from $z = 0$ to $z = z_E$. The right hand side then gives the horizontal divergence of the Ekman transport, and from the left hand side, we then obtain

$$w_E = -\nabla \cdot \mathbf{T} = h_E \zeta_0(x, y, t) \tag{5.3.8}$$

as the "suction velocity" which balances the mass budget in the boundary layer.

Strictly speaking, these vertical velocities should have been added to the momentum equations (5.2.1) in the form of $w\, \partial \mathbf{V}/\partial z$, but the latter are readily shown to have negligible magnitude compared to the Coriolis and viscous forces in the boundary layer. The suction velocities (5.3.8) will have an important dynamical effect, however, as they emerge from the boundary layer and enter the geostrophic interior region ($z \gg h_E$). Since the direct effect of viscosity in this region is negligible, the motions therein must satisfy the law of conservation of potential vorticity (2.3.13). To evaluate the $\partial w/\partial z$ term in the interior, we note that $w = dh/dt$ at the free surface $z = h$, w decreases linearly to $w = w_E$ at the bottom ($z = z_E \sim h_E$) of the "interior" region, and therefore

$$\partial w/\partial z = (dh/dt - w_E)/(h - h_E)$$

By expanding in the small quantity h_E, we have

$$\partial w/\partial z = h^{-1}\, dh/dt + (h_E/h^2)\, dh/dt - w_E/h + \ldots + O(h_E^2) \tag{5.3.9}$$

The w_E term in this relation is much larger than the preceding term because

$$\frac{h_E\, dh/dt}{h w_E} = \frac{dh/dt}{h\zeta_0} \sim \frac{|\partial w/\partial z|}{|\nabla \times \mathbf{V}_0|} \sim \frac{|\nabla \cdot \mathbf{V}_0|}{|\nabla \times \mathbf{V}_0|} \ll 1$$

where we have used the fact that the divergence of a quasi-geostrophic flow (5.3.3) is vanishingly small compared to the relative vorticity. Therefore, (5.3.9) becomes

$$\partial w/\partial z = h^{-1}\, dh/dt - w_E/h$$

When this is substituted into the potential vorticity equation (2.3.13), or

$$(d/dt)(\zeta_0) = (f + \zeta_0)\, \partial w/\partial z$$

we get

$$d\zeta_0/dt = (f + \zeta_0)[h^{-1}\, dh/dt - w_E/h] \tag{5.3.10}$$

Since the quasi-geostrophic approximation has been used previously, we will also discard ζ_0 (compared to f) in (5.3.10) and set $1/h = 1/H$. When (5.3.8) is used, the vorticity equation then becomes

$$d\zeta_0/dt = (f/H)(dh/dt - h_E\zeta_0)$$

or

$$(\partial/\partial t + \mathbf{V}_0(x, y, t) \cdot \nabla)(\zeta_0 - (fh/H)) = -(fh_E/H)\zeta_0 \qquad (5.3.11)$$

and Eqs. (5.3.3) and (5.3.4) complete the set for the determination of $h(x, y, t)$.

To illustrate the use of (5.3.11), we shall compute the decay of the geostrophic current $\mathbf{V}_0(x, t)$ shown in Fig. 5.1b. Since this current is entirely in the y direction, and since $\partial/\partial y = 0$, (5.3.11) reduces to

$$(\partial/\partial t)(\zeta_0 - (fh/H)) + (fh_E/H)\zeta_0 = 0$$

The geostrophic vorticity is $\zeta_0 = g/f\, \partial^2 h/\partial x^2$, and therefore we have

$$(\partial/\partial t)(\partial^2 h/\partial x^2 - f^2 h/gH) + (fh_E/H)\, \partial^2 h/\partial x^2 = 0$$

If the initial value of the field is given by $h = H + A \sin x/L$, then the solution of the above equation is

$$h - H = A \sin(x/L)e^{-t/\theta} \qquad \text{where} \qquad \theta = (H/h_E f)[1 + (f^2 L^2/gH)] \qquad (5.3.12)$$

is the characteristic time for the decay of the relative fluid motion, or the so-called "spin up time." From (5.3.12), we conclude that when the radius of deformation $(gH)^{1/2}f^{-1}$ is large compared to L, the geostrophic motion will decay to zero in a time of order $H/h_E f^{-1}$. On the other hand, a much larger decay time occurs when the radius of deformation is small, and in the limit $f^2 L^2/gh \gg 1$, we have

$$\theta = (f^2 L^2/gh_E)(1/f) = \sqrt{2}(f^{3/2} L^2/g\nu^{1/2})$$

The increase of the viscous decay time with L can readily be explained in terms of the corresponding increase in the ratio of available potential and kinetic energy (cf. Section 4.5). Thus, the potential energy serves to replenish the kinetic energy which is continually dissipated in the Ekman layer.

Reference

Greenspan, H. P. (1968). "The Theory of Rotating Fluids." Cambridge Univ. Press, London and New York.

CHAPTER VI

Shear Turbulence

6.1 Stability of Viscous Shear Flow

This chapter on the turbulence in a nonrotating system provides some background for the subsequent discussion of the oceanic Ekman layer, but it is not a prerequisite for Chapter VII, and therefore the reader may either skip this chapter or seek more information from the literature cited. We will start with a discussion of the breakdown of a laminar flow, and then proceed to Reynolds' formulation of the turbulence problem. A similarity argument for the drag law is given in Section 6.3, and the problem of homogeneous isotropic turbulence is discussed in the final section.

Consider a long wide channel having a uniform height H in the vertical (z) direction, and let ρG denote the uniform downstream (x) pressure gradient acting on the fluid of density ρ and viscosity ν. Whether the flow be laminar or turbulent, it is completely determined by these parameters, so that only one independent nondimensional number can be formed. Thus, if H is the length scale, and $(GH)^{1/2}$ the velocity scale, then all flows having the same value of the Reynolds number $(GH)^{1/2}H/\nu$ are dynamically similar. The mean speed U_0 at the channel center ($z = H/2$) can then be expressed as the product of with some universal function of the Reynolds number. We find it more convenient, however, to invert the problem by considering U_0 as "given," and to use

$$R = U_0H/\nu \tag{6.1.1}$$

as the Reynolds number, in which case we are required to determine G as a function of U_0^2/H and (6.1.1). The laminar Poiseulle flow, having a parabolic velocity profile, is observed only when R is less than some critical number $R_c \sim 10^3$, the prediction of which is supplied by the viscous stability problem surveyed below.

From the inviscid energy equation (1.5.11), we learned that a perturbation can gain energy from the mean flow only if its components are properly correlated (Reynolds stress), and it is readily shown [see (6.2.11)] that the corresponding viscous energy equation differs only insofar as the dissipation term (5.1.2b) is included. The rationale behind early theories of instability was that since viscosity acts to dissipate energy, one could use the inviscid problem (1.5.15) as a point of departure for the complete problem (5.1.5a).

Such an approach proved to be sound for the case of a jet flow and for other profiles having an inflection point. These satisfy Rayleigh's necessary condition (1.5.19) for instability, and the sufficiency condition for a free symmetric jet was established later (Lin, 1955). If H denotes the jet width and U_0 the maximum undisturbed velocity, then on dimensional grounds it follows that the maximum growth rate, with respect to the entire class of perturbations, must be proportional to U_0/H, and the associated wavelength must be proportional to H. Now we consider viscosity by noting that ν/H^2 is the free dissipation rate for a perturbation having scale H. Therefore, the maximum growth rate in a viscous shear flow reduces to zero when $U_0/H \sim \nu/H^2$, or $U_0H/\nu \sim 1$. Thus we see that the critical Reynolds number is $R_c \sim 1$ for the case of a free jet (Betchov and Criminale, 1967).

On the other hand, the Rayleigh condition is not satisfied for a parabolic velocity profile, and thus the Poiseulle flow is stable, according to the inviscid theory. But instability is *observed* at large R in flows having no inflection point, and therefore the effect of ν must be more subtle in these cases than in a jet flow. It seems that the small ν somehow overcomes those dynamical constraints which are reflected in the Rayleigh theorem and in the initial value solution (1.5.13) for the simple Couette flow. The latter calculation is instructive because it shows how the shear tends to distort or "disorganize" a perturbation, even if the initial value of the correlation coefficient (Reynolds stress) is favorable for the abstraction of energy from the mean flow. From this precedent, we may say that the shear of the basic flow has a dual "stabilizing–destabilizing" effect. The destabilizing aspect of the shear is manifest in the inviscid jet problem, whereas the stabilizing aspect is manifest in Couette flow.

The effect of the shear on the structure of a perturbation can be illustrated further by the following theorem. "If $\bar{U}''(z) \neq 0$, then (1.5.15) has no real c eigenvalues and no continuously differentiable eigenfunctions." We already know from Rayleigh's theorem that there are no *complex* eigenvalues, and the import of the present proposition is that there are no initial value perturbations which can preserve their form in the course of time. The proposition will be proved by assuming the contrary to be correct, and by then displaying a contradiction. Accordingly, a real eigenvalue c and a well behaved real solution $\bar{\psi}(z)$ is assumed for (1.5.15). Note that the function

$$f(z) \equiv \bar{\psi}(z)/(\bar{U}(z)-c) \tag{6.1.2}$$

is always well behaved, even if $\bar{U}(z) - c$ should vanish in $0 < z < H$, because (1.5.15) implies that $\bar{\psi}(z)$ must also vanish when $U(z) - c$ vanishes and when $\bar{U}''(z) \neq 0$. The equation for f which results when (6.1.2) is used in (1.5.15) is

$$(d/dz)[(\bar{U}(z)-c)^2\, df/dz] - k^2(\bar{U}-c)^2 f = 0$$

By multiplying this with the real function f, by integration over z, and by using the boundary condition $(\bar{U}-c)f = 0$ at $z = 0, H$, we have

$$-\int_0^H dz\, [(df/dz)^2(\bar{U}-c)^2 + k^2(\bar{U}-c)^2 f^2] = 0$$

Since the left hand side is negative definite, we have reached a contradiction, and thus there are no normal modes when $\bar{U}''(z) \neq 0$. This means that the initial value problem (1.5.8) must be solved without the help of the method of normal modes, and all initial perturbations will be continuously distorted by the shear in a flow without an inflection point.

Let us now consider the viscous vorticity equation (5.1.5a) for a two-dimensional motion. Thus, if u_1, w_1 denotes the x, z perturbation velocities, ψ the corresponding stream function, and $\zeta_1(x, z, t)$ the vorticity perturbation, then the linearization of (5.1.5a) with respect to the equilibrium basic flow $U(z)$ gives

$$(\partial/\partial t + U(z)\,\partial/\partial x)\zeta_1 + U''(z)\,\partial\psi/\partial x = \nu(\partial^2/\partial x^2 + \partial^2/\partial z^2)\zeta_1$$

$$\zeta_1 = -\nabla^2\psi, \qquad w_1 = \partial\psi/\partial x, \qquad u_1 = -\partial\psi/\partial z$$

This equation differs most notably from the inviscid vorticity equation (1.5.8) insofar as it contains fourth-order derivatives of the stream function with respect to z. The highest derivatives are multiplied by ν, and these terms are lost when we set $\nu = 0$. It is this singular behavior which accounts for the "lost" normal modes in the inviscid problem [with $U''(z) \neq 0$]. When the small ν is retained, and when the normal mode $\psi = \text{Re}\,\hat{\psi}(z)\exp ik(x-ct)$ is introduced above, we obtain the Orr–Sommerfeld equation

$$(U(z)-c)\hat{\zeta}(z) + U''\hat{\psi} = (\nu/ik)(d^2/dz^2-k^2)\hat{\zeta} \tag{6.1.3}$$

where $\hat{\zeta} = -(d^2/dz^2 - k^2)\hat{\psi}$. The boundary conditions for the viscous problem require that u as well as w vanish, and therefore

$$0 = \hat{\psi}(0) = \hat{\psi}(H) = \hat{\psi}'(0) = \hat{\psi}'(H)$$

This very difficult problem is solved by computing c as a function of k and R, and by then setting the imaginary part of c equal to zero. This gives R as a function of k, and the smallest possible $R = R_c$ gives the critical Reynolds number for the instability of the mean flow.

Let us consider the balance of terms in (6.1.3), when ν is small but finite ($R \gg 1$). Although the right hand side of (6.1.3) is then "generally" small, and although the inviscid differential equation is then satisfied almost everywhere, there may be *thin* layers in which $\hat{\psi}$ varies so rapidly with z that the viscous term is comparable with the inertial terms on the left hand side of (6.1.3). To illustrate this, we will confine attention to the neutral stability problem in which c is real. (Note that the minimum critical Reynolds number is, in fact, obtained from this class of neutral solutions.) Furthermore, attention will be further restricted to those wave numbers k for which

$$U(z_1)-c = 0 \tag{6.1.4}$$

for some $0 < z_1 < H$, because the first inertial term in (6.1.3) will then vanish at $z = z_1$, and therefore the viscous term will have to play an important role in balancing the remaining inertial term $U''(z_1)\hat{\psi} \neq 0$. For $|z - z_1| > 0$, the order of magnitude of $(U(z) - c)\hat{\zeta}$ increases and will become comparable with the other inertial term at some level

$$z_1 \pm \Delta z_1 \tag{6.1.5}$$

But within this thin *critical layer* of thickness

$$\Delta z_1/H \ll 1 \tag{6.1.6}$$

the viscous and inertial terms all have the same order of magnitude. At the "edge" of the critical layer, the order of magnitude of the first term in (6.1.3) is

$$(U(z)-c)\hat{\zeta}(z) = (z-z_1)U'(z_1)\hat{\zeta}(z_1) + \ldots + O(z-z_1)^2 \tag{6.1.7}$$

By equating this inertial term to the other one $(U''(z_1)\,\hat{\psi}(z_1))$, we have

$$\hat{\zeta}(z) \sim \frac{U''(z_1)\hat{\psi}(z_1)}{U'(z_1)(z-z_1)} \qquad \text{for} \qquad z-z_1 \sim \Delta z_1$$

The second derivative of this expression is

$$\frac{d^2\hat{\zeta}}{dz^2} \sim \frac{U''(z_1)\hat{\psi}(z_1)}{U'(z_1)(z-z_1)^3} \sim \frac{\hat{\zeta}}{(\Delta z_1)^2}$$

and therefore the order of magnitude of the right hand side of (6.1.3) is $(\nu/k)(\hat{\zeta}/(\Delta z_1)^2)$. By equating this viscous term to the inertial term (6.1.7) evaluated at $(z - z_1) \sim \Delta z_1$, we obtain $U'(z_1)\,\Delta z_1 \sim (\nu/k)(\Delta z_1)^{-2}$, or

$$(\Delta z_1)^3 = \nu/kU'(z_1) \tag{6.1.8}$$

The order of magnitude of $U'(z_1)$ is equal to the maximum value of $U(z)$ divided by H, and the relevant wavelength $1/k$ is of the same order as H. By inserting these values in (6.1.8), we then find that the relative thickness of the critical layer is

$$\Delta z_1/H \sim R^{-1/3} \ll 1 \tag{6.1.9}$$

Thus, we infer that a neutral disturbance can be maintained because of the thin (6.1.6) critical layer, and the Poiseulle flow in the channel can become unstable, when $R^{1/3} = H/\Delta z_1 \gg 1$. The important viscous forces in this layer serve to connect the essentially inviscid solution that exists outside the critical layer. Perhaps one may then be permitted to say that the viscosity in the critical layer provides the "glue" which allows a disturbance (the viscous normal mode) to preserve its form in the course of time, thereby offsetting the "disorganizing" effect of the shear. Thus, the viscosity also has a dual effect, since it allows the disturbance to organize itself in such a manner that it can draw upon the kinetic energy which is available in the basic flow.

6.2 Reynolds Equations

As mentioned above, the critical R_c is determined by the smallest possible value of the Reynolds number for which a neutral wave is possible. Thus, when $R > R_c$, there is some preferred wave number k, determined by the linear stability theory, whose amplitude will start to increase exponentially with time. The evolution is, however, determined by nonlinear effects which have been excluded from the stability theory. The basic consideration (6.2.11) now is that the growing perturbation will modify the basic state, so that a new equilibrium will be reached in which the transfer of energy from the mean flow to the perturbation is exactly balanced by the dissipation of the perturbation. We now proceed to consider the field equations that apply to this nonlinear problem and also to the problem of fully developed turbulence which occurs when $R \gg R_c$.

In the latter case, we have to determine the horizontally averaged velocity $\bar{u}(z)$ in the channel flow, as well as the statistical properties of the fluctuating field

$$\mathbf{V}'(x, y, z, t) \equiv \mathbf{V}(x, y, z, t) - \mathbf{i}\bar{u}(z) \tag{6.2.1}$$

where $\mathbf{V}$ is the total velocity at a point, $\mathbf{i}$ is a unit vector in the direction of mean flow (x), and the bar denotes an average over all x, y, t in a channel that is infinitely extensive in the x, y directions. If u', v', w' denote the x, y, z components of $\mathbf{V}'$, then their horizontal average must vanish by definition, or

$$\overline{u'} = \overline{v'} = \overline{w'} = 0 \tag{6.2.2}$$

and a similar decomposition will be made for the total pressure field ρp_*. Thus, if the mean horizontal pressure gradient in the x direction is denoted by ρG, then we have

$$p_* = F(z) - xG(z) + p'(x, y, t, z) \tag{6.2.3}$$

where the "fluctuating pressure" p' has the property $\bar{p}' \equiv 0$, and the significance of $F(z)$ and $G(z)$ appears below.

The substitution of (6.2.1) and (6.2.3) in the Navier–Stokes equation (5.1.2a) yields

$$\frac{\partial \mathbf{V}'}{\partial t} + \mathbf{V}' \cdot \nabla \mathbf{V}' + \bar{u}\frac{\partial \mathbf{V}'}{\partial x} + w'\frac{\partial \bar{u}}{\partial z}\mathbf{i} = -\nabla p' + \nu \nabla^2 \mathbf{V}' + \nu \frac{\partial^2 \bar{u}}{\partial z^2}\mathbf{i} - \nabla(F - xG) \tag{6.2.4}$$

This equation is now averaged horizontally to obtain mean momentum equations, and we consider first the x component of (6.2.4). The average of those terms that are linear (or bilinear) in a perturbation quantity $\mathbf{V}', p'$ will vanish according to (6.2.2). Note also that the horizontal average of the x derivative of a function [such as $\overline{\partial(u')^2/\partial x}$] vanishes because the distance over which the average is taken is infinite. Therefore, when the average of the x component of (6.2.4) is taken, the only nonvanishing term on the left hand side is

$$\overline{\mathbf{V}' \cdot \nabla u'} = \overline{\nabla \cdot (\mathbf{V}'u')} = (\partial/\partial z)\overline{(w'u')}$$

and thus we have

$$(\partial/\partial z)\overline{(w'u')} = \nu\, \partial^2\bar{u}/\partial z^2 + G \tag{6.2.5}$$

It is obvious that the transport of y momentum vanishes, or $\overline{w'v'} = 0$. When we take the horizontal average of the z component of (6.2.4), the result is

$$(\partial/\partial z)\overline{(w')^2} = -(\partial/\partial z)F(z) + x\, \partial G(z)/\partial z$$

Since $\overline{(w')^2}$ does not depend on x, this equation can only be satisfied if

$$\partial G/\partial z = 0$$

and therefore

$$(\partial/\partial z)\overline{(w')^2} = -\partial F/\partial z \tag{6.2.6}$$

Thus, we see that the mean pressure gradient G in the direction of flow is constant, and we also have $F = -\overline{(w')^2}$.

Equation (6.2.5) can be integrated immediately, and the constant of integration appearing therein is determined as follows. The velocity profile $\bar{u}(z)$ in a channel flow must clearly be symmetric about the center $z = H/2$, and since G is constant, Eq. (6.2.5) implies that $\overline{w'u'}$ is antisymmetric about $z = H/2$. Thus, we have $\overline{w'u'} = 0 = \partial\bar{u}/\partial z$ at $z = H/2$, and the integral of (6.1.5) that satisfies these conditions is

$$\nu\, \partial\bar{u}/\partial z - \overline{w'u'} = -(z - H/2)G \tag{6.2.7}$$

Since $\bar{u}$ must vanish at the $z = 0$ boundary, the integral of (6.2.7) gives

$$\bar{u}(z) = \nu^{-1} \int_0^z dz\, \overline{w'u'} - \nu^{-1} G(z^2/2 - Hz/2) \tag{6.2.8}$$

as the relation between the mean field and the fluctuating field. When Eqs. (6.2.5) and (6.2.6) are subtracted from (6.2.4), we have

$$\begin{gathered} \frac{\partial \mathbf{V}'}{\partial t} + \mathbf{V}' \cdot \nabla\mathbf{V}' - \frac{\partial}{\partial z}\overline{w'\mathbf{V}'} + \bar{u}\frac{\partial \mathbf{V}'}{\partial x} + w'\frac{\partial \bar{u}}{\partial z}\mathbf{i} = -\nabla p' + \nu \nabla^2 \mathbf{V}' \\ \nabla \cdot \mathbf{V}' = 0 \end{gathered} \tag{6.2.9}$$

and by using (6.2.8) to eliminate the mean field $\bar{u}$, we can obtain a closed set of integro-differential equations for the "fluctuating field" $\mathbf{V}'$.

Although these equations have not been solved, a most important relation can be obtained by taking the scalar product of $\mathbf{V}'$ with (6.2.9), and by averaging the result over z as well as x, y, t. In view of the boundary conditions, we then find that the only nonvanishing terms are

$$\frac{1}{H}\int_0^H \overline{w'u'}\frac{\partial\bar{u}}{\partial z}\, dz = \frac{1}{H}\int_0^H \overline{\mathbf{V}' \cdot \nu \nabla^2 \mathbf{V}'}\, dz \tag{6.2.10}$$

or

$$-\frac{1}{H}\int_0^H \overline{w'u'}\frac{\partial\bar{u}}{\partial z}\, dz = \frac{\nu}{H}\int_0^H \overline{(\nabla \times \mathbf{V}')^2}\, dz \tag{6.2.11}$$

where use has been made of the identity, $\nabla^2 \mathbf{V}' = -\nabla \times \nabla \times \mathbf{V}'$ (when $\nabla \cdot \mathbf{V}' = 0$), to obtain (6.2.11) from (6.2.10). The right hand side of (6.2.11) represents the average dissipation of energy by the fluctuating field of motion, and the left hand side is the work done by the Reynolds stress on the mean field

$\bar{u}$. From (6.2.11), it also follows that $\overline{u'w'}$ must be negatively correlated with $\bar{u}_z$, in consequence of which one says that the average momentum flux is down the mean gradient.

At $z = 0$, the Reynolds stress in (6.2.7) must vanish, and therefore the viscous stress τ at the wall is given by

$$\tau = \nu\ \partial\bar{u}(0)/\partial z = GH/2 \tag{6.2.12}$$

This equation merely states that the horizontal momentum lost at the boundaries is supplied by the downstream pressure drop inside the fluid. Accordingly, the value of G can be determined immediately from τ.

6.3 Rough Similarity Law in a Smooth Pipe

In a fully turbulent ($R \to \infty$) flow, $\mathbf{V}(x, y, z, t)$ is a stochastic field, having an infinite number of nonredundant statistical moments, but in any practical problem we are only interested in a very small number of the average properties of the flow. In the oceanic context, for example, the relationship between the surface stress and the atmospheric wind is obviously fundamental, and therefore we turn attention to the related problem of the wall stress in the channel flow of Section 6.2.

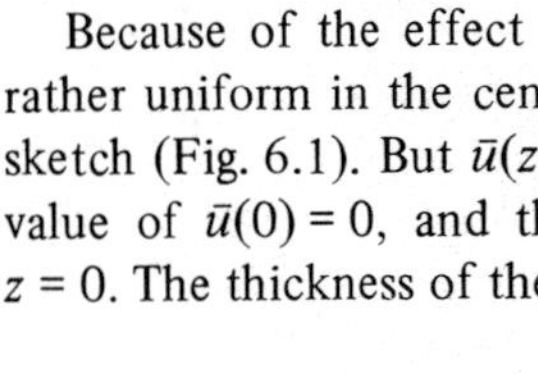

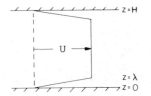

FIG. 6.1 Highly simplified diagram of the mean velocity profile $\bar{u}(z)$ in a turbulent channel flow.

Because of the effect of the turbulent eddies in mixing momentum, $u(z)$ is rather uniform in the center of the channel, as indicated in the highly simplified sketch (Fig. 6.1). But $\bar{u}(z)$ must decrease from its maximum $\bar{u}(H/2) = U_0$ to the value of $\bar{u}(0) = 0$, and therefore $\bar{u}(z)$ must have a relatively large gradient at $z = 0$. The thickness of the boundary layer may be defined by

$$\lambda = U_0 [\partial\bar{u}(0)/\partial z]^{-1} \tag{6.3.1}$$

and is schematically indicated by a straight line profile in Fig. 6.1.

When (6.3.1) is used to eliminate $\partial\bar{u}(0)/\partial z$ in (6.2.12), the wall stress can be expressed as

$$\tau = U_0^2/R_\lambda \qquad \text{where} \qquad R_\lambda = U_0\lambda/\nu \tag{6.3.2}$$

The quantity R_λ represents a boundary layer Reynolds number based upon the thickness λ and the velocity difference U_0 across the layer. In the following similarity theory, we ask whether there is any asymptotic relationship between R_λ and $R \to \infty$. We will argue that λ approaches a finite value in a sequence of experiments with U_0 and ν held constant and with $H \to \infty$. If this be true, then R_λ must approach a constant as $R \to \infty$, and (6.3.2) then implies that the stress is proportional to U_0^2. After giving a physical basis for the limit assumption, and after estimating the value of R_λ, we shall mention the discrepancy between this (6.3.2) drag law and measurements in smooth pipes.

The formal definition of λ in (6.3.1) can be applied to the parabolic velocity profile $\bar{u} = 4U_0(H - z)zH^{-2}$ which occurs in a channel of height H, when $R < R_c$ and when the flow is laminar. In that case, (6.3.1) gives $\lambda = H/4$, and we now proceed to consider how λ varies when we increase H with U_0 and ν held constant. As mentioned previously, the effect of the eddies (for $R > R_c$) is to remove the destabilizing velocity gradient from the interior and to concentrate it near the boundaries, thereby reducing λ below the value of $H/4$. In some sense the overall effect of the instabilities is to bring about a new and "more" stable $\bar{u}(z)$ profile, and the question then arises as to how one measures the "stability" of a turbulent flow (Malkus, 1956). Suppose we focus attention on the mean flow $\bar{u}(z)$, and particularly upon the energy generating shear layer λ. We might then make a *formal* stability calculation, by inserting any $\bar{u}(z)$ into the Orr–Sommerfeld equation and by computing the growth rates of the perturbations. These growth rates will depend on the value of the boundary layer Reynolds number which is associated with the given $\bar{u}(z)$, and if this R_λ is supercritical, then we would expect that such a profile will not be realized. We would expect to realize a different and more stable $\bar{u}(z)$, having a smaller R_λ. On the other hand, the R_λ of the realized state cannot be too small, because a finite shear layer λ is required to maintain the turbulent eddies which, in turn, maintain the real $\bar{u}(z)$. Since the real R_λ can be neither too large nor too small as $R \to \infty$, we argue that it must approach a limit R_c. The realized $\bar{u}(z)$ is then said to be marginally stable, in the sense indicated above, and R_c is to be estimated from the Orr–Sommerfeld equation.

The lower half of the mean velocity profile (Fig. 6.1) in fully turbulent channel flow has a crude resemblance to the boundary layer profile formed in the (Blausius) flow over a semi-infinite plate, and the critical Reynolds number in the latter case is known from both theory and experiment to be of the order $R_c \sim 10^3$ For the purpose of the estimation mentioned above, we may, therefore, use the estimate $R_\lambda \sim 10^3$, and when this is introduced in (6.3.2), we obtain fair agreement with the measured (Schlichting, 1960) range of drag coefficients. The experiments show, however, a small but systematic variation of τ/U_0^2 with R in *smooth* pipes, and therefore our basic similarity assumption (constant R_λ) is, in fact, not correct. We do believe, however, that the disagreement is not such as to discourage further development along this line.

6.4 Homogeneous Isotropic Turbulence

Because the problem of turbulence is so important, and so difficult, we must first try to understand the examples that offer minimum geometrical and parametrical complexity. The problem discussed previously is of this type because the statistical properties are the same (homogeneous) at all points in a horizontal plane, and because there is only one external nondimensional number upon which all the average properties depend. Another basic problem occurs when a screen or grid is placed crosswise to the flow in a wind tunnel, and we will discuss the turbulent motions that occur in the wake of the grid. The reader is also referred to the literature (Hinze, 1959; Gibson and Schwarz, 1963) for more comprehensive discussion.

The grid is located at $x = 0$ in the y, z plane (Fig. 6.2), and the distance between individual wires or bars in the grid is denoted by M. The uniform fluid velocity upstream of the grid is denoted by U, and the "grid Reynolds number" by UM/ν. In the immediate wake $(x \sim M)$ of the grid, the *energy bearing* eddies of scale M are formed, and these eddies proceed to decay as they are carried downstream with the mean current U. We assume that UM/ν is sufficiently large, so that the motion is still turbulent at the downstream position $x = x_{\mathrm{I}} \gg M$,

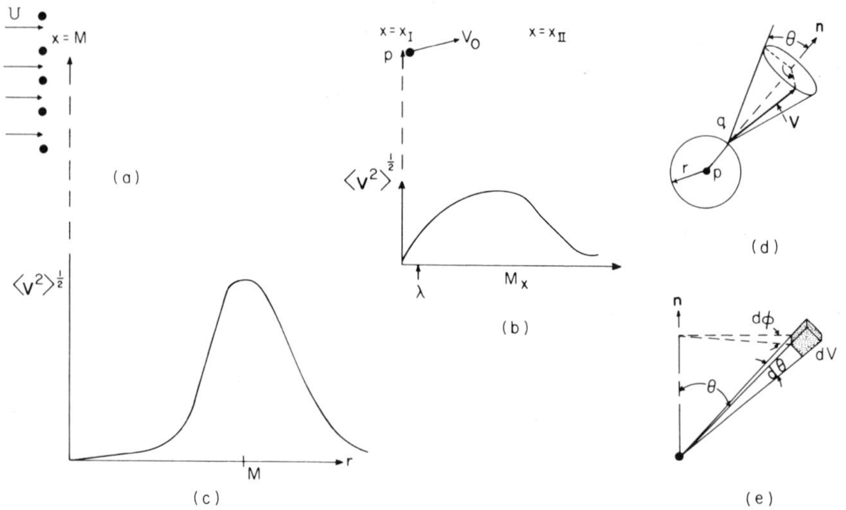

FIG. 6.2 (a) The uniform flow U through a grid whose wires are separated by distance M. (b) Sketch of the rms velocity difference $\langle \mathbf{V}^2 \rangle^{1/2}$ between two points separated by the distance $r \ll M_x \ll M \ll x_{\mathrm{I}}$ at $x = x_{\mathrm{I}}$. (c) Sketch of the rms velocity difference in the immediate wake $(x \sim M)$ of the grid. (d) A particular realization of the velocity difference $\mathbf{V}$ between points q and p is plotted on a conical surface having q as its vertex. (e) A volume element in "velocity space" is stippled and we ask for the probability that $\mathbf{V}$ lies therein. The polar axis $\mathbf{n}$ is the same unit vector that appears in the physical space (c).

which is shown in Fig. 6.2. The position $x_{\mathrm{II}} \gg x_{\mathrm{I}}$ corresponds to the final stage of decay, wherein the velocity fluctuation and typical eddy size are so small that the associated Reynolds number is of order unity.

The scale of the eddies in the wake $(x \sim M)$ is determined by the mesh size of the grid, and therefore the rms velocity difference $\langle \mathbf{V}^2 \rangle^{1/2}$ between two points separated by the distance r will have a maximum at $r \sim M$, as shown in Fig. 6.2c. Since the eddies decay downstream, the amplitude of the $\langle \mathbf{V}^2 \rangle^{1/2}$ curve is smaller at $x = x_{\mathrm{I}}$ (Fig. 6.2b). But this decay does not occur as a result of the *direct* action of ν on the large or "energy bearing scale" M_x, because the associated Reynolds number is too large. Instead, the larger scale motions transfer energy to the smaller scales

$$r \ll M_x \tag{6.4.1}$$

and the corresponding vorticity of the microstructure is then large enough to account for the dissipation (5.1.2b) of the energy. Because of the strong nonlinear interactions which give rise to the small scales (6.4.1), and because of the kinematical complexity, we assume that the properties of the small scale motion at x_{I} are independent of the history of the flow through the grid. Moreover, these scales are assumed to be *isotropic* because of the extent to which they are "whirled around" by the larger scale motions. If closely spaced observers move with the fluid, they will be unable to tell which direction is downstream, and they will have no way of knowing what the absolute velocity of the fluid is from their observations of each other. Thus, we will assume that the relative velocity

$$\mathbf{V} = \mathbf{V}_q{}^* - \mathbf{V}_p{}^* \tag{6.4.2}$$

is *statistically independent* of the absolute velocities $\mathbf{V}_q{}^*$ and $\mathbf{V}_p{}^*$, between two points q and p, separated by the distance (6.4.1) at $x = x_{\mathrm{I}}$, The first and most interesting statistical moment is the *structure function*

$$\langle \mathbf{V}^2 \rangle = \langle (\mathbf{V}_q{}^* - \mathbf{V}_p{}^*)^2 \rangle \tag{6.4.3}$$

where the brackets denote an average (for given r, x) with respect to an ensemble of similarly prepared experiments.

The eddies march downstream under the influence of the uniform mean current $\langle \mathbf{V}_p{}^* \rangle = \langle \mathbf{V}_q{}^* \rangle$, and consequently the eddy kinetic energy is transported by this motion, rather than by "pressure work." The smaller scales of motion are generated by the breakdown or instability of the energy bearing eddies, and the energy transferred thereby is dissipated, as stated above. Thus the total flux of kinetic energy across a y, z plane decreases downstream, and the average divergence of this flux at any value of x is denoted by

$$\nabla \cdot \langle \tfrac{1}{2} \mathbf{V}^* (\mathbf{V}^*)^2 \rangle \equiv -\epsilon < 0 \tag{6.4.4}$$

(where $\epsilon = -\nu\langle \mathbf{V}^* \cdot \nabla^2 \mathbf{V}^* \rangle$ [cf. (5.1.2a)] is also the viscous dissipation per unit volume). We will assume that the eddies decay slowly downstream, so that the variation of ϵ with x can be neglected in the relatively small distance (6.4.1). Therefore, if point p be surrounded by a sphere of radius r, and if (6.4.4) be integrated over the spherical volume, then we obtain

$$\tfrac{1}{2} \oiint d\mathbf{A} \cdot \langle \mathbf{V}_q^*(\mathbf{V}_q^*)^2 \rangle = -\tfrac{4}{3}\pi r^3 \epsilon \tag{6.4.5}$$

where $\mathbf{V}_q^*$ is the absolute velocity of a point q on the surface of the sphere, $d\mathbf{A} = \mathbf{n}\, dA$ is the element of surface area, and $\mathbf{n}$ is an outward pointing unit vector at q. The neglect of the ϵ variation is consistent with the assumption of statistical homogeneity around the sphere, and the isotropy assumption implies that $\mathbf{V}$ has statistics which depend on $r = |\mathbf{r}|$ but not on the orientation of $\mathbf{r}$ or on the "absolute" velocity $\mathbf{V}^*$.

The vector $\mathbf{V}$ drawn in Fig. 6.2d corresponds to a particular realization of the relative velocity between the two points p and q. This vector is specified by its magnitude V, by the angle θ which it makes with the normal $\mathbf{n}$ to the sphere at point q, and by the azimuthal angle, $0 < \phi < 2\pi$, which corresponds to the location of $\mathbf{V}$ on the conical surface having $\mathbf{n}$ as its axis. Because of isotropy, all positions of $\mathbf{V}$ on this cone are indistinguishable and equiprobable. The end point of $\mathbf{V}$ lies in the stippled volume element

$$dV(V\, d\theta)(V \sin\theta\, d\phi) \tag{6.4.6}$$

in a velocity space (Fig. 6.2e) having V, θ, ϕ as its spherical polar coordinates. Each stippled point corresponds to a particular realization in the ensemble, and the total number of realizations of $\mathbf{V}$ is denoted by

$$P(r, V, \theta)\, dV(V\, d\theta)(V \sin\theta\, d\phi) \tag{6.4.7}$$

where the density function P equals the number of realizations in each unit of the velocity space. This density function does not depend on azimuthal angle ϕ because of the isotropy assumption. Also, P does not depend on the "absolute" velocity, since $\mathbf{V}$ and $\mathbf{V}_p^*$ are statistically independent. Consequently, when the product of a function of $\mathbf{V}$ and a function of $\mathbf{V}_p^*$ is averaged, the result is equal to the product of the individual averages. Thus, when (6.4.2) is used to eliminate $\mathbf{V}_q^*$ in (6.4.5), we obtain

$$\begin{aligned} -\epsilon(8\pi/3\, r^3) &\doteq \oiint d\mathbf{A} \cdot \langle (\mathbf{V} + \mathbf{V}_p^*)(\mathbf{V}^2 + 2\mathbf{V}\cdot\mathbf{V}_p^* + (\mathbf{V}_p^*)^2) \rangle \\ &= \oiint d\mathbf{A} \cdot \langle \mathbf{V}\mathbf{V}^2 \rangle + H \end{aligned} \tag{6.4.8}$$

where

$$\begin{aligned} H = \oiint d\mathbf{A} \cdot \{ & 2\langle \mathbf{V}\mathbf{V} \rangle \cdot \langle \mathbf{V}_p^* \rangle + \langle \mathbf{V} \rangle \langle (\mathbf{V}_p^*)^2 \rangle \\ & + \langle \mathbf{V}_p^* \rangle \langle \mathbf{V}^2 \rangle + 2\langle \mathbf{V} \rangle \cdot \langle \mathbf{V}_p^* \mathbf{V}_p^* \rangle + \langle \mathbf{V}_p^*(\mathbf{V}_p^*)^2 \rangle \} \end{aligned} \tag{6.4.9}$$

The kinematics of the following paragraph will show that $H = 0$, and (6.4.8) then implies that the divergence of the flux of kinetic energy can be computed by replacing the absolute velocities with the relative velocities, as indicated in (6.4.12).

The vector **V** in Fig. 6.2d can also be resolved into a component $u_r\mathbf{n}$, which is normal to the sphere, and into a tangential component $\mathbf{u}_t$, or

$$V = u_r\mathbf{n} + \mathbf{u}_t$$

where

$$u_r = V\cos\theta, \qquad \text{and} \qquad |\mathbf{u}_t| = V\sin\theta \tag{6.4.10}$$

Since the average value of u_r is proportional to the mean mass flux through the sphere, the continuity equation implies $\langle u_r \rangle = 0$. We also note that for any given u_r the values of $+\mathbf{u}_t$ and $-\mathbf{u}_t$ are equiprobable, according to the conical symmetry or isotropy. In general, $\mathbf{u}_t$ is independent of u_r, and $\mathbf{u}_t$ has zero average, or

$$\langle \mathbf{V} \rangle = 0, \qquad \text{and} \qquad \langle u_r\mathbf{u}_t \rangle = 0 \tag{6.4.11}$$

We are now prepared to show that each of the five terms in (6.4.9) vanishes. The two terms containing $\langle \mathbf{V} \rangle$ vanish because of the first relation in (6.4.11). The term $\langle \mathbf{V}_p{}^*(\mathbf{V}_p{}^*)^2 \rangle$ has the same value at all points on the sphere, and therefore its integral in (6.4.8) is proportional to the surface integral of $d\mathbf{A} = \mathbf{n}\,dA$. The latter integral clearly vanishes, and thus the term containing $\langle \mathbf{V}_p{}^*(\mathbf{V}_p{}^*)^2 \rangle$ in (6.4.8) vanishes. The vector $\langle \mathbf{V}_p{}^* \rangle \langle \mathbf{V}^2 \rangle$ is also constant for all points on the surface of the sphere, and therefore the integral of that term also vanishes. The only term which remains in (6.4.9) is

$$2 \oiint d\mathbf{A} \cdot \langle \mathbf{V}\mathbf{V} \rangle \cdot \langle \mathbf{V}_p{}^* \rangle = 2 \oiint dA\, \langle u_r(u_r\mathbf{n} + \mathbf{u}_t) \rangle \cdot \langle \mathbf{V}_p{}^* \rangle$$

$$= 2\langle u_r^2 \rangle \oiint d\mathbf{A} \cdot \langle \mathbf{V}_p{}^* \rangle + 2\langle \mathbf{V}_p{}^* \rangle \cdot \oiint dA\, \langle u_r\mathbf{u}_t \rangle = 0$$

because of the second relation in (6.4.11), and also because $\oiint d\mathbf{A} = 0$. Thus, we have shown that $H = 0$, and therefore (6.4.8) becomes

$$-\epsilon\left(\frac{8\pi}{3}\right) r^3 = \oiint dA\, \langle u_r\mathbf{V}^2 \rangle = 4\pi r^2 \langle u_r\mathbf{V}^2 \rangle \qquad \text{or} \qquad \langle u_r\mathbf{V}^2 \rangle = -\tfrac{2}{3}\epsilon r \tag{6.4.12}$$

These ensemble averages can also be expressed in terms of integrals of the density function P over the volume of the velocity space. For example, by multiplying $u_r = V\cos\theta$ with (6.4.7), by integrating the result, and by dividing with the volume integral of (6.4.7), we obtain $\langle u_r \rangle$. Since this expression must vanish because of continuity, we then have

$$\int_0^\infty dV \int_0^\pi d\theta\, V^3 \sin\theta \cos\theta\, P(r, V, \theta) = 0 \tag{6.4.13}$$

Since $u_r \mathbf{V}^2 = V^3 \cos\theta$, a similar procedure for (6.4.12) gives

$$\frac{\int_0^\infty dV \int_0^\pi d\theta\, V^5 \sin\theta \cos\theta\, P(r, V, \theta)}{\int_0^\infty dV \int_0^\pi d\theta\, V^2 \sin\theta\, P(r, V, \theta)} = -\frac{2}{3} r\epsilon \tag{6.4.14}$$

We would like to compute the mean of $\mathbf{V}^2 = V^2$, or

$$\langle \mathbf{V}^2 \rangle = \frac{\int_0^\infty dV \int_0^\pi d\theta\, V^4 \sin\theta\, P(r, V, \theta)}{\int_0^\infty dV \int_0^\pi d\theta\, V^2 \sin\theta\, P(r, V, \theta)} \tag{6.4.15}$$

as a function of r.

By introducing the transformation

$$V = (\epsilon r)^{1/3} v \qquad \text{and} \qquad P(r, (\epsilon r)^{1/3} v, \theta) = F(v, \theta) \tag{6.4.16}$$

Eq. (6.4.14) becomes

$$\frac{\int_0^\infty dv \int_0^\pi d\theta\, v^5 (\sin\theta \cos\theta)\, F(v, \theta)}{\int_0^\infty dv \int_0^\pi d\theta\, v^2 (\sin\theta) F} = -\frac{2}{3}$$

and Eq. (6.4.15) gives the Kolmogoroff law

$$\langle \mathbf{V}^2 \rangle = C(\epsilon r)^{2/3} \tag{6.4.17}$$

where

$$C \equiv \frac{\int_0^\infty dv \int_0^\pi d\theta\, v^4 (\sin\theta) F}{\int_0^\infty dv \int_0^\pi d\theta\, v^2 (\sin\theta) F} \tag{6.4.18}$$

is a *nondimensional* constant.

At this point, one would like to close the statistical problem (and determine C) by imitating the procedure used in the molecular theory of gases. Such a procedure involves a determination of the dynamically relevant constraints on the distribution function, and a determination of which states (realizations) are equiprobable. With regard to the first question, we note that the average work done by the pressure (p^*) force must vanish, or $\langle \mathbf{V}^* p^* \rangle = 0$. By taking the divergence of the Navier-Stokes equations, and by using $\nabla \cdot \mathbf{V}^* = 0$, we then obtain $\nabla^2 p^* = -\nabla \cdot (\mathbf{V}^* \cdot \nabla \mathbf{V}^*)$. By using the method of potential theory, we can readily solve this equation for the pressure $p^*(\mathbf{r})$ at point q in Fig. 6.2d. Thus, if $\mathbf{r}_1$ denotes the distance from the origin to any other point in the field, then the vanishing of the pressure work is given by

$$\left\langle \mathbf{V}^*(\mathbf{r}) \cdot \iiint dx_1\, dy_1\, dz_1 \frac{\nabla \cdot (\mathbf{V}^*(\mathbf{r}_1) \cdot \nabla \mathbf{V}^*(\mathbf{r}_1))}{|\mathbf{r}_1 - \mathbf{r}|} \right\rangle = 0 \tag{6.4.19}$$

Like (6.4.12) and (6.4.11), Eq. (6.4.19) represents a statistical constraint, and the latter is the only point of contact with the Navier–Stokes equations which has been made so far. But (6.4.19) cannot be expressed in a form such as (6.4.15), because it depends on the simultaneous velocity distribution at multiple points, and thus we now require the introduction of a *joint* density distribution function. Here we have to describe the probability of simultaneous velocities $\mathbf{V}(\mathbf{r})$, $\mathbf{V}(\mathbf{r}_1)$ throughout the field, and consequently a greater number of geometrical dimensions must appear in the distribution function. But no additional physical dimensions (such as ϵ) need appear here, or in the eventual specification of those realizations which are equiprobable. According to the program of statistical mechanics, we should then seek the most probable distribution subject to the constraints mentioned above. The result can only depend on ϵ and certain geometric quantities like $\mathbf{r}$, and therefore (6.4.17) can only depend on ϵ and r. Since no nondimensional number can be formed from ϵ and r alone, it follows that C must be a universal numerical constant. Thus, the content of the *Kolmogoroff law* (6.4.17) is that the rms velocity difference increases as the $\frac{1}{3}$ power of the separation distance between two points.

The validity of the law is limited, on one hand, by the consideration that r must be small compared to the scale of the energy bearing eddies. The law is also inapplicable for very small r because a formal Taylor series expansion of $\mathbf{V}$ implies that $\langle \mathbf{V}^2 \rangle$ is proportional to r^2 as $r \to 0$. Thus, there must be additional dynamical constraints which have not been considered above, and which will reflect the importance of viscosity ν for the smallest scales of motion. In order to estimate the region of validity of (6.4.17), we form the Reynolds number

$$R_r \equiv r(\epsilon r)^{1/3}/\nu \tag{6.4.20}$$

from the separation distance and the relative velocity. Equation (6.4.17) is valid for those r which make $R_r \gg 1$, whereas viscosity assumes direct importance for those separation distances $r \sim \lambda$ at which $R_r \sim 1$. Setting (6.4.20) equal to unity and solving for r gives the Kolmogoroff length

$$\lambda = (\nu^3/\epsilon)^{1/4} \tag{6.4.21}$$

at which (6.4.17) becomes inapplicable.

References

Betchov, R., and Criminale, W. O., Jr. (1967). "Stability of Parallel Flows." Academic Press, New York.

Gibson, C. H., and Schwarz, W. H. (1963). *J. Fluid Mech.* **16,** 365-384.

Hinze, J. O. (1959). "Turbulence." McGraw-Hill, New York.

Lin, C. C. (1955). "The Theory of Hydrodynamic Stability." Cambridge Univ. Press, London and New York.

Malkus, W. V. R. (1956). *J. Fluid Mech.* **1,** 521–539.

Schlichting, H. (1960). "Boundary Layer Theory." 4th ed. McGraw-Hill, New York.

CHAPTER VII

Wind Driven Circulation

7.1 The Turbulent Ekman Layer

The solution (5.2.2) for the *laminar* Ekman layer has two components of velocity, these being given by linear combinations of the real and imaginary parts of $\exp \pm i^{1/2}(f/\nu)^{1/2}z$, where z denotes the vertical coordinate, f the Coriolis parameter, and ν the viscosity. The particular solution (5.2.4) is maintained by the pressure gradient of a geostrophic flow, and another particular solution can readily be obtained for the case in which a horizontally uniform wind stress of $\rho\tau$ (dynes/cm) acts on the free surface of a semi-infinite liquid of density ρ. Since there is now no horizontal pressure gradient, we set $G = 0$ in Eqs. (5.2.1), and since the velocities must be finite at $z = -\infty$, we set $A = 0$ in (5.2.2). The remaining integration constant B is then determined by setting τ/ν equal to the derivative of (5.2.2) at $z = 0$. The resulting solution is called an Ekman spiral, because of the way in which the vector formed from u, v rotates with depth.

Both of the laminar Ekman layers mentioned above become unstable and turbulent as the boundary layer Reynolds number increases. This number is based on the Ekman depth $(\nu/f)^{1/2}$ and for the wind driven problem, the velocity scale is $\tau(\nu f)^{-1/2}$, where $\tau = |\tau|$. Thus, the Reynolds number is

$$R = \tau(\nu/f)^{1/2}/(\nu f)^{1/2}\nu = (\tau/\nu f) \qquad (7.1.1)$$

The minimum critical value of this boundary layer Reynolds number varies between 10 and 100 (Tatro and Mollo-Christensen, 1967; Lilly, 1966; Faller and Kaylor, 1967), depending on whether the Ekman layer is driven by a wind or by a geostrophic pressure gradient. Most experiments have been performed under the latter conditions, and at $R = 500$, the flow is rather turbulent. The value of (7.1.1) is 10^6 for a typical wind stress of 1 dyne/cm^2, so that the oceanic boundary layer is turbulent, even in the absence of other mechanisms (Assaf *et al.*, 1971). Since so little is known about turbulence in nonrotating shear flows, it is both fortunate and remarkable that the integral property discussed below is independent of the detailed mechanics of the eddies.

Consider a semi-infinite ocean subjected to a uniform wind stress $\boldsymbol{\tau}$, and let $\mathbf{V}(z)$ denote the horizontally averaged velocity in the water. The fluctuating component $\mathbf{V}'(x, y, z, t)$ is associated with the turbulent eddies, and is schematically indicated by the loops in Fig. 7.1. Although a detailed description of the eddies is not necessary for the derivation of Eq. (7.1.6) the nomenclature introduced below will be useful later on. We shall use the terms "small scale" and "momentum transporting eddies" synonymously, to denote those horizontal wavelength components of the turbulence that are predominant in the downward transport of the momentum supplied by the wind. The space–time scale of these eddies is such that they are neither in geostrophic nor hydrostatic balance. The turbulent velocities decrease with depth, and we denote the characteristic depth of the turbulent Ekman layer by $z = -h_e$. The value of h_e is much less than the total depth of the ocean, and is estimated by the similarity theory given in Section 8.1. When $\boldsymbol{\tau}$ is uniform, $\mathbf{V}(z)$ is also confined to a depth h_e, and the term "pure boundary layer" will be used to refer to the case in which *all* of the wind induced velocities vanish at great depths. For the case (Section 7.3) in which $\boldsymbol{\tau}$ varies horizontally, we will find, on the other hand, that wind driven motions are induced at great depths.

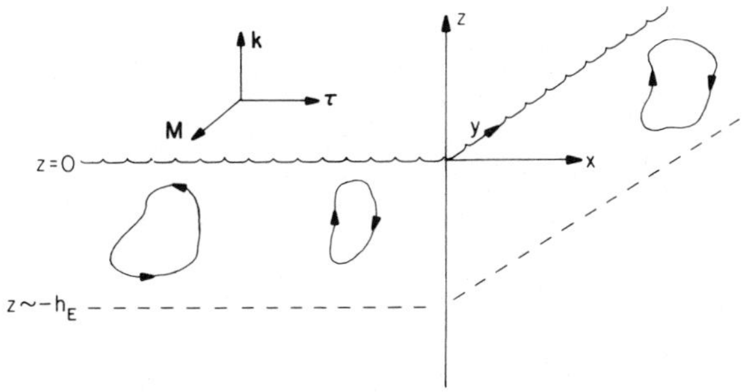

FIG. 7.1 Perspective diagram of a uniform x-directed wind stress $\boldsymbol{\tau}$ acting on the free surface ($z = 0$) of a semi-infinite ocean. $\mathbf{k}$ is the unit vertical direction and $\mathbf{M}$ is the vertically integrated value of the mean current. The small scale and turbulent eddies are schematically indicated by a loop.

We now consider the statistically steady field in the pure boundary layer, and form a momentum budget for the infinite horizontal strip between $z < 0$ and $z - dz$. Let $\rho\boldsymbol{\theta}(z)$ denote the average downward transport of horizontal momentum, or the stress, acting on the top of this strip. The stress on the bottom of the strip is then given by $-\rho\boldsymbol{\theta}(z - dz)$, and therefore $\rho\ \partial\boldsymbol{\theta}/\partial z\ dz$ is the average horizontal force due to the action of the turbulence. The average Coriolis force is $\rho\ dz\ f\mathbf{k} \times \mathbf{V}$, and since there is no horizontal pressure gradient in this case, the relation

$$f\mathbf{k} \times \mathbf{V}(z) = \partial\boldsymbol{\theta}/\partial z \tag{7.1.2}$$

expresses an exact momentum balance for the turbulent boundary layer.

Since the Reynolds stress $\boldsymbol{\theta}$ must vanish at great depths, and since $\rho\boldsymbol{\theta}$ must equal the given wind stress at $z = 0$, we have the boundary conditions

$$\boldsymbol{\theta}(0) = \boldsymbol{\tau} \qquad \text{and} \qquad \boldsymbol{\theta}(-\infty) = 0 \tag{7.1.3}$$

Therefore, the vertical integral of (7.1.2) is

$$f\mathbf{k} \times \mathbf{M} = \boldsymbol{\tau} \tag{7.1.4}$$

where

$$\mathbf{M} = \int_{-\infty}^{0} \mathbf{V}\, dz \tag{7.1.5}$$

By sketching the orthogonal triad $(\mathbf{M}, \mathbf{k}, \boldsymbol{\tau})$, it is readily seen that (7.1.4) is equivalent to

$$\mathbf{M} = (\boldsymbol{\tau} \times \mathbf{k})/f \tag{7.1.6}$$

This beautifully simple relation, when extended to the case of variable $\boldsymbol{\tau}, f$ in Section 7.3, provides the basis for the theory of the wind driven ocean circulation.

7.2 Where Does the Momentum Go?

In the steady state model considered above, there is a downward flux of horizontal momentum at the free surface, but no flux at great depths. Since the model is also horizontally homogeneous, it may then appear that the momentum supplied by the wind is disappearing in the boundary layer. The paradox is resolved in the following instructive generalization.

Let $\bar{u}_\phi(r)$ denote the *absolute* azimuthal velocity of an undisturbed axisymmetric vortex in a semi-infinite liquid, where r denotes the radial distance from the vertical axis of the vortex. The vortex is in equilibrium because of the balance of the centrifugal force with the radial pressure gradient, and u_ϕ^2/gr is the slope of the free surface. We suppose, however, that g is sufficiently large so that this slope can be neglected in that which follows, and thus the free surface is taken at $z = 0$. The problem then is to determine the velocities induced by the application of an axisymmetric wind stress $\tau(r)$ on $z = 0$. In particular, we want to construct the circular analog of Section 7.1, and to consider the budget of absolute *angular* momentum, this being the appropriate dynamical invariant. We will take into account the vertical transport of angular momentum by the turbulence and the lateral transport of angular momentum by the wind induced mean motion, but we will neglect the relatively small viscous stress that arises from the lateral variation in $\bar{u}_\phi(r)$, should the vortex not be in a state of solid body rotation.

The radial ($\tau_r(r)$) and the azimuthal ($\tau_\phi(r)$) components of τ will induce a mean radial velocity $\hat{u}_r(r, z)$ in the water, and the mean azimuthal component is denoted by $\bar{u}_\phi(r) + \hat{u}_\phi(r, z)$. We now ask for that distribution of τ which will produce a "pure" boundary layer, in the sense that $\hat{u}_r$, $\hat{u}_\phi$ vanish at great depths. A necessary condition for a pure boundary layer is that the radial volume flux

$$m = 2\pi \int_{-\infty}^{0} dz\, r\hat{u}_r(r, z) \tag{7.2.1}$$

be independent of r, for otherwise the divergent flow will induce motion at great depths (cf. Sections 5.3 and 7.3). Should the undisturbed vortex be contained in a circular annulus of *finite* radial width, then a constant value of (7.2.1) could be realized by having the same kind of clearance space at the vertical boundaries of the annulus as in the boundary layer flow of Fig. 5.1a.

Another necessary condition for a pure boundary layer is provided by the conservation of absolute angular momentum. Consider a control volume located between radii r and $r + dr$, and having semi-infinite vertical boundaries. The torque (in consistent units) of the wind on the top of this ring is

$$(2\pi r\, dr) r\rho\tau_\phi(r) \tag{7.2.2}$$

In the steady state, this torque is balanced for by the difference between the fluxes of angular momentum of the fluid at the vertical boundaries of the control volume. The absolute angular momentum (per unit volume) at any radius is given by $r\rho(\bar{u}_\phi + \hat{u}_\phi)$, and therefore the radial flux of angular momentum is

$$2\pi\rho \int_{-\infty}^{0} dz\, (r\hat{u}_r) r(\bar{u}_\phi + \hat{u}_\phi) = \rho r\bar{u}_\phi m + O(\hat{u}_r \hat{u}_\phi) \tag{7.2.3}$$

where the last term indicates the part of the integral that is quadratic in the amplitude of the velocity components induced by the wind. The $\rho r\bar{u}_\phi m$ term, on the other hand, is linear in this amplitude, and therefore dominant for sufficiently small m (or τ). The value of (7.2.3) at $r + dr$ minus its value at r

must equal (7.2.2), according to the angular momentum principle, and therefore we have

$$2\pi r^2 \tau_\phi = (\partial/\partial r)[r\bar{u}_\phi m + O(\hat{u}_r \hat{u}_\phi)]$$

As mentioned above, the quadratic $(\hat{u}_r \hat{u}_\phi)$ term can be neglected for sufficiently small τ, and since m is independent of r, we obtain

$$r^2 \tau_\phi = (m/2\pi)(\partial/\partial r)(r\bar{u}_\phi) \qquad \text{or} \qquad \frac{m}{2r\pi} = \frac{\tau_\phi(r)}{(\partial \bar{u}_\phi/\partial r) + (\bar{u}_\phi/r)} \tag{7.2.4}$$

From (7.2.4), we conclude that a necessary condition for a pure boundary layer is that the torque of the wind $(r\tau_\phi)$ must be proportional to the *absolute* vorticity

$$(\partial \bar{u}_\phi/\partial r) + (\bar{u}_\phi/r) \tag{7.2.5}$$

of the undisturbed vortex at all r. (If τ does not satisfy this proportionality, then the divergences in the Ekman layer will generate motions at great depths, as indicated by the theory of the following section.) The coefficient of proportionality between wind torque and (7.2.5) is the mass transport function m.

The correspondence between (7.2.4) and (7.1.6) can readily be made by considering the special case in which the undisturbed vortex is in solid body rotation with angular velocity $f/2$, so that $\bar{u}_\phi = fr/2$. For simplicity, we also take $\tau_r(r) = 0$, and Eq. (7.2.4) then reduces to

$$m/2r\pi = \tau_\phi/f \tag{7.2.6}$$

Since the mass transport per unit azimuthal distance $(m/2\pi r)$ equals the vertical integral of the radial velocity, and since the wind is directed in the azimuth $(\tau_r = 0)$, we see that (7.2.6) is in exact correspondence with (7.1.6) when the radius of curvature r is large. The momentum paradox in the latter problem is thereby removed by the fact that the divergence of the *absolute* momentum flux balances the wind stress in the horizontally homogeneous model of Section 7.1.

The preceding discussion, moreover, need not be restricted to the case in which the original vortex is in a state of solid body rotation. Equation (7.2.4) implies that a pure boundary layer occurs whenever $r\tau_\phi$ is proportional to the absolute vorticity, or to the sum of the Coriolis parameter and the relative vorticity. In this case, we see that the Ekman transport is inversely proportional to the *absolute* vorticity, and this provides an introduction to an effect discussed further in the next chapter.

7.3 Sverdrup Theory

We now want to consider the case in which the τ, f of Section 7.1 varies slowly with latitude, in accord with the global distribution of the mean atmospheric wind stress. For maximum simplicity, we will use the Cartesian β-plane approximation (Section 2.5), in which the x axis points eastward and the y axis points northward. For reasons mentioned previously, we shall also neglect the slight tilt of the free surface ($z = 0$) which is associated with the large scale pressure gradient computed below.

The *global* scale variation of τ can hardly influence the $\boldsymbol{\theta}$ of the *small* scale turbulence, and therefore the *local* balance of the eddy stress with the Coriolis force is still given by (7.1.6). But the slow variation of τ, f with latitude will produce a divergence in the Ekman transport $\mathbf{M}$, and the resulting "suction velocity" (Section 5.3) will induce an additional field underneath the turbulent boundary layer. We now give two derivations of this effect, and the reader may prefer the second [following (7.3.8)] because it is mathematically simpler. The first derivation, however, is more revealing and useful later on.

Consider the two layer model of Fig. 7.2, wherein the depth h_e of the turbulent layer is a small fraction of the thickness h of the upper layer, and the local value of the vertically integrated boundary layer flow is given by (7.1.6).

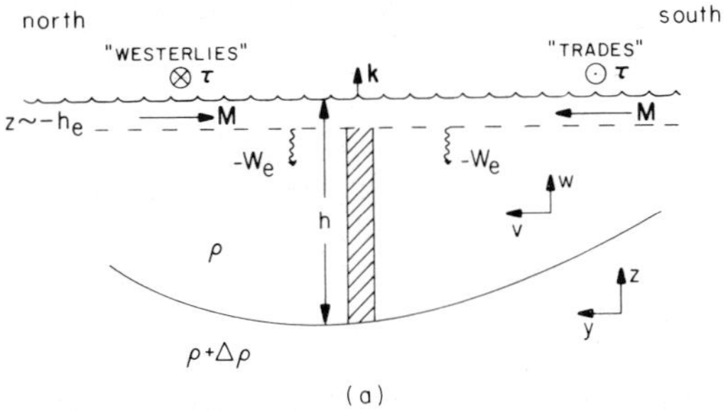

FIG. 7.2 A north–south vertical section through a two layer ocean driven by a variable zonal wind stress $\tau(y)$. y is positive in the north direction and the x direction is into the page. $h(x, y)$ is the steady state depth of the interface which separates light fluid (density ρ) and heavy fluid ($\rho + \Delta\rho$). The region of small scale turbulence is confined to the depth $h_e \ll h$, and $-w_e$ is the Ekman suction velocity.

By integrating the continuity equation with the boundary condition $w(x, y, 0) = 0$, we then find that the upward vertical velocity $w = w_e$ at the bottom of the Ekman layer is given by

$$w_e = \nabla \cdot \mathbf{M} \qquad \text{or by} \qquad w_e = \nabla \cdot (\tau \times \mathbf{k}/f) \tag{7.3.1}$$

when (7.1.6) is used. The wind stress assumed in the model (Fig. 7.3) acts in the zonal direction, and has magnitude $\tau(y)$. Thus, the northward Ekman transport is

$$M_y = -\tau(y)/f(y) \tag{7.3.2}$$

where $f(y)$ is the local Coriolis parameter, and $\beta = \partial f/\partial y$ in that which follows.

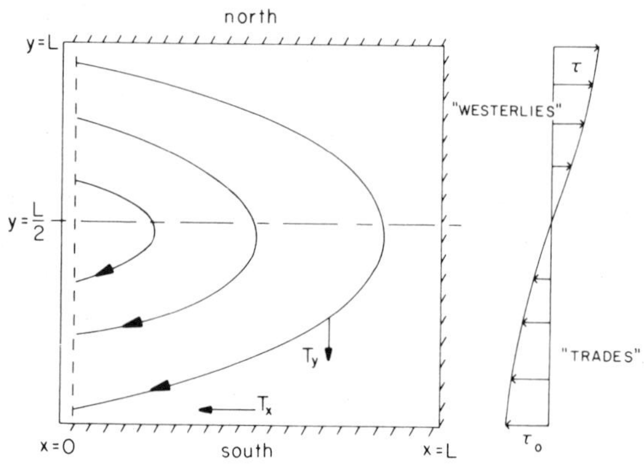

FIG. 7.3 Schematic diagram of total mass transport streamlines in a square ocean basin. The assumed wind stress distribution is shown on the right. T_x, T_y are the components of the total transport vector. The anticyclonic Sverdrup gyre extends only up to the western boundary layer.

The corresponding value of (7.3.1), or

$$w_e = -(\partial/\partial y)(\tau(y)/f) \tag{7.3.3}$$

is sketched in Fig. 7.2, and thus we see that downward directed velocities are forced into the fluid beneath the turbulent boundary layer at latitude $y = L/2$.

Consider now the effect of (7.3.3) on the marked material column (Fig. 7.2) whose top is located at $z \sim -h_e$ and whose bottom is located at the interface $z = -h(x, y)$ which separates the upper layer from a very deep bottom layer of density $\rho + \Delta\rho$. The downward directed Ekman suction velocity tends to "squash" the cylinder in the vertical direction, thereby increasing its cross-sectional area. The product of this cross-sectional area and the vertical component of the earth's vorticity f thereby tends to increase. But the motion in the region $z < -h_e$ must satisfy the circulation theorem (conservation of potential vorticity), and consequently a relative velocity must develop which will conserve the product of cross-sectional area with the normal component of absolute vorticity. This can be accomplished by a southward movement of the column, since $f(y)$ decreases southward, and the circulation around the "squashed" cylinder can thereby be conserved. Thus we see how geostrophic velocities are generated in the water beneath the Ekman layer, and the concomitant pressure gradient will then cause the interface $h(x, y)$ to adjust, as indicated in Fig. 7.2.

These quantities will now be computed for the case in which $\tau(y)$ and its gradient are small. The suction velocities and the induced geostrophic velocities are then small, and the nonlinear terms can be neglected in the equation for conservation of potential vorticity (3.7.3a). Thus, we have the linearized version (3.7.4), or

$$v\beta = f\,\partial w/\partial z \tag{7.3.4}$$

where v is the northward component of geostrophic velocity. Since $w = w_e$ at $z = -h_e$, and $w = -dh/dt$ at $z = -h$, the linear variation of w with z implies

$$\partial w/\partial z = (w_e + \mathbf{V}\cdot\nabla h)/(h - h_e) \simeq w_e/h \tag{7.3.5}$$

because $dh/dt = \mathbf{V}\cdot\nabla h$ is quadratic in $\mathbf{V}$, and also because $h_e \ll h$. The approximations made in obtaining (7.3.4) and (7.3.5) are

$$\zeta \ll f, \qquad |\mathbf{V}\cdot\nabla\zeta| \ll |\beta v|, \qquad |\mathbf{V}\cdot\nabla h| \ll |(\partial/\partial y)(\tau/f)| \tag{7.3.6}$$

The result of combining (7.3.5), (7.3.4), and (7.3.3) is

$$vh = -(f/\beta)(\partial/\partial y)(\tau/f) = -\beta^{-1}\,\partial\tau/\partial y + \tau/f \tag{7.3.7}$$

This transport beneath the Ekman layer is comparable with (7.3.2), but the typical value of v is much smaller than the typical boundary layer velocity M_y/h_e. Thus, we see that when the northward velocity is integrated from the interface to the free surface, the *total* transport function T_y is given by the sum of (7.3.7) and (7.3.2), or

$$T_y = -\beta^{-1}\,\partial\tau/\partial y \tag{7.3.8}$$

This is a particular case of the Sverdrup relation (Robinson, 1963; Stommel, 1965), which states that the total northward transport is equal to β^{-1} multiplied by the curl of the wind stress.

In the second and mathematically simpler derivation of the Sverdrup relation, we do not separate the total transport $\mathbf{T}$ into components (e.g., $\mathbf{M}$) but directly integrate the linearized momentum equation in the vertical. Thus, the integral of the Coriolis force from $z = -h$ to $z = 0$ is written as $f\mathbf{k}\times\mathbf{T}$, and the integral of the wind induced force $\partial\boldsymbol{\theta}/\partial z$ is $\boldsymbol{\tau}$. Since the lower layer (Fig. 7.2) is not in direct contact with the wind stress, we may assume the motion therein to be at rest in the steady state. Therefore, $g^*\nabla h$ is the horizontal pressure gradient force in the upper layer where g^* is the reduced value of gravity, and the vertically integrated pressure force then becomes $g^*h\,\nabla h$. Therefore, $f\mathbf{k}\times\mathbf{T} = -\nabla(g^*h^2/2) + \boldsymbol{\tau}$, and the vertically integrated continuity equation is $\nabla\cdot\mathbf{T} = 0$. By taking the curl of the first equation and by utilizing $\nabla\cdot\mathbf{T} = 0$, one readily obtains the Sverdrup relation for the northward component of $\mathbf{T}$ as a function of curl $\boldsymbol{\tau}$. The reader will also find it instructive to perform this calculation in spherical coordinates, so as to examine the validity of the β-plane approximation used in deriving (7.3.8).

In addition to the northward component T_y, there is also an eastward component T_x of the transport vector **T**. The conservation of mass requires that **T**(x, y) be nondivergent, and therefore

$$\partial T_x/\partial x + \partial T_y/\partial y = 0$$

There can be no transport normal to the eastern boundary, or

$$T_x(L, y) = 0$$

and consequently the integration of the previous equation gives

$$T_x = (x - L)(\partial/\partial y)\beta^{-1}\, \partial\tau/\partial y \tag{7.3.9}$$

In Fig. 7.3, we have $\partial\tau/\partial y = 0$ and $\partial^2\tau/\partial y^2 < 0$ at the northern boundary, and therefore $T_x > 0$ for $y > L/2$, as shown by the clockwise circulation of the streamlines of **T**.

We note, however, that (7.3.9) does not satisfy the boundary condition $T_x(0\, y) = 0$ at the western wall, and the deep significance of this is discussed subsequently. Since T_x changes sign near $y = L/2$, the eastward component of geostrophic velocity T_x/h must also change sign, and the associated north–south slope of the interface (Fig. 7.2) can be computed as follows. Since the lower layer is at rest, the slope of the interface is given by Margules' equation (4.4.5), and by using (7.3.7), we then find that the east–west slope of the interface is

$$\frac{\partial h}{\partial x} = \frac{fv}{g(\Delta\rho/\rho)} = -\frac{f^2}{\beta hg(\Delta\rho/\rho)}\frac{\partial}{\partial y}\left(\frac{\tau}{f}\right) \tag{7.3.10}$$

Therefore, the slope is upward to the east along the central latitude ($\tau = 0$).

Let us now examine the reason why the solution given above is unable to satisfy the western boundary condition $T_x(0, y) = 0$. The approximate vorticity equation (7.3.4) was based on (7.3.6), and the first two of these approximations can be written as

$$1 \gg |\mathbf{V}\cdot\nabla\zeta|/|v\beta| \sim |\zeta|/f \sim |v|/fL \sim \tau/f^2Lh \tag{7.3.11}$$

wherein we assume the lateral scale L of the wind to be the same as the radius of the earth, and (7.3.7) has also been used in (7.3.11). When the typical oceanic values

$$\tau = 1\ (\text{cm/sec})^2, \quad f = 10^{-4}\ \text{sec}^{-1}, \quad h = 10^5\ \text{cm}, \quad L = 5 \times 10^8\ \text{cm}$$

are inserted, we see that the right hand side of (7.3.11) is 10^{-5}, so that there is little doubt about the validity of the asymptotic solution. But the $\mathbf{V}\cdot\nabla\zeta$ term contains the *highest* horizontal derivatives in the exact vorticity equation. Therefore, the neglect of that term lowers the order of the differential equation, and thereby removes some of the solutions which are necessary for the satisfaction of all the inviscid boundary conditions. Accordingly, we now look for one of these "lost" solutions whose horizontal scale is much less than L, and for which the relations (7.3.11) and (7.3.4) are not applicable.

7.4 Inertial Western Boundary Layers

The procedure to be used in modifying the general circulation picture (Fig. 7.3), is suggested by the problem considered in Section 2.6. We showed that a westward current (Fig. 2.3) having small relative vorticity ζ tends to intensify into a thin jet as it is forced toward high latitudes by the western wall, and the nonlinear terms $\mathbf{V} \cdot \nabla\zeta$ are essential in the dynamics. Therefore, we expect a similar effect to occur as the flow in the southwest corner of Fig. 7.3 approaches the western boundary.

An expanded view of the southwest corner ($x = 0$, $y = 0$) is given in Fig. 7.4, in which the solid curves labeled h = constant are also the isobars or streamlines of the geostrophic flow above the interface. We now "stop" the linear solution of Section 7.3 at some small longitude $x = \delta$, and then proceed to compute the flow further downstream by including the previously neglected nonlinear vorticity terms (Charney, 1955; Morgan, 1956).

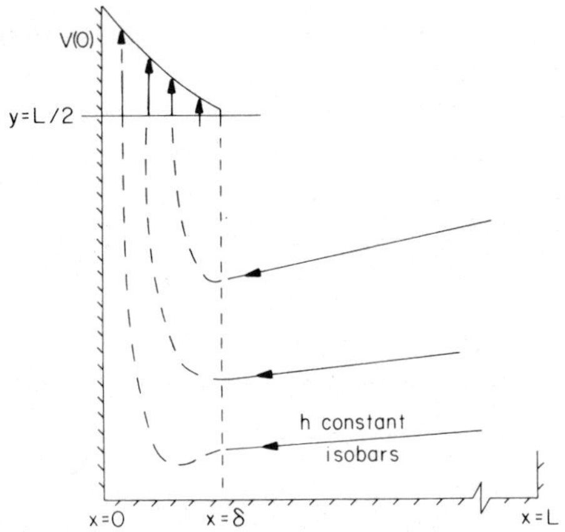

FIG. 7.4 The western boundary current and its connection with the geostrophic interior solution. See Figs. 7.2 and 7.3 and the text.

The present problem only differs from Fig. 2.3, because the model now includes an interface and the associated buoyancy $g\,\Delta\rho/\rho$ between two layers. Since the geostrophic velocities (7.3.7) at $x = \delta$ are fixed, the Margules equation, or (7.3.10), indicates that the gradients in h will be negligibly small for an asymptotic case in which g is very large. Therefore, if we start the discussion with this case, then the variations in h can be neglected, and the rigid bottom calculation (Fig. 2.3) can then be applied to the upper layer in Fig. 7.4. Thus, h is constant, and absolute vorticity $f + \zeta$ is conserved, as columns are deflected northward by the western boundary. If $\bar{f}$ denotes the Coriolis parameter at some high latitude ($y \sim L/2$), then the fluid on the boundary ($x = 0$) will have a relatively large vorticity $-\bar{f}$, and the east–west width is then given by (2.6.3), or

$$\delta \sim (\hat{T}/\bar{f})^{1/2} \tag{7.4.1}$$

where $h\hat{T}$ is total volume transport of the north-going boundary jet. Continuity requires this transport to equal the total southbound Sverdrup transport, as given by the longitudinal integral of (7.3.8), from $x = \delta$ to $x = L$, and thus we have

$$\hat{T}h \simeq \int_0^L dx\,(-T_y) = \beta^{-1} \int_0^L (\partial\tau/\partial y)\,dx \sim \tau L/\bar{f}$$

because $\partial\tau/\partial y \sim \tau/L$ and $\beta \sim \bar{f}/L$. The width (7.4.1) of the jet then becomes

$$\delta \sim (\bar{\tau}L/\bar{f}^2 h)^{1/2} \tag{7.4.2}$$

Since this width approaches zero, and since the right hand side of (7.3.11) also approaches zero as $\tau \to 0$, it is most appropriate to refer to the jet as an inertial *boundary* layer.

Let us now estimate the modification of the width of the western boundary current caused by a finite value of g. The variations in $h(x, y)$ are now dynamically significant, and we let Δh denote the typical variation in $h(x, y)$ across the width of the jet. From the geostrophic relation $fV(x) = g(\Delta\rho/\rho)\,\partial h/\partial x$, we obtain the order of magnitude relation $fV \sim g(\Delta\rho/\rho)\,\Delta h/\delta$, where $V \sim \hat{T}/\delta$. For large but finite g, we can use (7.4.1) as a first approximation in these relations, and thus we obtain

$$\frac{\Delta h}{h} \sim \frac{\bar{f}^2\delta^2}{g(\Delta\rho/\rho)h} \sim \frac{\tau}{g(\Delta\rho/\rho)h}\left(\frac{L}{h}\right) \tag{7.4.3}$$

If the right hand side of (7.4.3) is small, or

$$\frac{\tau}{g(\Delta\rho/\rho)h}\left(\frac{L}{h}\right) \ll 1 \tag{7.4.4}$$

then Δh is small, the variation of h along the western boundary is also small, and (7.4.2) will indeed determine the width of the jet. Equation (7.4.2) can also be used for order of magnitude purposes, even when the left hand side of (7.4.4) is of order unity. But for such values of the parameters, the buoyancy effect† must be considered in any quantitative theory, and the reader is referred to the literature cited previously. The significant nondimensional number (7.4.3) equals the square of the ratio of (7.4.2) to the Rossby radius of deformation. This number is nearly equal to unity for the Gulf Stream and Kuroshio boundary currents.

† See Section 4.1 for an example of the inertial boundary layer when no β effect is present.

The considerations of this section show that the Sverdrup solution (Section 7.3) can be accepted for the "interior" of the ocean, even though it does not satisfy the proper boundary conditions at the southwest boundary. The situation at the northwest boundary is discussed below.

7.5 Comparison with Observations

The interface in our model will be identified with some constant density surface at mid depth in the main thermocline of the ocean. Each of these density surfaces is observed to reach maximum depth near the latitude where the zonal wind stress changes sign, and thus we have qualitative agreement with the variation of the interface shown in Fig. 7.2 (Von Arx, 1962; Fuglister, 1960).

The total clockwise circulation of the North Atlantic and North Pacific Oceans has been determined by a geostrophic calculation of the southbound flow, using observed east–west density gradients. The transports also have been determined from measurements of the compensating northbound flow in the Gulf Stream and Kuroshio boundary currents. By using realistic wind stresses, Munk (1950) obtained favorable comparison with the Sverdrup transport. Munk also gave a completely closed circulation theory by assuming a viscously dominated boundary layer, in contrast with the inertial western boundary current expounded above. The precise dynamics of the western boundary, however, has negligible effect on the Sverdrup interior solution, and the total amount of water circulating clockwise is obtained by multiplying the maximum value of (7.3.8) with the width of the ocean. Thus Munk obtained 36×10^{12} gm/sec for the North Atlantic and 39×10^{12} for the North Pacific. Subsequent direct measurements of the northbound transport in the Gulf Stream gave a value of 33×10^{12} gm/sec at 30° N, but the transport continues to increase and reaches 147×10^{12} gm/sec, 2000 km further downstream (Knauss, 1969). This large downstream increase in transport has not been satisfactorily explained. For this reason, and for others given below, we have not attempted to continue the western boundary flow northward in Fig. 7.4, and thus our circulation pattern is not yet closed.

Reference is made to Hansen's (1970) description of the synoptic variations in the position of the axis of the Gulf Stream jet. South of Cape Hatteras (35° N) the axis is relatively straight and steady, but meanders of increasing amplitude develop downstream. Typical wavelengths are 200–400 km, and phase speeds are about 8 cm/sec (as compared to maximum jet velocities of 100–200 cm/sec). Occasionally, the amplitude of a meander wave becomes so large that a ring of water detaches from the main stream (Fuglister, 1972) and thereby deposits a large volume of cold fresh coastal water on the seaward side of the mean stream or, inversely, a large volume of warm salty ocean water on the shoreward side of the mean stream (Saunders, 1971). These meander waves may be related to the geostrophic instabilities discussed in Sections 4.3 and 4.4 and also to the large variations in the depth of the ocean bottom along the path of the Gulf Stream. The exchange of heat and salt that occurs when the meanders detach from the western boundary current may be important in the thermodynamical budgets of the tropical and polar oceans. Thus, the observations indicate the existence of important processes that are beyond the scope of the previous model, and therefore the dynamics of the northwest territory is left "open". The circulation pattern in Fig. 7.3 is not only kinematically open in the upper left corner, but the overall ocean energetics is also open, as indicated below.

The energy which is pumped downward at the bottom of the turbulent boundary layer is given by the product of the pressure $\phi(x, y, -h_e)$ and the vertical velocity

$$-w_e = -\nabla \cdot \mathbf{M} \tag{7.5.1}$$

when integrated over some constant level surface ($z = -h_e$) that lies beneath the Ekman layer. If dA denotes an area element on that surface, then the pressure work is

$$\oiint dA\,(-w_e)\phi(x, y, -h_e) = \oiint \mathbf{M} \cdot \nabla_2 \phi\, dA \tag{7.5.2}$$

since the normal component of $\mathbf{M}$ at the vertical coastal boundaries must, in fact, vanish. Since the Sverdrup theory is linear, the kinetic plus potential energy which develops in the interior of the ocean must be entirely due to the pressure work (7.5.2). Since the geostrophic relation

$$f\mathbf{k} \times \mathbf{V} = -\nabla_2 \phi$$

applies at $z = -h_e$, the value of (7.5.2) is

$$\oiint dA\, \tau \times \mathbf{k} \cdot \mathbf{V} \times \mathbf{k} = \oiint dA\, \tau \cdot \mathbf{V} \tag{7.5.3}$$

Thus, we see that the rate at which energy is supplied to the *deep* ocean is given by the product of wind stress and the *geostrophic* velocity, or by $\tau T_x/h$.

The geostrophic transport streamlines in the interior of the ocean (Fig. 7.3), imply a positive contribution to (7.5.3), and therefore energy is "still" being pumped downward into the steady Sverdrup interior! This energetic paradox is due to the fact that no mechanism has been provided for the transformation and dissipation of the work done by the wind, and a purely *inertial* western boundary current cannot be made to connect with the Sverdrup solution in the northwest corner of the basin. Munk (1950) solved this problem for the case of a homogeneous ocean by introducing lateral friction in the western boundary. But for a stratified ocean (Part II), we will need to utilize some of the wind work to mix waters of different temperature (salinity) and to thereby satisfy the global heat budget requirements of the ocean. Consequently, we will neither dispose of the prime energy source (wind work) nor close the circulation, until the thermodynamics is settled.

References

Assaf, G., Gerard, R., and Gordon, A. L. (1971). *J. Geophys. Res.* **76**, No. 27, 6550.

Charney, J. G. (1955). The Gulf Stream as an Inertial Boundary Layer. *Proc. Nat. Acad. Sci. Wash.* pp. 731–740.

Faller, A. J., and Kaylor, R. (1967). Instability of the Ekman Spiral with Applications to the Planetary Boundary Layers. *Phys. Fluids* **10**, Suppl. 212–219.

Fuglister, F. C. (1960). "Atlantic Ocean Atlas." Woods Hole Oceanographic Institution, Woods Hole, Massachusetts.

Fuglister, F. C. (1972). Cyclonic Rings Formed by the Gulf Stream 1965–1966. *In* "Studies in Physical Oceanography" (A. L. Gordon, ed.), Vol. I (Wust Birthday Volume). Gordon and Breach, New York.

Hansen, D. V. (1970). Gulf Stream Meanders between Cape Hatteras and the Grand Banks. *Deep Sea Res. Oceanogr. Abstr.* **17**, 495–511.

Knauss, J. A. (1969). A Note on the Transport of the Gulf Stream. *Deep Sea Res. Oceanogr. Abstr.* **16**, Suppl. 117–123.

Lilly, D. K. (1966). On the Instability of Ekman Boundary Layer Flow. *J. Atmos. Sci.* **23**, 481–494.

Morgan, G. W. (1956). On the Wind Driven Ocean Circulation. *Tellus* **8**, No. 3, 301–320.

Munk, W. H. (1950). On the Wind Driven Ocean Circulation. *J. Meteorol.* 7, No. 2, 79–93.

Robinson, A. R. (ed.) (1963). "Wind Driven Ocean Circulation." Ginn (Blaisdell), Boston, Massachusetts.

Saunders, P. M. (1971). Anticyclonic Eddies Formed from Shoreward Meanders of the Gulf Stream. *Deep Sea Res. Oceanogr. Abstr.* **18**, 1207–1219.

Stommel, H. M. (1965). "The Gulf Stream." 2nd ed. Univ. of California Press, Berkeley.

Tatro, P., and Mollo-Christensen, E. L. (1967). Experiments on Ekman Layer Instability. *J. Fluid Mech.* **28**, 531–543.

Von Arx, W. (1962). "An Introduction to Physical Oceanography." Addison-Wesley, Reading, Massachusetts.

CHAPTER VIII

Wind Driven Appendix

8.1 Depth of the Turbulent Ekman Layer

In Section 7.1, we determined the mass transport (7.1.6) in a "pure boundary layer" of vertical extent $z \sim -h_E$. If τ denotes the magnitude of the horizontally uniform wind stress, then a mass transport velocity can be defined by

$$U = |\mathbf{M}|/h_E = \tau/fh_E \tag{8.1.1}$$

The integrated theory of the turbulent boundary layer does not determine the separate values of U or h_E, and one may even ask whether a statistically steady boundary layer having finite depth h_E is possible. The point at stake here is illustrated by the nonrotating counterpart, wherein a steady state is *not* achieved because the momentum supplied by the uniform wind to the semi-infinite fluid *continually* diffuses downward by turbulent transport. Thus, we must now determine the factors which lead to the finite thickness h_E of the boundary layer in the rotating case.

The only externally imposed parameters that enter this problem are the uniform values of τ and f and the kinematic molecular viscosity ν. Only one length scale

$$h_E \equiv \tau^{1/2}/f \tag{8.1.2}$$

can be formed from τ and f, but the viscosity will introduce another length scale $(\nu/f)^{1/2}$, and the ratio of the two gives the Reynolds number (7.1.1). When the latter is large, we have the turbulent case for which the following similarity theory (Caldwell *et al.*, 1972) applies. If one assumes, somewhat arbitrarily, that the depth of the turbulent boundary layer is independent of ν, then that depth must be equal to (8.1.2) multiplied by some universal constant. We note that the value of (8.1.2) is about 100 m for typical ocean parameters, and this figure is a relevant upper bound for the thickness of the oceanic mixed layer. This similarity argument does not, however, provide any insight into the mechanism by which the boundary layer is limited, and the latter point is considered in the remainder of this section. [It should also be pointed out that the stratification of the underlying thermocline plays an important role in determining the depth of the turbulent boundary layer (Pollard *et al.*, 1973) of the ocean.]

The tendency of the eddies is to mix momentum downward and to increase the vertical extent of the turbulent boundary layer. We shall now develop the idea that this tendency is limited by the spontaneous formation of inertia oscillations when the layer depth exceeds a certain value. These inertia oscillations will have a larger space–time scale than the turbulence, and the waves will be generated as a new instability of the shear flow. The effect of such an instability will be to radiate energy to $z = -\infty$, thereby removing part of the source which would otherwise be available to the smaller scales of turbulent motion. Thus, the onset of the new instability mode provides a mechanism for limiting the downward spread of the turbulent layer (and the amplitude of the inertial oscillations is likewise so limited).

Suppose the turbulent boundary layer to be in equilibrium with the horizontally averaged momentum equation given by (7.1.2), and let us inquire into the stability with respect to a new kind of small perturbation $\mathbf{V}'$. This problem is extraordinarily difficult because the interaction of $\mathbf{V}'$ with the field of turbulence, as well as with the mean field $\mathbf{V}(z)$, should be taken into account. But we assume that $\mathbf{V}'$ will derive its energy from $\mathbf{V}(z)$, and therefore the interaction of the infinitesimal $\mathbf{V}'$ with the small scale turbulence is neglected. Likewise, we also neglect the infinitesimal change in the small scale stress $\boldsymbol{\theta}(z)$ when $\mathbf{V}'$ is applied, and therefore the linearized equation of motion becomes

$$\partial \mathbf{V}'/\partial t + \mathbf{V}' \cdot \nabla \mathbf{V}(z) + \mathbf{V} \cdot \nabla \mathbf{V}' + f\mathbf{k} \times \mathbf{V}' = -\nabla p', \qquad \nabla \cdot \mathbf{V}' = 0$$

For any given $\mathbf{V}(z)$, this leads to a linear eigenvalue equation for $\mathbf{V}' \propto e^{i\omega t}$, and we now pose the following question. Under what conditions do modes $\mathbf{V}'$ exist such that they continually propagate (ω = real) energy away from the boundary layer flow $\mathbf{V}(z)$ and into the far field ($z = -\infty$)? As a first orientation into the solution of this stability problem, and because the form of $\boldsymbol{\theta}(z)$ is not known, we will consider the (unrealistic) case in which $\partial\boldsymbol{\theta}/\partial z$ is directed along the y axis with constant magnitude in a shear layer $0 > z > -h$. The upper surface ($z = 0$) is assumed to be dynamically rigid ($w' = 0$), and $\boldsymbol{\theta}(z) = 0$ beneath the $z = -h$ interface. Since $\partial\boldsymbol{\theta}/\partial z$ is balanced by $f\mathbf{k} \times \mathbf{V}$, we see that the basic flow $[\mathbf{V} = (U(z+h)/h,\ 0)]$ is directed along the x axis with uniform vertical shear and with U denoting the surface current. For $z < -h$, we have $\mathbf{V} = 0$. By further restricting the analysis to the case of longitudinal modes, we can express the vertical velocity w' by the real part of

$$w(z)e^{ily}e^{i\omega t} \qquad (l > 0)$$

and because these modes have no x variation, the perturbation equation for $z > -h$ reduces to

$$\begin{aligned} i\omega u - fv + w(U/h) &= 0 \\ i\omega v + fu &= -ilp \\ i\omega w &= -\partial p/\partial z \\ ilv + \partial w/\partial z &= 0 \end{aligned} \tag{8.1.3}$$

where u and v are the x, y perturbation velocities and p is the pressure term. The corresponding equations for the lower layer can be obtained from (8.1.3) by setting $U = 0$.

The result of eliminating u from the first two equations in (8.1.3) is

$$v(f^2 - \omega^2) - fwU/h = lp\omega$$

and the use of the last equation in (8.1.3) then gives

$$(i/l)(f^2 - \omega^2)\,\partial w/\partial z - fwU/h = lp\omega \tag{8.1.4}$$

When the third equation in (8.1.3) is used to eliminate p, Eq. (8.1.4) becomes

$$\frac{\partial^2 w}{\partial z^2} + \frac{iflU}{h(f^2 - \omega^2)}\frac{\partial w}{\partial z} + \left(\frac{l^2\omega^2}{f^2 - \omega^2}\right)w = 0 \tag{8.1.5}$$

This is to be solved subject to $w(0) = 0$ and subject to the boundary condition obtained by connecting the solution with the field of vertical velocity $w_-(z)$ in the region of zero shear ($z < -h$). The eigenfunction for the latter region can be obtained by setting $U = 0$ in (8.1.5), and thus we have

$$\frac{\partial^2 w_-}{\partial z^2} + \frac{l^2\omega^2 w_-}{f^2 - \omega^2} = 0 \tag{8.1.6}$$

The eigenvalue relation for ω is now obtained from the solutions of (8.1.5) and (8.1.6) by matching the vertical velocities and the pressures at $z = -h$. We will show that when h is sufficiently large, there is a solution having the real frequency

$$0 < \omega < f \tag{8.1.7}$$

which *continually* pumps energy downward to $z = -\infty$ from the shear layer. Thus, the basic state is unstable at this critical value of h because of the monotonic loss of energy produced by such (neutral) modes.

When (8.1.7) is satisfied, there are two sinusoidal solutions of (8.1.6), one of which is

$$w_- = \exp + il\omega(f^2 - \omega^2)^{-1/2} z \tag{8.1.8a}$$

and the corresponding pressure field obtained by setting $U = 0$ in (8.1.4) is

$$p_- = \frac{i(f^2 - \omega^2)}{l^2 \omega} \frac{\partial w_-}{\partial z} = -(1/l)(f^2 - \omega^2)^{1/2} w_- \tag{8.1.8b}$$

Note that $(\mathrm{Re}\, p-)(\mathrm{Re}\, w-) < 0$ for $l > 0$, and this negative pressure work implies *downward* energy propagation by (8.1.8a). The other solution of (8.1.6) propagates energy upward (from $z = -\infty$), and it is therefore discarded because the only possible energy source is the shear layer *above* $z = -h$.

By equating the pressures (8.1.4) above and below $z = -h$, and by using

$$w(-h) = w_-(-h)$$

we obtain the matching condition

$$\frac{\partial w(-h)}{\partial z} + \frac{1}{h}\left(\frac{iflU}{f^2 - \omega^2}\right) w(-h) = \frac{\partial w_-(-h)}{\partial z} = \left[\frac{il\omega}{(f^2 - \omega^2)^{1/2}}\right] w(-h)$$

or

$$\frac{\partial w(-h)}{\partial z} + i\lambda w = 0 \qquad \text{where} \qquad \lambda = \frac{fUl}{h(f^2 - \omega^2)} - \frac{l\omega}{(f^2 - \omega^2)^{1/2}} \tag{8.1.9}$$

Equation (8.1.9) is the lower boundary condition for the solution of (8.1.5). Since $w(0) = 0$ is the upper boundary condition, the solution becomes

$$w = A\left[e^{i\beta_1 z} - e^{i\beta_2 z}\right] \tag{8.1.10}$$

where A is an integration constant and β_1 and β_2 are the two roots of

$$\beta^2 + \beta \frac{fUl}{h(f^2 - \omega^2)} - \frac{l^2 \omega^2}{f^2 - \omega^2} = 0$$

From the solution of this quadratic equation, we have

$$\beta_2-\beta_1=\left[\left(\frac{flU}{h(f^2-\omega^2)}\right)^2+\frac{4l^2\omega^2}{f^2-\omega^2}\right]^{1/2} \quad \text{and} \quad \beta_2+\beta_1=-\frac{flU}{h(f^2-\omega^2)} \tag{8.1.11}$$

The eigenvalue equation obtained by substituting (8.1.10) in (8.1.9) then becomes

$$(\beta_2+\lambda)/(\beta_1+\lambda)=e^{i(\beta_2-\beta_1)h} \tag{8.1.12}$$

Since $\beta_2-\beta_1>0$ is real, and since λ is also real. Eq. (8.1.12) can be satisfied only if

$$h(\beta_2-\beta_1)=(2m-1)\pi \qquad \text{and} \qquad (\beta_2+\lambda)/(\beta_1+\lambda)=-1$$

where m is any integer. When (8.1.11) is used, the first of these two equations becomes

$$\frac{(2m-1)^2\pi^2}{h^2}=\frac{f^2l^2U^2}{h^2(f^2-\omega^2)^2}+\frac{4l^2\omega^2}{f^2-\omega^2} \tag{8.1.13}$$

and the second equation becomes

$$-\tfrac{1}{2}(\beta_2+\beta_1)=\lambda$$

When (8.1.11) and (8.1.9) are substituted, the latter equation becomes

$$\frac{flU}{2h(f^2-\omega^2)}=\frac{flU}{h(f^2-\omega^2)}-\frac{l\omega}{(f^2-\omega^2)^{1/2}}$$

and simplification gives

$$U/fh=2(\omega/f)(1-\omega^2/f^2)^{1/2} \tag{8.1.14}$$

The maximum value of the right hand side of (8.1.14) is unity (corresponding to $\omega/f=1/\sqrt{2}$), and therefore if

$$hf/U<1 \tag{8.1.15}$$

there will be no neutral solution. On the other hand, if $h>U/f$, then the frequency of the neutral solution is determined by (8.1.14), and the possible values of the wavelength are determined by (8.1.13). The neutral wave in the lower region is an inertial oscillation which *continually transfers energy* downward via the pressure-work term mentioned previously, and this energy flux is supplied by the mean flow above $z=-h$.

We have suggested that if $h>U/f$, then the pure boundary layer of the equilibrium state is not stable because it will spontaneously radiate inertial oscillations to the far field. The nonlinear modifications brought about by the latter effect would probably reduce the turbulence and the thickness of the boundary layer, because of the energy removed from the latter region. Thus, we argue that an excessively deep boundary layer would be "driven back" to a smaller h by the onset of such new modes as discussed above. For sufficiently small values of (8.1.15), on the other hand, all the modes are trapped near the boundary layer and are therefore available for the maintenance of the turbulence in that layer.

This calculation provides a basis for assuming that the equilibrium depth of the turbulent Ekman layer is determined by (8.1.15), or $h_E = U/f$. A second relation between h_E and U is given by (8.1.1), and from the combination, we then obtain $h_E \sim \tau^{1/2}/f$, in agreement with the similarity assertion (8.1.2). The calculation also suggests that the statistical fluctuations in the depth of the oceanic boundary layer may generate inertia waves which then radiate downward into the thermocline (see Chapter IX).

8.2 Ekman Transport Generalized

The purpose of this section is to consider the transport **M** of the turbulent boundary layer (Section 7.1) from a more general point of view, and the class of problems to which the results apply will be indicated in the following sections.

The Reynolds method for separating the mean and fluctuating components of a turbulent field is unambiguous in all previously considered problems. In Section 6.2, for example, the average velocity is obtained by integrating the total velocity over the entire horizontal plane, and the fluctuation is then given by the residual. Likewise, the mean motion in Section 7.3 may be unambiguously defined at any point by forming the *time* average of the total velocity. A problem arises, however, in those problems for which the "mean" motion of interest varies in space and time. In order to use Reynolds' method to separate these from the smaller scales of turbulent eddies, there must exist a spectral gap which separates the two different scales. Accordingly, we specify that the "mean" motion to be discussed herein has a horizontal scale which is so large that the hydrostatic approximation applies to it. Such scales correspond to those inertial motions which were investigated earlier in this book, and for which the law of conservation of potential vorticity applied. We now seek an augmentation of these hydrostatic equations which takes into account the average effect of the small scale eddies that arise when a wind stress τ is present.

The result of averaging the velocity field over x, y, t intervals that are large compared to the scale of the eddies but small compared to the "mean" motion scale is denoted by

$$\mathbf{V}(x, y, z, t) + w(x, y, z, t)\mathbf{k} \tag{8.2.1}$$

where w is the vertical component and **V** the horizontal component. The average vertical transport of horizontal momentum (Reynolds stress) produced by the small scale turbulence has a local value denoted by $\boldsymbol{\theta}(x, y, z, t)$, and $\partial\boldsymbol{\theta}/\partial z$ is the corresponding force which must be inserted into the horizontal equation of motion for $\mathbf{V}$. Although there are, in general, nine stress components, it is clear that the forces other than $\partial\boldsymbol{\theta}/\partial z$ will be negligible when the vertical dimension h_E of the turbulent boundary layer is small compared to the lateral scale of the mean motion. Therefore, the equations of motion for the mean (hydrostatic) field are

$$\partial\mathbf{V}/\partial t + (\mathbf{V}\cdot\nabla_2 + w\,\partial/\partial z)\mathbf{V} + f\mathbf{k}\times\mathbf{V} = -\rho^{-1}\nabla_2 p + (\partial/\partial z)\boldsymbol{\theta} \qquad (8.2.2)$$

$$\partial p/\partial z = -\rho g \qquad (8.2.3)$$

$$\nabla_2\cdot\mathbf{V} + \partial w/\partial z = 0 \qquad (8.2.4)$$

where p denotes pressure, and the symbol ∇_2 is again used for the horizontal component of nabla.

Although the detailed structure of $\boldsymbol{\theta}$ is not known, the boundary value

$$\boldsymbol{\theta}(x, y, 0, t) = \tau \qquad (8.2.5)$$

is given, and we also know that $\boldsymbol{\theta}$ must vanish for $z \ll -h_e$. We assume that the interface depth $z = -h(x, y, t)$ in Fig. 8.1 is much greater than h_e, so that the lower boundary condition on $\boldsymbol{\theta}$ becomes

$$\boldsymbol{\theta}(x, y, -h, t) = 0 = \partial\boldsymbol{\theta}(x, y -h, t)/\partial z \qquad (8.2.6)$$

In the use of (8.2.5), and also in the subsequent use of the kinematical boundary condition

$$w(x, y, 0, t) = 0 \qquad (8.2.7)$$

we neglect the small tilt of the free surface ($z = 0$) relative to the level surface (but the horizontal pressure gradient at $z = 0$ will not be neglected).

Since the turbulence is confined to the upper half of the upper layer

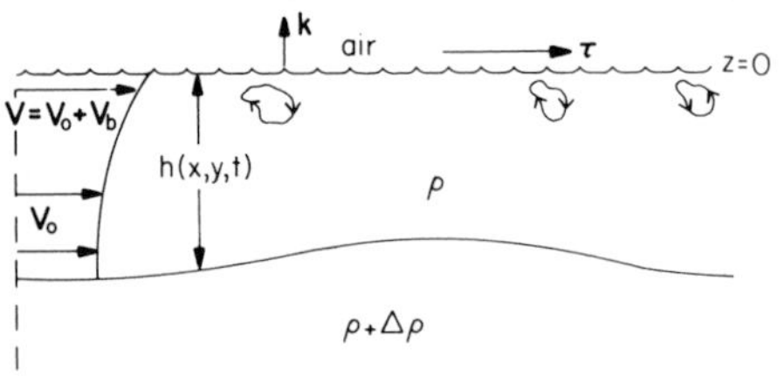

FIG. 8.1 Schematic diagram used for the small stress expansion. The stress vector τ acts on the free surface at $z = 0$, producing small scale eddies in the upper part of a homogenous layer of density ρ. The latter extends to the interface at $z = -h(x, y, t)$, below which is another homogenous layer of density $\rho + \Delta\rho$. The total horizontal velocity $\mathbf{V}(x, y, z, t)$ in the upper layer equals the sum of the two components discussed in the text.

(Fig. 8.1), the last term in (8.2.2) will vanish in the lower part of the upper layer and also in the fluid of density $\rho + \Delta\rho$. In these two regions, (8.2.2) will then reduce to the ideal fluid equations. Thus, $\partial \mathbf{V}/\partial z = 0$ in the region *immediately* above $z = -h$, but $\mathbf{V}$ increases with z in the upper part of the upper layer as shown in Fig. 8.1. Therefore, it is convenient to separate the velocity field (8.2.1) into two fields, denoted by $[\mathbf{V}_0(x, y, t), w_0(x, y, z, t)]$ and $[\mathbf{V}_b(x, y, z, t), w_b]$. The first field will be required to satisfy the equation (8.2.16) which is formally obtained from (8.2.2) by setting $\partial \boldsymbol{\theta}/\partial z = 0$, and the second field will be required to satisfy the residual momentum balance. Since the latter is nonlinear, the coupling of the two fields will be reflected therein. But (8.2.4) is linear, and therefore if the first field be required to satisfy the continuity equation, then the second field must also satisfy it, or

$$\partial w_0/\partial z = -\nabla_2 \cdot \mathbf{V}_0(x, y, t) \tag{8.2.8}$$

$$\partial w_b/\partial z = -\nabla_2 \cdot \mathbf{V}_b(x, y, z, t) \tag{8.2.9}$$

The total field (8.2.1) is given by the sums

$$\mathbf{V} = \mathbf{V}_0 + \mathbf{V}_b \qquad \text{and} \qquad w = w_0 + w_b \tag{8.2.10}$$

and we require that $(\mathbf{V}, w)$ must approach $(\mathbf{V}_0, w_0)$ in the region of the interface. Thus, the second field is required to satisfy the lower boundary conditions

$$0 = \mathbf{V}_b(x, y, -h, t) = w_b(x, y, -h, t) \qquad \text{and} \qquad 0 = \partial \mathbf{V}_b/\partial z = \partial w_b(x, y, -h, t)/\partial z \tag{8.2.11}$$

One of the coupling conditions between the two fields, obtained from (8.2.7) and (8.2.10), is

$$w_0(x, y, 0, t) = -w_b(x, y, 0, t)$$

By integrating (8.2.9) in z, and by using the boundary conditions (8.2.11), we can express the previous equation as

$$w_0(x, y, 0, t) = \nabla \cdot \mathbf{M} \tag{8.2.12}$$

where

$$\mathbf{M} = \int_{-h(x,y,t)}^{0} \mathbf{V}_b \, dz \tag{8.2.13}$$

The subscript 2 in (8.2.12) has been discarded because it is superfluous or redundant, and the same rule is followed below. By integrating (8.2.8) in z, and by using the interfacial boundary condition, $w_0(x, y, -h, t) = -dh/dt$, we also have

$$w_0(x, y, 0, t) + dh/dt = -h \, \nabla \cdot \mathbf{V}_0(x, y, t) \tag{8.2.14}$$

The substitution of (8.2.12) then gives

$$\partial h/\partial t + \nabla \cdot (\mathbf{V}_0 h) = -\nabla \cdot \mathbf{M} \tag{8.2.15}$$

In the region just above $z = -h$, $\mathbf{V} = \mathbf{V}_0, \partial\boldsymbol{\theta}/\partial z = 0$, and therefore (8.2.2) reduces to

$$\partial \mathbf{V}_0/\partial t + (\mathbf{V}_0 \cdot \nabla)\mathbf{V}_0 + f\mathbf{k} \times \mathbf{V}_0 = -\rho^{-1} \nabla_2 p \tag{8.2.16}$$

But Eqs. (8.2.15) and (8.2.16) do not give a closed system for $\mathbf{V}_0, h$ because $\mathbf{M}$ is not yet determined.

To obtain this relation for $\mathbf{M}$, we first substitute (8.2.10) in (8.2.2), and by subtracting (8.2.16) we then obtain

$$\frac{\partial \mathbf{V}_b}{\partial t} + (\mathbf{V}_b \cdot \nabla_2)\mathbf{V}_0 + (\mathbf{V}_0 \cdot \nabla_2)\mathbf{V}_b + (\mathbf{V}_b \cdot \nabla_2)\mathbf{V}_b + (w_0 + w_b)\frac{\partial \mathbf{V}_b}{\partial z} + f\mathbf{k} \times \mathbf{V}_b = \frac{\partial \boldsymbol{\theta}}{\partial z} \tag{8.2.17}$$

When this equation is integrated from $z = -h$ to $z = 0$, the last term becomes τ, and the next to last term becomes $f\mathbf{k} \times \mathbf{M}$ [cf. (8.2.13)]. The vertical integral of the first term in (8.2.17) equals $\partial\mathbf{M}/\partial t$, because $\mathbf{V}_b(x, y, -h, t)$ vanishes at the lower limit, $z = -h(x, y, t)$, of integration. Likewise, the vertical integral of $(\mathbf{V}_0 \cdot \nabla_2)\mathbf{V}_b$ equals $(\mathbf{V}_0 \cdot \nabla_2)\mathbf{M}$, and the vertical integral of $(\mathbf{V}_b \cdot \nabla_2)\mathbf{V}_0$ equals $(\mathbf{M} \cdot \nabla)\mathbf{V}_0(x, y, t)$. Consider next the vertical integral of the term

$$\begin{aligned}(w_0 + w_b)\, \partial \mathbf{V}_b/\partial z &= (\partial/\partial z)[(w_0 + w_b)\mathbf{V}_b] - \mathbf{V}_b(\partial/\partial z)(w_0 + w_b) \\ &= (\partial/\partial z)[(w_0 + w_b)\mathbf{V}_b] + \mathbf{V}_b(\nabla \cdot \mathbf{V}_0) + \mathbf{V}_b(\nabla_2 \cdot \mathbf{V}_b)\end{aligned}$$

Since $w_0 + w_b = 0$ at $z = 0$, and since $\mathbf{V}_b = 0$ at $z = -h$, the vertical integral is

$$\int_{-h}^{0} (w_0 + w_b)(\partial \mathbf{V}_b/\partial z)\, dz = \mathbf{M}(\nabla \cdot \mathbf{V}_0) + \int_{-h}^{0} \mathbf{V}_b(\nabla_2 \cdot \mathbf{V}_b)\, dz$$

Therefore, the vertical integral of (8.2.17) can be written as

$$\partial \mathbf{M}/\partial t + (\mathbf{M} \cdot \nabla)\mathbf{V}_0 + (\mathbf{V}_0 \cdot \nabla)\mathbf{M} + \mathbf{M}(\nabla \cdot \mathbf{V}_0) + f\mathbf{k} \times \mathbf{M} + \mathbf{Q} = \tau \tag{8.2.18a}$$

where

$$\mathbf{Q} \equiv \int_{-h}^{0} dz(\mathbf{V}_b \cdot \nabla_2 \mathbf{V}_b + \mathbf{V}_b\, \nabla_2 \cdot \mathbf{V}_b) = \nabla_2 \cdot \int_{-h}^{0} \mathbf{V}_b \mathbf{V}_b\, dz \tag{8.2.18b}$$

and the dyadic notation $(\mathbf{V}_b\mathbf{V}_b)$ merely means that the x component of $\mathbf{Q}$ is to be computed by multiplying $\mathbf{V}_b$ with its x component and by taking the divergence of the vector so formed.

In the case where $\mathbf{V}_0 = 0$ and τ is constant, we have $\mathbf{Q} = 0$ because the vertical integral of the momentum transport term (8.2.18b) is horizontally homogeneous, and therefore (8.2.18a) reduces to the familiar Ekman relation (7.1.6) under steady state conditions.

Let us consider next the inverse case in which the wind stress is "small" and the $\mathbf{V}_0(x, y, t)$ component is large, or

$$|\tau|^{1/2}/|\mathbf{V}_0| \ll 1 \tag{8.2.19}$$

The amplitude of the $\mathbf{V}_b$ component and the boundary layer transport $\mathbf{M}$ will then be small in the same sense as $|\tau|$. But the $\mathbf{Q}$ term is quadratically small in $\mathbf{V}_b$ (or $|\tau|$), and therefore Q is negligible compared to those terms in (8.2.18) which are either linear in $\mathbf{M}$ or bilinear in M and $\mathbf{V}_0$. Therefore, in the limit of a small wind stress, Eq. (8.2.18) reduces to

$$\partial\mathbf{M}/\partial t + (\mathbf{M} \cdot \nabla)\mathbf{V}_0 + (\mathbf{V}_0 \cdot \nabla)\mathbf{M} + \mathbf{M}(\nabla \cdot \mathbf{V}_0) + f\mathbf{k} \times \mathbf{M} = \tau \tag{8.2.20}$$

This equation describes the evolution of the boundary layer transport $\mathbf{M}$ as a functional of $\mathbf{V}_0$, and the joint solution with (8.2.15) and (8.2.16) determines $\mathbf{V}_0$ and M.

The nature of the coupling can be made more explicit by considering the two layer model in Fig. 8.1, and by assuming that the lower layer is sufficiently deep so that the motions therein can be neglected at all times. Thus, the horizontal pressure gradient just above the interface is $\rho g^* \nabla h$, where $g^* = g\,\Delta\rho/\rho$ (see Section 3.6), and therefore (8.2.16) becomes

$$\partial\mathbf{V}_0/\partial t + (\mathbf{V}_0 \cdot \nabla)\mathbf{V}_0 + f\mathbf{k} \times \mathbf{V}_0 = -g^* \nabla h \tag{8.2.21}$$

Equations (8.2.21), (8.2.20), and (8.2.15) are now seen to constitute a complete set of five scalar equations in five scalar unknowns h, $\mathbf{V}_0$, and $\mathbf{M}$. This generalized system contains no arbitrary turbulence parameter, and therefore it may be used to explore new effects which arise from the fact that typical ocean velocities are much larger than the time-averaged velocities in Chapter VII.

8.3 Example of Wind Stress–Vortex Interaction

Suppose that the upper layer contains a geostrophically balanced current at time $t < 0$, and let $(0, v(x))$ denote the (x, y) components of this initial flow. Suppose that a horizontally uniform wind stress having components $(0, \tau \sin \omega t)$ is then applied at $t > 0$, where ω is the given frequency of the wind stress. We will show that the boundary layer transport $(\mathbf{M} = M_x, M_y)$ is divergent and that the field, $\mathbf{V}_0(x, t) = (u_0, v_0(x, t))$ will then evolve from the initial state $(0, v(x))$ according to Eqs. (8.2.20) and (8.2.21).

Let us also assume that the initial state is cyclic (periodic) in the x direction, and averages in that direction will be denoted by a bar. Thus, we define an average transport function by

$$\mathbf{T} = \overline{h\mathbf{V}_0} \tag{8.3.1}$$

Since the fields are independent of y and cyclic in x, a term such as

$\overline{(g^* \nabla h^2/2)}$ will vanish, and therefore the product of h with (8.2.21) will have an average given by

$$0 = \overline{h\, \partial \mathbf{V}_0/\partial t} + \overline{h\mathbf{V}_0 \cdot \nabla \mathbf{V}_0} + \overline{f\mathbf{k} \times \mathbf{V}_0 h}$$

or

$$0 = \partial \mathbf{T}/\partial t - \overline{\mathbf{V}_0\, \partial h/\partial t} - \overline{\mathbf{V}_0\, \nabla \cdot (\mathbf{V}_0 h)} + f\mathbf{k} \times \mathbf{T}$$

When (8.2.15) is used, we then obtain

$$\partial \mathbf{T}/\partial t + f\mathbf{k} \times \mathbf{T} = -\overline{\mathbf{V}_0\, \nabla \cdot \mathbf{M}} \tag{8.3.2}$$

Let us now assume that τ is so small, and therefore $\mathbf{M}$ is so small, that we can linearize the equations of motion with respect to this small quantity. Thus, the linearized value of the y component of the last term in (8.3.2) is $-v(x)\, \partial M_x/\partial x$, where $v(x)$ is the undisturbed component of $\mathbf{V}_0$. On the other hand, the x component, or $-u_0\, \partial M_x/\partial x$, is neglected below because $u_0 \to 0$ as $\tau \to 0$. Thus, the linearization of (8.3.2) gives

$$\partial T_x/\partial t - fT_y = 0 \qquad \text{and} \qquad \partial T_y/\partial t + fT_x = -[\overline{v(x)\, \partial M_x/\partial x}]$$

where T_x and T_y are the x, y components of $\mathbf{T}$. By elimination and simplification, we then obtain

$$f^{-1}((\partial^2/\partial t^2) + f^2)T_x = \overline{(M_x\, \partial v/\partial x)} \tag{8.3.3}$$

The field equation for $\mathbf{M}$ is given by (8.2.20), and it is readily found that the linearization yields

$$\partial M_x/\partial t - fM_y = 0 \qquad \text{and} \qquad \partial M_y/\partial t + M_x\, \partial v/\partial x + fM_x = \tau \sin \omega t \tag{8.3.4}$$

By eliminating M_y, we then have

$$(\partial^2/\partial t^2 + f^2 + f\, \partial v/\partial x)M_x = \tau f \sin(\omega t)$$

and for $t > 0$, we have the solution

$$M_x = \frac{\tau f \sin \omega t}{f^2 - \omega^2 + f\, \partial v/\partial x} \tag{8.3.5}$$

When this is substituted in (8.3.3), the solution for $T_x(t)$ is

$$T_x = \frac{\tau f^2 \sin \omega t}{f^2 - \omega^2} \overline{\left\{ \frac{\partial v/\partial x}{f^2 - \omega^2 + f\, \partial v/\partial x} \right\}}$$

$$= \frac{\tau f^2 \sin \omega t}{(f^2 - \omega^2)^2} \left[-\frac{f}{f^2 - \omega^2} \overline{\left(\frac{\partial v}{\partial x}\right)^2} + \ldots + O\overline{\left(\frac{\partial v}{\partial x}\right)^4} \right] \tag{8.3.6}$$

When the basic vorticity $\partial v/\partial x$ vanishes, (8.3.6) vanishes, and the entire motion then consists of an oscillating Ekman transport (8.3.5), whose amplitude becomes infinite at the resonant frequency $\omega = f$. In the more general case, however, the boundary layer transport (8.3.5) will vary on a horizontal scale which is equal to that of the basic geostrophic velocity. Moreover, the amplitude of the oscillations in $\mathbf{V}_0(x, t)$, or (8.3.6), will increase more rapidly than the amplitude of (8.3.5) as $\omega \to f$. [For further examples of the way in which the vertical component of geostrophic vorticity influences the wind driven motion, see Stern (1966) and Niiler (1969)]. The effect discussed herein must also be distinguished from those motions which are directly forced by the *horizontal variation* in the wind (see Pollard, 1970). All of these considerations bear on the problem of explaining the strength of the horizontal velocity fluctuations which are observed in the surface layer of the sea.

8.4 Equatorial Undercurrent

There are important features of the mean atmospheric wind stress and the mean ocean circulation that are not indicated in Fig. 7.3. Thus, a relatively strong "equatorial countercurrent" is located between 6°N and 9°N in the Atlantic Ocean (Montgomery and Stroup, 1962). This eastward flowing current may be related to the fact that the maximum value of the trade winds occurs somewhat north of the equator (Munk, 1950). On the other hand, the "equatorial undercurrent" is observed on the equator (of all the oceans), and is more difficult to reconcile with the Sverdrup theory of Chapter VII, especially since Eq. (7.3.7) becomes infinite if we extend the southern boundary to the equator where $f = 0$.

Observations (Tsuchiya, 1968) of the Pacific undercurrent indicate that surfaces of constant density in the thermocline slope upward in the eastward direction, and a relatively strong eastward current is found within a zone of $1\frac{1}{2}°$ of latitude on either side of the equator. The largest value of the eastward directed transport (40×10^{12} cm^3/sec) (Knauss, 1966) occurs in the center (140°W) of the Pacific, and the maximum eastward speed (150 cm/sec) occurs at a depth of 150 m on the equator. The upper 20 m, however, is often found to be moving in the same direction (westward) as the prevailing equatorial wind. The eastward transport of the Pacific undercurrent also increases slowly from the western boundary to the center of the ocean, and then decreases slowly toward the eastern boundary (Tsuchiya, 1968; Montgomery, 1962; Knauss, 1964, 1966). A similar equatorial undercurrent occurs in the Atlantic (Montgomery, 1962), and also in the Indian Ocean (Taft and Knauss, 1967) when the monsoon blows westward. The seasonal reversal of the monsoon does not, however, produce a simple reversal of the current, but the normal longitudinal pressure gradient is weakened and stronger eddies also appear. Let us now set some of these facts into the context of the same two layer model (Fig. 8.2) that was previously used in the exposition of the extratropical circulation.

Since the Coriolis parameter vanishes on the equator, it is interesting to consider first the wind driven circulation in a nonrotating model, such as applies to a lake. As the superficial water is pushed downwind, water is piled up on the lee shore, a horizontal pressure gradient is established beneath the superficial layer, and therefore the subsurface water moves in the opposite direction to the superficial water. In a two-layer fluid (such as Fig. 8.2), we then obtain a closed vertical circulation cell for the wind driven motion in the upper layer. The lower layer has no horizontal pressure gradients because it is assumed to be deep and resting. Consequently, the slope of the interface must be opposite that of the free surface, and the relatively large value of the latter provides a basis for computing the wind stress, as follows. The steady state momentum principle implies that the wind stress times the width of the lake is equal to the difference between the vertically integrated pressure force on the two shores of the lake.

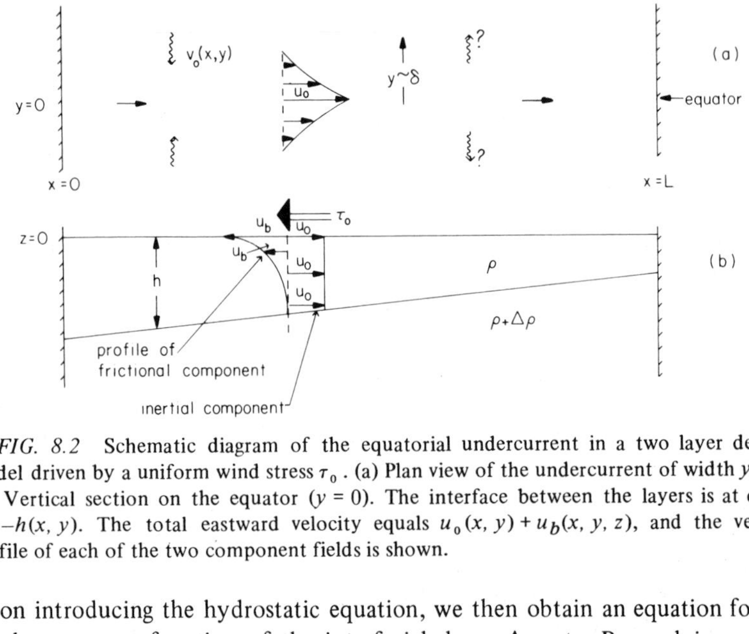

FIG. 8.2 Schematic diagram of the equatorial undercurrent in a two layer density model driven by a uniform wind stress τ_0. (a) Plan view of the undercurrent of width $y \sim \delta$. (b) Vertical section on the equator ($y = 0$). The interface between the layers is at depth $z = -h(x, y)$. The total eastward velocity equals $u_0(x, y) + u_b(x, y, z)$, and the vertical profile of each of the two component fields is shown.

Upon introducing the hydrostatic equation, we then obtain an equation for the wind stress as a function of the interfacial slope, $\Delta\rho$, etc. By applying such a method to density measurements made on the equator, it is possible to determine the mean wind stress, and the value used by Munk (1950) was 0.3–0.5 dyn/cm^2. The nonrotating model fails, however, to explain why the transport of the undercurrent is so much greater than the transport of the superficial water. Thus, we now turn to the equatorial dynamics on a rotating sphere, the discussion again being confined to order of magnitude relations, and the reader is also referred to the mathematical theories cited below.

The horizontal plane shown in Fig. 8.2a is tangent to the sphere at the equator, and we use the Cartesian β-plane approximation with the y axis pointing northward and with the x axis pointing eastward. The wind stress is directed westward with uniform magnitude τ_0, and we also assume that the heavy fluid beneath the interface at $z = -h(x, y)$ is deep and resting. Thus, Eqs. (8.2.21), (8.2.20), and (8.2.15) apply to the problem of determining the steady values of the boundary layer transport $\mathbf{M}$ and the velocity $\mathbf{V}_0\ (x, y)$ immediately above the interface. The eastward component $u_0(x, y)$ of $\mathbf{V}_0$ is indicated on one half of the profile in Fig. 8.2b, and the x component u_b of $\mathbf{V}_b(x, y, z, t)$ is indicated on the other half of the profile. The sum $u_0 + u_b$ gives the total eastward velocity, and the total eastward transport is given by $hu_0 + M_x$, where M_x and M_y are the x, y components of $\mathbf{M}$.

Although we shall not attempt to solve these complicated equations here, the most important point to note is that (8.2.20) is nonsingular at the equator ($f = 0$), as compared to (7.3.2), and the new system of equations reduces to the Sverdrup theory at high latitudes. In that which follows, we give a similarity theory which is based upon the new system at the equator, and its connection with the Sverdrup theory in extratropical latitudes.

In Fig. 8.2a, the northward component v_0 of $\mathbf{V}_0$ is sketched, and the jet-like profile of the eastward component u_0 anticipates the fact that the relatively strong undercurrent is confined within a small distance $y \sim \delta$ from the equator ($y = 0$). We want to determine the way in which the magnitudes of u_0 and δ depend on the external parameters

$$\tau_0, \quad L, \quad g^*H, \quad \beta = \partial f(y)/\partial y$$

where H is the given mean depth of the interface and L is the width of the ocean.

For reasons of symmetry, we have

$$v_0(x, 0) = 0 \tag{8.4.1}$$

and since $f(0) = 0$, the only terms that appear in the x component of (8.2.21) at the equator are

$$u_0(x, 0)\ \partial u_0/\partial x = -g^*(\partial/\partial x)h(x, 0) \tag{8.4.2}$$

This equation states that the pressure gradient just above the interface is balanced by a downstream acceleration. Equation (8.4.2) can be immediately integrated, and if we tentatively assume L to be so large that

$$|h(0, 0) - h(L, 0)| \sim H \tag{8.4.3}$$

then we obtain

$$u_0 \sim (g^*H)^{1/2} \tag{8.4.4}$$

as the maximum eastward velocity in the undercurrent. This result, being independent of the driving force τ_0, can only make sense in some asymptotic context relating L and τ_0. Before turning to this question, we list below the typical values of the oceanic parameters:

$$\tau_0 = \tfrac{1}{2}(\text{cm/sec})^2, \qquad H = 10^4 \text{ cm}$$

$$\beta = 2 \times 10^{-13} \text{ cm}^{-1} \text{ sec}^{-1}, \qquad g^* = (g\, \Delta\rho/\rho) = 3 \text{ cm}^2/\text{sec}$$

Thus, the value of (8.4.4) is $u_0 \sim \sqrt{3} \times 10^2$ cm/sec, and this is in reasonably good agreement with the previously cited value of the maximum velocity in the undercurrent.

The simplest way to estimate the order of magnitude of the width L of the basin which is required for the validity of (8.4.4) is to formulate a momentum balance for a strip of unit width which is centered on the equator and which extends from $x = 0$ to $x = L$. Thus, the wind stress exerts a force (consistent units) of $\tau_0 L$ on the surface of this strip, and for order of magnitude purposes, this must be balanced by the difference between the vertical integral of the pressure forces at the $x = 0$, L boundaries. From the hydrostatic equation, we find that this integrated pressure force equals

$$\tfrac{1}{2}g^*h^2(0,0) - \tfrac{1}{2}g^*h^2(0,L) \sim g^*H^2$$

where (8.4.3) has been used. Therefore, the momentum budget for the strip requires

$$\tau_0 L \sim g^*H^2 \qquad \text{or} \qquad L \sim g^*H^2/\tau_0 \sim 6 \times 10^3 \text{ km} \tag{8.4.5}$$

Therefore, (8.4.3) or (8.4.4) only applies when the ocean width is equal to, or larger than, (8.4.5). Since the numerical value given in (8.4.5) is, in fact, compatible with the equatorial width of the oceans, we can proceed.

Let us now consider the trajectory of an element of fluid which is initially located in the western half of the ocean, at latitude $y \sim \delta$, and at a depth which is located just above the interface. At this point, the velocity and the relative vorticity ζ correspond to, and can be computed from, the Sverdrup theory (Chapter VII) of the extratropical ocean. Let us recall that $\zeta \ll f(y)$ at such "high" latitudes, and therefore the vertical component of absolute vorticity of the aforementioned fluid element is $f(\delta) \sim \beta\delta$. As the element continues to move southward with the flow, the potential vorticity must be conserved (Fofonoff and Montgomery, 1955; Charney, 1960). Since the layer thickness h changes only by a factor of 2 (or so), we see that when the fluid reaches the equator ($f = 0$) the *relative* vorticity ζ must be of the same order $\beta\delta$ as the Coriolis parameter at $y \sim \delta$. The relative vorticity on the equator is $-\partial u_0/\partial y \sim u_0/\delta$, since $v_0(x, 0) = 0$, and therefore $u_0/\delta \sim \beta\delta$, or

$$\delta \sim (u_0/\beta)^{1/2} \sim (g^*H/\beta^2)^{1/4} \sim 3 \times 10^2 \text{ km} \tag{8.4.6}$$

This theoretical estimate of the width of the jet is also in reasonable agreement with the observations cited previously. The unrealistic cusp which appears in the u_0 profile of Fig. 8.2a is due to the fact that no dissipative mechanism has been included in this model, and therefore parcels of fluid arrive on either side of the equator with absolute vorticities of opposite sign (Charney and Spiegel, 1971; Gill, 1971).

The eastward transport of the undercurrent is given by the product of $u_0 H$ and δ, or

$$u_0 H\delta \sim (g^* H)^{1/2} H (g^* H/\beta^2)^{1/4} \tag{8.4.7}$$

and this is also independent of the magnitude of the wind stress. In order to obtain the total eastward transport, it is necessary to subtract from (8.4.7) the meridional integral of M_x, as mentioned previously, but the following short discussion will indicate that the order of magnitude of the integrated M_x is smaller than (8.4.7).

The Ekman theory of Section 7.1 is applicable at the relatively high ($y \sim \delta$) latitude where f is much larger than ζ. At the edge of the jet, we therefore have $M_y \sim \tau/(\beta\delta)$ and $M_x = 0$. At the lower latitudes, the Coriolis parameter should be replaced by the absolute vorticity in the Ekman transport relation (cf. Section 7.2), but the transport M_x *parallel* to the wind still vanishes. This kind of successive approximation to **M** breaks down, however, very close to the equator where the high order derivatives in (8.2.20) become important. Thus M_x can be large within a "sublayer" centered on $y = 0$. But even if M_x/h becomes as large as (8.4.4) on $y = 0$, the small ($\ll\delta$) meridional extent of the sublayer implies that the meridional integral of M_x is small compared to (8.4.7). Thus, we conclude that (8.4.7) is the order of magnitude of the total eastward transport of the undercurrent.

From the picture sketched above, it is easy to see (Fig. 8.2a) how the southbound Sverdrup velocities v_0 in the *western* half of the ocean couple into the equatorial undercurrent and thereby provide the observed downstream increase of transport. The situation in the eastern half of the ocean is, however, most puzzling from both the observational and theoretical point of view. Thus, the poleward motions indicated by the wiggly arrows (with the question mark) have been inserted only because of the imperative demands of the continuity equation, and we must emphasize that such poleward motions are *not* consistent with the southward motions shown in the lower right hand part of Figure 7.3. Thus, the observed decrease in the transport of the undercurrent from the center to the eastern boundary has not been qualitatively explained, and the nature of the coupling with higher latitudes is also not understood. It seems most appropriate, therefore, to leave the wind driven circulation at this point, and to begin a consideration of the effects associated with the continuous density variation in the ocean.

References

Caldwell, D. R., VanAtta, C. W., and Helland, K. N. (1972). A Laboratory Study of the Turbulent Ekman Layer. *Geophys. Fluid Dynamics* **3**, 125–160.

Charney, J. G. (1960). Non-linear Theory of a Wind Driven Homogeneous Layer near the Equator. *Deep Sea Res.* **6**, 303–310.

Charney, J. G., and Spiegel, S. L. (1971). Structure of Wind Driven Equatorial Currents in Homogeneous Oceans. *J. Phys. Oceanogr.* **1**, No. 3, 149–160.

Fofonoff, N. D., and Montgomery, R. B. (1955). The Equatorial Undercurrent in the Light of the Vorticity Equation. *Tellus* **7**, 518–521.

Gill, A. E. (1971). The Equatorial Current in a Homogeneous Ocean. *Deep Sea Res. Oceanogr. Abstr.* **18**, 421–431.

Knauss, J. A. (1964). Subsurface Ocean Currents. *Res. Geophys.* **2**, 271–290.

Knauss, J. A. (1966). Further Measurements of the Cromwell Current. *J. Mar. Res.* **24**, 205–240.

Montgomery, R. B. (1962). Equatorial Undercurrent Observations in Review. *J. Oceanogr. Soc. Japan.* 20th Anniv. Vol. 287–398.

Montgomery, R. B. and Stroup, E. D. (1962). Equatorial Waters and Currents at 150° W in July–August, 1962. *Johns Hopkins Oceanogr. Stud.* **1**, 68 pp.

Munk, W. H. (1950). On the Wind Driven Ocean Circulation. *J. Meteorol.* **7**, No. 2, 79–93.

Niiler, P. P. (1969). On the Ekman Divergence in an Oceanic Jet. *J. Geophys. Res.* **74**, No. 28, 7048–7052.

Pollard, R. (1970). On the Generation by Winds of Inertial Waves in the Ocean. *Deep Sea Res. Oceanogr. Abstr.* **17**, 795–812.

Pollard, R., Rhines, R., and Thompson, R. (1973). The Deepening of the Wind-Mixed Layer. *Geophys. Fluid Dynamics* **3**, 381–404.

Stern, M. E. (1966). Interaction of a Uniform Wind Stress with Hydrostatic Eddies. *Deep Sea Res. Oceanogr. Abstr.* **13**, 193–203.

Taft, B. A., and Knauss, J. A. (1967). The Equatorial Undercurrent of the Indian Ocean, etc. *Bull. Scripps Inst. Oceanogr.* **9**, 1–163.

Tsuchiya, M. (1968). "The Upper Waters of the Inter Tropical Pacific Ocean," Vol. 4. Johns Hopkins Univ. Press, Baltimore, Maryland.

CHAPTER IX

Stratification

9.1 Equation of State and the Parcel Method

The density ρ of sea water depends on pressure p, temperature T, and salinity S, the latter being expressed in grams solute per gram of solution. The equation of state $\rho = \rho(p, T, S)$ can be expressed in the differential form

$$\rho^{-1}\, d\rho = +\gamma_T\, dp - \alpha\, dT + \beta\, dS \tag{9.1.1}$$

where $\alpha = -\rho^{-1}\, \partial\rho/\partial T$ is the coefficient of thermal expansion, $\gamma_T = +\rho^{-1}\, \partial\rho/\partial p$ is the *isothermal* compressibility, and $\beta = \rho^{-1}\, \partial\rho/\partial S$ is the fractional increase in density per unit increase in salinity when T and p are held constant.

If a stratified fluid is at rest, then the hydrostatic equation is

$$\partial\bar{p}/\partial z = -g\bar{\rho}(z) \tag{9.1.2a}$$

The *static stability* s is defined by the negative of $\rho^{-1}\, \partial\bar{\rho}(z)/\partial z$, and by using (9.1.1) and (9.1.2a) we have

$$\rho^{-1}\, \partial\bar{\rho}/\partial z = -\gamma_T\bar{\rho}g - \alpha\, \partial\bar{T}/\partial z + \beta\, \partial\bar{S}/\partial z \tag{9.1.2b}$$

We will now give a simple description of the nature of the restoring forces that arise when a small parcel of this stratified fluid is displaced vertically, and a more rigorous treatment appears in Section 9.3.

Suppose we isolate a small parcel of the fluid by means of a cellophane wrapper which is impervious to the flow of heat and salt, but which may expand freely if the pressure of the surrounding fluid be decreased. Let $\rho_0(z)$ denote the density at the center of the parcel in the undisturbed position z, T_0 the temperature, and S_0 the salinity. Suppose that an external force is applied so as to raise the parcel slowly to height $z + \delta z$, without disturbing the horizontal stratification of the surrounding fluid. The hydrostatic pressure acting on the parcel is then changed by $\delta p = -\rho_0 g\, \delta z$ (for small δz), and the adiabatic expansion will change the density of the parcel to

$$\rho_0(z + \delta z) = \rho_0(z) + \gamma_a \rho_0\, \delta p = \rho_0(z) - \gamma_a \rho_0{}^2 g\, \delta z \tag{9.1.3}$$

where

$$\gamma_a \equiv \rho_0^{-1}(\partial \rho / \partial p)_a \tag{9.1.4}$$

is called the *adiabatic* compressibility. From thermodynamics, we know that the volume of a fluid will decrease by a greater amount for isothermal compression than for adiabatic compression, and therefore $\gamma_a < \gamma_T$. We shall show that the dynamics of the displaced parcel depends upon the difference of the two compressibilities, or on the quantity

$$c_1 \equiv (\rho\gamma_T - \rho\gamma_a)^{-1/2} \tag{9.1.5}$$

This quantity has the same dimensions, and order of magnitude, as the speed of sound in the liquid.

According to Archimedes (213 B.C.), the upward directed pressure force acting on the parcel at $z + \delta z$ is equal to the local volume of the parcel multiplied by the local density $\bar{\rho}(z + \delta z)$ of the *surrounding* fluid. The downward force of gravity is equal to the product of the volume and the density $\rho_0(z + \delta z)$ of the parcel. Thus the resultant force per unit mass in the upward direction is

$$(g/\rho_0)[\bar{\rho}(z + \delta z) - \rho_0(z + \delta z)] = g[\rho_0^{-1}\, \partial\bar{\rho}/\partial z + \gamma_a \rho_0 g]\, \delta z$$

where (9.1.3) has been used. This determines the magnitude of the external force which is required to hold the parcel in equilibrium at $z + \delta z$, or the upward directed acceleration $d^2(\delta z)/dt^2$, if the external force is removed. In the latter case, we have the equation of motion

$$(d^2/dt^2)(\delta z) = -N^2\, \delta z \tag{9.1.6}$$

where

$$N^2 = -g[\rho_0^{-1}\, \partial\bar{\rho}/\partial z + \gamma_a \rho_0 g] = g[\alpha\, \partial\bar{T}/\partial z - \beta\, \partial\bar{S}/\partial z + (g/c_1{}^2)] \tag{9.1.7}$$

and where (9.1.2b) and (9.1.5) have been used. From (9.1.6), we see that the parcel will oscillate with the *Brunt–Väisälä frequency N.*

The restoring force on a displaced parcel vanishes ($N^2 = 0$) in an isohaline ($\partial\bar{S}/\partial z = 0$) fluid when the basic temperature gradient is

$$\partial\bar{T}/\partial z = -g/\alpha c_1^2 \tag{9.1.8}$$

and in such a stratification, the temperature $T_0(z + \delta z)$ of the displaced parcel must equal the temperature $\bar{T}(z + \delta z)$ of the surrounding fluid, because the densities (as well as the pressures) are equal when $N = 0$. Therefore, the adiabatic temperature change δT_0 of a parcel, is equal to $(\partial\bar{T}/\partial z)\,\delta z$, and when (9.1.8) is used, we have

$$\delta T_0 = -(g/\alpha c_1^2)\,\delta z \qquad \text{or} \qquad \delta(T_0 + (zg/\alpha c_1^2)) = 0 \tag{9.1.9}$$

Thus, we see that the *potential temperature* $(T_0 + zg/\alpha c_1^2)$ of a parcel is conserved, and such an adiabatic invariant is clearly independent of the stratification (9.1.8). The latter was used as an artifice to obtain this invariant, whereas a more rigorous thermodynamic derivation would start from the entropy. Also note that the salinity, like the mass, is conserved.

From (9.1.9), we see that the parcel temperature T_0 is approximately conserved when the speed of sound [or (9.1.5)] is large, and this approximation will be used in all that follows (corrections can readily be obtained by merely replacing T_0 with the potential temperature). The variability in the coefficients α and β will also be neglected subsequently, but first we discuss the "caballing" effect which is due to this variation in α at low temperature.

In a process which takes place at constant pressure ($dp = 0$) and density ($d\rho = 0$), the equation of state (9.1.1) implies an isopycnal relation $T = T(S)$, the slope of the curve being

$$dT/dS = \beta/\alpha \tag{9.1.10}$$

In pure water, we know that the coefficient of thermal expansion α vanishes at 4° C and increases with temperature. This temperature of maximum density is depressed by the salt content, and α increases with temperature in the entire range of sea water. Consequently, the value of (9.1.10) and the slope of the T–S curve (Fig. 9.1) increases as temperature decreases. Suppose we take m gm of water, having coordinates T_1, S_1, and mix it thoroughly with $(1 - m)$ gm of the T_2, S_2 type. The resulting 1-g mixture contains $mS_1 + (1 - m)S_2$ gm of salt, and thus the salinity of the mixture is

$$S = mS_1 + (1-m)S_2$$

Likewise, the conservation of internal energy implies that the temperature of the mixture is

$$T = mT_1 + (1-m)T_2$$

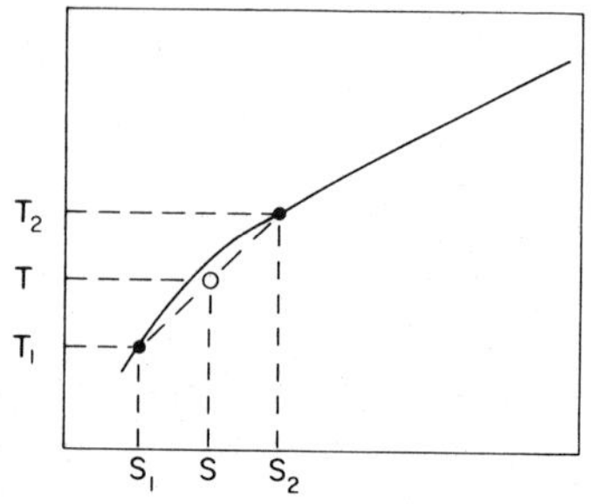

FIG. 9.1 The curve indicates the variation in temperature and salinity at constant density and pressure. An isopycnal curve of greater density would lie below the curve shown. Point (T, S) represents a mixture of two water samples with properties (T_1, S_1) and (T_2, S_2).

Thus, the T, S point representing the mixture lies somewhere on the straight line (Fig. 9.1) that connects T_1, S_1 and T_2, S_2. The precise location of the T, S point depends on the value of m, but all points on the connecting line lie *beneath* the relevant isopycnal curve, and therefore the mixture will have a density greater than that of the original component. The dynamical consequences of this effect can be appreciated by considering the stability of a fluid stratified with compensating temperature–salinity gradients. Suppose, for example, that the density is uniform, and that the vertical variation of (T, S) corresponds to the curve in Fig. 9.1. A small perturbation of this equilibrium stratification will cause some mixing, and consequently the mixture will have a greater density than basic state. The fluid so formed will proceed to sink and thereby generate new motions and additional mixing. Thus, we see that the basic state of uniform density is unstable because of the variable α. If the T–S gradients in the basic state are not quite compensating so that the density decreases upward, then the basic state may still be unstable with respect to large amplitude perturbations which bring about the mixing effect mentioned above.

9.2 Rotating Stratified Fluids

With the exception of Sections 9.2 and 9.9, this chapter is concerned with small scale effects in a nonrotating fluid, and therefore the reader may prefer to proceed immediately to the example given in Section 9.3. In this section, we give some general properties of rotating stratified fluids, an example of which can be found in Section 9.9. Other examples are implicit in the layer models of Part I, and thus the reader may generalize the theory of Kelvin waves and Rossby waves to the case of a uniformly stratified fluid.

(a) *Statics.* If a stratified fluid be in a state of solid body rotation ω, then the centrifugal acceleration is $-\nabla(\omega^2 R^2/2)$, where the notation of Section 2.1 has been used. Therefore, the force per unit volume due to the centrifugal and gravity accelerations can be written as $\rho(x, y, z)\,\nabla\psi$ where ψ is the total potential function. The static equilibrium is then described by

$0 = -\nabla p - \rho \nabla \psi$, where p is the pressure field. Thus the component of $\nabla \psi$ along an isobaric surface must vanish, and the level surfaces of constant ψ coincide with the isobaric surfaces. We also have $\nabla \times \rho^{-1} \nabla p = 0 = \nabla \rho \times \nabla p$, and this implies that the isopycnal surfaces coincide with the isobaric surfaces (and also with the level surfaces). Figure 9.2 is a plane section passing through the axis of rotation (not shown), and the isopycnals are indicated by dashed lines. The sum of the gravity and centrifugal accelerations acts in the $-z$ direction, and g denotes the effective magnitude. The projection of 2ω on the local normal (z axis) is denoted by the Coriolis parameter f, and the symbol f_H will be used to denote the horizontal component of 2ω. The $z = 0, H$ boundaries are parts of level surfaces (e.g., spheres with the y axis pointing north).

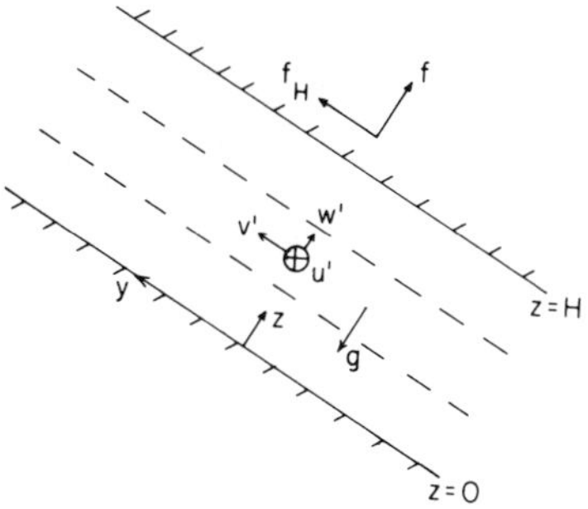

FIG. 9.2 A plane section passing through the axis of rotation (not shown) of a stratified fluid contained between two rigid level surfaces ($z = 0, H$). The projection of the rotation vector 2ω on the vertical axis is denoted by f. See Sections 9.2 and 9.9.

Let us now consider an axisymmetric equilibrium state in which the isopycnal surfaces are inclined to the level surfaces, so that a steady geostrophic velocity $\mathbf{V}' = (u'(y, z), 0, 0)$ appears in the x, y, z directions of the rotating Cartesian coordinate system. The Coriolis force and its x, y, z components are

$$2\rho\omega \times \mathbf{V}' = \rho(0, fu', -f_H u') \tag{9.2.1}$$

An additional centrifugal acceleration $(u')^2/R$ also appears because of the relative azimuthal velocity, but this acceleration will be neglected, compared to (9.2.1), because the distance from the axis of rotation R is much larger than u'/f. By equating the y component of (9.2.1) to the horizontal pressure gradient force, we have the geostrophic equation

$$fu'\rho = -\partial p/\partial y \tag{9.2.2}$$

and the vertical force balance is

$$-f_H u'\rho = -\partial p/\partial z - g\rho \tag{9.2.3}$$

The elimination of p between (9.2.2) and (9.2.3) gives the *thermal wind* equation

$$f\,\partial(u'\rho)/\partial z = g\,\partial\rho/\partial y - (\partial/\partial y)(f_H u'\rho)$$

or

$$f\,\partial(u'\rho)/\partial z \simeq g\,\partial\rho/\partial y \tag{9.2.4}$$

when

$$\partial(u'\rho)/\partial z \gg (f_H/f)\,\partial(u'\rho)/\partial y \tag{9.2.5}$$

The condition (9.2.5) is satisfied, and the approximation (9.2.4) is valid, when the horizontal scale of the current is much larger than the vertical. In this case, the vertical component of the Coriolis force is negligible in (9.2.3).

(b) *Vortex Laws* (*Inertial Reference Frame*). If $\mathbf{V}$ denotes the *absolute* velocity of a particle of density ρ in an incompressible ($d\rho/dt = 0$) and stratified fluid, then the general continuity equation (1.1.2) again becomes

$$0 = \nabla \cdot \mathbf{V}$$

and the Euler equation of motion is given by

$$d\mathbf{V}/dt + \rho^{-1}(x, y, z, t)\,\nabla p = \nabla G \tag{9.2.6}$$

where G is the gravitational potential. Let us derive a "circulation theorem" for a stratified fluid by constructing a closed contour $C(t)$ on the isopycnal surface $\rho(x, y, z, t) = \text{constant}$ (Fig. 9.3). Following the procedure of Fig. 1.1, we allow each arc element $d\boldsymbol{\lambda}$ on C to move with the local $\mathbf{V}$. Since each fluid parcel conserves density, the curve C will always remain on the same isopycnal surface. The closed line integral of ∇G obviously vanishes, and we also have

$$\oint_C \rho^{-1}\,\nabla p \cdot d\boldsymbol{\lambda} = \rho^{-1} \oint_C \nabla p \cdot d\boldsymbol{\lambda} = 0$$

since ρ is constant on C. Thus, (9.2.6) implies that the line integral of $d\mathbf{V}/dt$

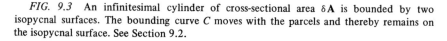

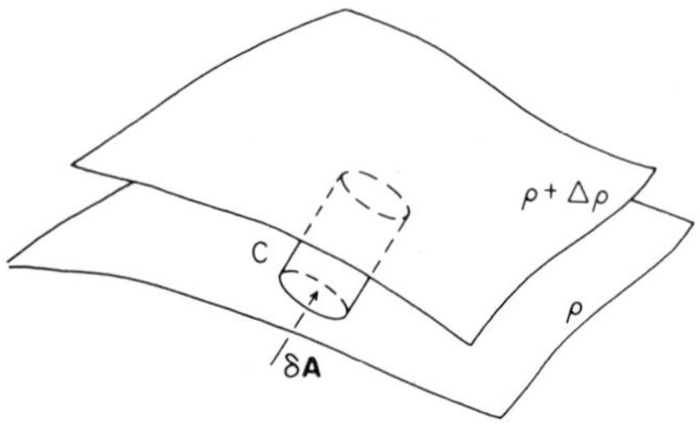

FIG. 9.3 An infinitesimal cylinder of cross-sectional area $\delta\mathbf{A}$ is bounded by two isopycnal surfaces. The bounding curve C moves with the parcels and thereby remains on the isopycnal surface. See Section 9.2.

vanishes, and by applying the kinematics of Fig. 1.1, we then conclude that

$$(d/dt) \oint_C \mathbf{V} \cdot d\boldsymbol{\lambda} = 0 \tag{9.2.7}$$

Therefore, the circulation on an isopycnal surface is conserved. [See Eq. (9.8.21) for the rate of change of circulation on nonisopycnal surfaces.]

If C be a small curve (Fig. 9.3) and $|\delta \mathbf{A}|$ denote the infinitesimal area enclosed on the ρ = constant surface, then (9.2.7) implies

$$(d/dt)(\boldsymbol{\zeta} \cdot \delta \mathbf{A}) = 0 \qquad \text{where} \qquad \boldsymbol{\zeta} = \nabla \times \mathbf{V} \tag{9.2.8}$$

is the absolute vorticity, and the vector area element $\delta \mathbf{A}$ points normal to the isopycnal surface. The density gradient $\nabla\rho$ also points normal to the isopycnal surface, and therefore we now consider the scalar coefficient of proportionality between $\delta \mathbf{A}$ and $\nabla\rho$. The cylinder having $\delta \mathbf{A}$ as its base in Fig. 9.3 is constructed by erecting normals to the ρ = constant surface and by terminating these normals at the isopycnal surface $\rho(x, y, z, t) + \Delta\rho$ = constant, where $\Delta\rho$ is an arbitrarily assigned infinitesimal constant. The height of this cylinder multiplied by $|\nabla \rho|$ equals $\Delta\rho$, and therefore the volume of the material cylinder is

$$|\delta \mathbf{A}|\ |\nabla\rho|^{-1}\ \Delta\rho$$

Since this volume is conserved, and since $\delta \mathbf{A}$ is parallel to $\nabla\rho$, we can replace $\delta \mathbf{A}$ by $(\Delta\rho)^{-1}\nabla\rho$ in (9.2.8). Since $\Delta\rho$ is a constant, we then obtain *Ertel's theorem*

$$(d/dt)(\boldsymbol{\zeta} \cdot \nabla\rho/\rho) = 0 \tag{9.2.9}$$

This form of the law of conservation of potential vorticity provides the basis for extending many of the calculations in Part I to the stratified regime. For an example, we refer to (7.3.4) and generalize the idea therein to the continuously stratified case. Thus we note that a downward directed velocity field tends to "squeeze" the isopycnals together or to increase the local $|\partial\rho/\partial z|$. Equation (9.2.9) then implies a corresponding increase in the vertical component of absolute vorticity, and thus southward movement occurs in accord with (7.3.4).

9.3 Boussinesq Approximation

The parcel method (9.1.6) indicates that the dynamically significant effect of the density variation lies in the modification of the gravity force, rather than in the modification of the inertia of a parcel. This observation provides the basis for simplifying the equations of motion when the departures of the density $\rho(x, y, z, t)$ from the mean ρ_0 are small, or

$$\left|\frac{\rho(x, y, z, t) - \rho_0}{\rho_0}\right| \ll 1 \tag{9.3.1}$$

Before making use of this approximation in the momentum equation, let us note that the continuity equation is exactly given by

$$\nabla \cdot \mathbf{V} = 0 \tag{9.3.2}$$

because $d\rho/dt = 0$ for an incompressible fluid.

In a stratified fluid, the gravity force [cf. (1.1.3)]can be rewritten as

$$-\rho_0 g\mathbf{k} - (\rho - \rho_0) g\mathbf{k}$$

where ρ_0 is the constant mean density of the fluid. The first of these terms can be incorporated with the pressure term, but the *buoyancy force* $(\rho - \rho_0)g\mathbf{k}$ is not necessarily irrotational, and thus small density variations can generate small accelerations. In this case, the inertial acceleration in (1.1.3) can be approximated by

$$\rho_0\, d\mathbf{V}/dt + (\rho - \rho_0)\, d\mathbf{V}/dt \simeq \rho_0\, d\mathbf{V}/dt$$

and the neglected term $(\rho - \rho_0)d\mathbf{V}/dt$ is small compared to both $\rho_0\ d\mathbf{V}/dt$ and $(\rho - \rho_0)g\mathbf{k}$. When this approximation is made in (1.1.3), we have the Boussinesq equation

$$\rho_0\, d\mathbf{V}/dt = -\nabla p - g\rho\mathbf{k} \tag{9.3.3}$$

which states that (in a nonrotating fluid) the density may be considered constant except where it multiplies the gravitational acceleration. (For the case of a rapidly rotating fluid, one must also take into consideration the modification of the centrifugal force by the density variations. See Section 9.2.)

In the remainder of this section, we shall separate the density into mean and fluctuating fields using Reynolds method (Section 6.2), but the full utility of this procedure will appear only in the convection problem (Chapter X). As in Section 6.2, the bar symbol is used to denote the horizontal x, y average of a field, and the prime is used for the fluctuation from the mean. Thus, the respective values of pressure, density, temperature, and salinity are given by

$$p = \bar{p}(z, t) + p'(x, y, z, t), \qquad T = \bar{T}(z, t) + T'(x, y, z, t)$$

$$\rho = \bar{\rho}(z, t) + \rho'(x, y, z, t), \qquad S = \bar{S}(z, t) + S'(x, y, z, t) \tag{9.3.4}$$

The horizontal average of all fluctuating fields, including that of the velocity $\mathbf{V} = u, v, w$, vanishes by definition, or

$$0 = \overline{p'} = \overline{\rho'} = \bar{w}$$

In taking the average of a nonlinear term like

$$\overline{dw/dt} = \partial \bar{w}/\partial t + \overline{\nabla \cdot \mathbf{V}w} = \partial \overline{w^2}/\partial z$$

we assume vertical bounding walls on which the normal component of $\mathbf{V}$

vanishes, or else an unbounded fluid having cyclic variations of **V** in (x, y). In the latter case, for example, the average of $\partial(uw)/\partial x + \partial(vw)/\partial y$ vanishes because its horizontal integral is finite and because the distance over which the average is taken is infinite. Thus, the average of $\nabla \cdot \mathbf{V}w$ equals the average of $\partial w^2/\partial z$, as indicated above. By writing the momentum equation (9.3.3) as

$$\rho_0\, d\mathbf{V}/dt = -\nabla(p' + \bar{p}) - g\rho'\mathbf{k} - g\bar{\rho}\mathbf{k} \tag{9.3.5}$$

and by averaging the vertical component of (9.3.5), we then obtain

$$(\partial/\partial z)(\rho_0\overline{w^2} + \bar{p}) = -g\bar{\rho} \tag{9.3.6}$$

When this generalization of the hydrostatic equation is subtracted from (9.3.5), we get the fluctuation equation

$$d\mathbf{V}/dt = -\nabla\phi - g(\rho'/\rho_0)\mathbf{k} \qquad \text{where} \qquad \phi = (p'/\rho_0) - \overline{w^2} \tag{9.3.7}$$

When the nondiffusive conservation laws for heat and salt

$$0 = dT/dt = \partial T/\partial t + \mathbf{V}\cdot\nabla T \qquad \text{and} \qquad 0 = dS/dt = \partial S/\partial t + \mathbf{V}\cdot\nabla S$$

are subjected to a similar procedure, we obtain

$$0 = (d/dT)(S' + \bar{S}) = \partial S'/\partial t + \mathbf{V}\cdot\nabla S' + \partial\bar{S}/\partial t + w\,\partial\bar{S}(z, t)/\partial z$$

$$0 = (d/dt)(T' + \bar{T}) = \partial T'/\partial t + \mathbf{V}\cdot\nabla T' + \partial\bar{T}/\partial t + w\,\partial\bar{T}/\partial z \tag{9.3.8}$$

and the horizontal averages of these are

$$0 = \partial\bar{S}/\partial t + (\partial/\partial z)(\overline{wS'}) \qquad \text{and} \qquad 0 = \partial\bar{T}/\partial t + (\partial/\partial z)(\overline{wT'}) \tag{9.3.9}$$

By subtracting (9.3.8) from the last equation in (9.3.9), we obtain the fluctuation equations

$$\partial S'/\partial t + \mathbf{V}\cdot\nabla S' - (\partial/\partial z)\,\overline{wS'} + w\,\partial\bar{S}/\partial z = 0 \tag{9.3.10a}$$

$$\partial T'/\partial t + \mathbf{V}\cdot\nabla T' - (\partial/\partial z)\,\overline{wT'} + w\,\partial\bar{T}/\partial z = 0$$

From the linear combination of these, and with the use of the equation of state

$$\rho'/\rho_0 = \beta S' - \alpha T' \tag{9.3.10b}$$

we obtain the fluctuation density equation

$$\partial/\partial t(\rho'/\rho_0) + \mathbf{V}\cdot\nabla(\rho'/\rho_0) - (\partial/\partial z)\overline{(\rho' w/\rho_0)} + w(\rho_0^{-1}\,\partial\bar{\rho}/\partial z) = 0 \tag{9.3.11}$$

where

$$\rho_0^{-1}\,\partial\bar{\rho}/\partial z = \beta\,\partial\bar{S}/\partial z - \alpha\,\partial\bar{T}/\partial z$$

A similar procedure applied to (9.3.9) gives

$$\rho_0^{-1}\,\partial\bar{\rho}/\partial t = -(\partial/\partial z)\overline{(w\rho'/\rho_0)} \tag{9.3.12}$$

An equation for the time rate of change of potential energy can be obtained from (9.3.12) by multiplying it with z and by averaging the result in the vertical. The latter average is denoted by brackets, and we assume that $\mathbf{V}$ is cyclic in an unbounded vertical direction, or else that $\overline{w}$ vanishes at finite horizontal boundaries. In the latter case, for example, $\{z\, \partial \overline{w\rho'}/\partial z\} = -\{\overline{w\rho'}\}$, and therefore the potential energy equation obtained from (9.3.12) is

$$(\partial/\partial t)\langle zg\bar{\rho}/\rho_0\rangle = (g/\rho_0)\langle \overline{w\rho'}\rangle \tag{9.3.13}$$

The kinetic energy equation obtained by taking the scalar product of $\mathbf{V}$ with (9.3.7), and by averaging the result, is

$$\tfrac{1}{2}(\partial/\partial t)\langle \overline{\mathbf{V}^2}\rangle = -(g/\rho_0)\langle \overline{\rho' w}\rangle \tag{9.3.14}$$

When (9.3.13) and (9.3.14) are added, the result

$$(\partial/\partial t)\langle \tfrac{1}{2}\mathbf{V}^2 + gz\bar{\rho}/\rho_0\rangle = 0 \tag{9.3.15}$$

expresses the conservation of total energy. We now apply the theory of this section to the propagation of internal waves in a continuously stratified thermocline.

9.4 Internal Waves

Surprisingly large amplitude internal waves in the ocean are often observed (Webster, 1968) near the Coriolis frequency, but we shall defer a consideration of rotational effects until Section 9.9. The higher frequency waves are of interest because of their possible role in the mixing of the thermocline.

One mechanism for generating *internal* waves is provided by the reflection of the body tides from the continental shelves (Rattray, 1960). Internal waves can also be generated locally in the central ocean by the back and forth movement of the tides over the irregular bottom topography (Cox and Sandstrom, 1962). The variable wind stress in a storm can produce surface oscillations which are then transmitted to greater depths (Chapters VII and VIII; also Pollard, 1970). The following considerations, however, apply mainly to the propagation problem, and therefore we shall parameterize the generating mechanism in an unrealistic but convenient way.

The moving boundary employed in Section 2.2 will again be used to generate internal waves in the stratified fluid. Accordingly, we consider an upper rigid boundary having sinusoidal undulations of wavelength $2\pi/k$ in the x direction. At $t > 0$, we translate this boundary with speed c, thereby generating the infinitesimal vertical velocity

$$w'(x, 0, t) = \mathrm{Re}[\hat{w}^{ik(x-ct)}] \tag{9.4.1}$$

just beneath the upper surface ($z = 0$). The amplitude $\hat{w}$ is determined by c, k, and the vertical amplitude of the undulations on the moving upper surface. We assume that the undisturbed static stability

$$s = -\rho_0^{-1}\,\partial\bar{\rho}/\partial z > 0 \tag{9.4.2}$$

is uniform, and that the fluid is of semi-infinite extent in $z < 0$.

If u' denotes the horizontal perturbation velocity, then the linearization of (9.3.7) and (9.3.11) gives

$$\begin{aligned} \partial u'/\partial t &= -\partial\phi'/\partial x, \qquad (\partial/\partial t)(\rho'/\rho_0) - w's = 0 \\ \partial w'/\partial t &= -\partial\phi'/\partial z - g\rho'/\rho_0, \quad \partial u'/\partial x + \partial w'/\partial z = 0 \end{aligned} \tag{9.4.3}$$

and by eliminating u', ϕ', and ρ'/ρ_0, one obtains

$$(\partial^2/\partial x^2 + \partial^2/\partial z^2)\,\partial^2 w'/\partial t^2 + gs\,\partial^2 w'/\partial x^2 = 0 \tag{9.4.4}$$

Before proceeding further, let us form an energy equation by multiplying the first equation in (9.4.3) with u', the second with w', and the third with $gs^{-1}(\rho'/\rho_0)$, and by adding the results. Simplification by means of the last equation in (9.4.3) then gives

$$\tfrac{1}{2}(\partial/\partial t)[(u')^2 + (w')^2 + (g/s)(\rho'/\rho_0)^2] = -(\partial/\partial x)(u'\phi') - (\partial/\partial z)(w'\phi') \tag{9.4.5}$$

This equation states that the sum of wave kinetic energy and wave potential energy $(g/2s)(\rho'/\rho_0)^2$ can only change because of a net convergence of "pressure work." The rate of increase of the total energy *beneath* some z level is now computed by integrating (9.4.5) from $z = -\infty$, and by averaging horizontally (such averages, again, being denoted by a bar). Since energy can only propagate downward at a finite rate, we have $w'(x, -\infty, t) = 0$, for any finite time after the motion of the upper boundary is initiated. When this boundary condition is used in the integration of (9.3.5), we obtain

$$\text{rate of increase of energy beneath } z = -\overline{w'\phi'} \qquad \text{evaluated at } z \tag{9.4.6}$$

At any finite value of z, however, the solution of (9.4.4) approaches the "steady state" value given by the product of $\exp ik(x - ct)$ and a function of z. These steady state solutions are now examined.

If the excitation frequency exceeds the Brunt–Väisälä frequency, or $k^2c^2 > gs$, then the only finite solution of (9.3.4) is

$$w'(x, z, t) = \mathrm{Re}[\hat{w}(\exp ik(x - ct))(\exp (z/c)(k^2c^2 - gs)^{1/2})]$$

This solution decreases exponentially with depth, and the corresponding value of (9.4.6) vanishes because w' and ϕ' are in quadrature [cf. the first and last equations in (9.4.3)].

On the other hand, if $k^2c^2 < gs$, then there are two bounded solutions of (9.4.4), these being

$$w'(x, z, t) = \text{Re}\, \hat{w} \exp[ik(x-ct) \pm izl] \qquad \text{where} \qquad l = c^{-1}(gs-k^2c^2)^{1/2} > 0 \tag{9.4.7}$$

Both solutions in (9.4.7) satisfy the $z = 0$ boundary condition, and both are finite at $z = -\infty$. It may then appear that there are an infinite number of linear combinations, each of which is an acceptable solution. In order to resolve this ambiguity we must now introduce more explicit information about the conditions at $z = -\infty$. Note that the solutions (9.4.7) apply equally well to the (different) problem in which energy is propagated *upward* from a wave generating source located at great depths. But no such motion (at $z = -\infty$) exists in the present problem, and we must make explicit use of this fact. One way is to solve the transient problem and then let $t \to \infty$, but it is much simpler to use a *radiation boundary condition* at $z = -\infty$. According to this, we consider each ($\pm$) solution in (9.4.7), and show that the (+) sign corresponds to a wave that propagates energy *downward*, whereas the ($-$) sign corresponds to the case in which energy propagates *upward*. It will be shown that the downward propagating wave is given by

$$w'(x, z, t) = \text{Re}\, \hat{w} \exp[ik(x-ct) + izc^{-1}(gs-k^2c^2)^{1/2}] \tag{9.4.8}$$

To prove (9.4.8), we now examine the value of flux (9.4.6) for each of the two ($\pm$) solutions. The pressure eigenfunction ϕ' that corresponds to either w' eigenfunction can easily be computed from the first and last equation in (9.4.3). Thus, we obtain the two relations: $\phi' = cu'$ and $ku' \pm lw' = 0$, or $-\phi' = \pm(lc/k)w'$, where the ($\pm$) convention corresponds to that used in (9.4.7). Therefore, the downward propagation of wave energy, as given by (9.4.6), is

$$-\overline{w'\phi'} = \pm\, c\, \overline{(w')^2}\,(l/k)$$

and since this must be positive in the present problem, the (+) sign must be used, as indicated in (9.4.8).

When the (+) sign is introduced in the continuity equation, we have $ku' + lw' = 0$, and thus we see that there is a Reynolds stress

$$\overline{u'w'} = -(l/k)\overline{(w')^2} < 0$$

or a downward transport of x momentum by the steady state wave. This stress arises because of the horizontal force which must be continually supplied to the moving upper boundary in order to supply the wave energy which appears at successively deeper strata as time increases. The mean horizontal force on the upper boundary must be transmitted to greater depths in order to avoid a pile-up of mean momentum, and in order to obtain a steady state at finite z. Although the Reynolds stress given above is independent of z, the problem considered next shows how the energy and momentum can converge, thereby suggesting a mechanism for the generation of mean flow and the dissipation of wave energy.

9.5 Critical Layer in a Stratified Shear Flow

We shall first derive a more general wave equation by considering a basic state consisting of an arbitrary mean current $U(z)$ in the x direction and also a vertically varying field of static stability $s(z)$. The infinitesimal vertical component of perturbation velocity is again denoted by

$$w'(x, z, t) = \text{Re}\, \hat{w}(z) \exp ik(x-ct) \tag{9.5.1}$$

and therefore the operator $\partial/\partial t$ may be replaced by $-c\, \partial/\partial x$ wherever it appears in the linearized equations of motion. The total horizontal velocity is now given by $U(z) + u'(x, z, t)$, and therefore the linearized acceleration is

$$\frac{d}{dt}(u' + U(z)) = \frac{du'}{dt} + w' \frac{\partial U}{\partial z} = \frac{\partial u'}{\partial t} + U \frac{\partial u'}{\partial x} + w \frac{\partial U}{\partial z} = (U-c) \frac{\partial u'}{\partial x} + w \frac{\partial U}{\partial z}$$

Thus, the two-dimensional perturbation equations now become

$$\begin{aligned}(U(z)-c)\, \partial u'/\partial x + w'\, \partial U/\partial z &= -\partial \phi'/\partial x \\ (U-c)\, \partial w'/\partial x &= -\partial \phi'/\partial z - g\rho'/\rho_0 \\ (U-c)(\partial/\partial x)(\rho'/\rho_0) - w's(z) &= 0 \\ \partial u'/\partial x + \partial w'/\partial z &= 0\end{aligned} \tag{9.5.2}$$

By replacing $\partial u'/\partial x$ with $-\partial w'/\partial z$ in the first equation of (9.5.2), and by subsequent elimination of ϕ' and ρ', we obtain the wave equation

$$(U-c)^2 [\partial^2 w'/\partial x^2 + \partial^2 w'/\partial z^2] + [gs - (U-c)\, \partial^2 U/\partial z^2] w' = 0 \tag{9.5.3}$$

These are called *transverse* modes, since they are independent of distance y normal to the mean flow. The *longitudinal* modes, on the other hand, have $u' = 0$ and a nonvanishing velocity component $v'(y, z, t)$ in the y direction. For the longitudinal modes, we have $\partial v'/\partial t = -\partial \phi'/\partial y$, and the basic current $U(z)$ also does not appear in the other equations of motion. Thus, the longitudinal modes are *decoupled* from the basic shear flow, and consequently the equation for $w'(y, z, t)$ may be obtained from (9.5.3) by setting $U = 0$ and x equal to y, or

$$c^2(\partial^2 w'/\partial y^2 + \partial^2 w'/\partial z^2) + gsw' = 0 \qquad \text{(longitudinal mode)} \tag{9.5.4}$$

Let us now examine the effect of a shear current on the internal waves studied in Section (9.4). Thus, we retain a constant s but add a basic current having the form

$$U(z) = \begin{cases} U_0 & (z < 0) \\ U_0 - z(gs/R)^{1/2} & (z > 0) \end{cases} \tag{9.5.5}$$

where

$$R \equiv gs/(\partial U/\partial z)^2 \tag{9.5.6}$$

is a nondimensional constant that measures the shear above $z = 0$. As in Section 9.4, we generate downward propagating waves having horizontal speeds

$$c > U_0 > 0 \tag{9.5.7}$$

by translating a sinusoidal boundary, the latter being located at some level above $z = 0$. Thus, $\partial^2 U/\partial z^2$ in (9.5.3) vanishes in the regions above and below $z = 0$. We shall also confine the discussion to small values of k, so that when the long wave approximation

$$\partial^2 w'/\partial x^2 \ll \partial^2 w'/\partial z^2$$

is made, Eq. (9.5.3) becomes

$$(U(z) - c)^2\, \partial^2 w'/\partial z^2 + gsw' = 0 \tag{9.5.8}$$

for $z \lessgtr 0$.

The current $U(z) = U_0$ is constant in the region $z < 0$, and therefore the solutions of (9.5.8) are $\exp[\pm iz(gs)^{1/2}(c - U_0)^{-1}_{-}]$. The particular solution (see Section 9.4) that corresponds to a *downward* propagation of energy is

$$w' = B \exp\,[+iz(gs)^{1/2}(c - U_0)^{-1}] \tag{9.5.9}$$

where B is a constant of integration. In the region $z > 0$, we have a constant $\partial U/\partial z$, and by direct substitution, one readily shows that $w = (c - U(z))^q$ is a solution of (9.5.8), provided the exponent satisfies the indicial equation $q(q - 1) + R = 0$. The two roots of this quadratic equation are

$$q_+ = \tfrac{1}{2} + i(R - \tfrac{1}{4})^{1/2} \quad \text{and} \quad q_- = \tfrac{1}{2} - i(R - \tfrac{1}{4})^{1/2} \tag{9.5.10}$$

and therefore the solution of (9.5.8) is given by

$$w = A_+(c - U(z))^{q_+} + A_-(c - U)^{q_-} \qquad (z > 0) \tag{9.5.11}$$

where the integration constants A are determined by matching solutions at $z = 0$. The character of the solution depends on the value of the *Richardson number* (9.5.6), and the case

$$R > \tfrac{1}{4} \tag{9.5.12}$$

is discussed below. (The case of low Richardson number is discussed in the next section.)

At the $z = 0$ surface, $U(z)$ is continuous, and the discontinuity in $\partial U/\partial z$ requires that the solutions (9.5.9)–(9.5.11) be connected in such a way as to insure the continuity of w' and $\partial\phi'/\partial x$. By referring to the first equation in (9.5.2), we see that the pressure continuity condition implies

$$-(U-c)\,\partial w'/\partial z + w'\,\partial U/\partial z \qquad \text{is continuous at} \quad z = 0 \tag{9.5.13}$$

The continuity of (9.5.9) and (9.5.11) at $z = 0$ implies

$$B = A_+(c - U_0)^{q_+} + A_-(c - U_0)^{q_+} \tag{9.5.14}$$

and (9.5.13) implies

$$-i(gs)^{1/2}B = A_+q_+(\partial U/\partial z)(c-U_0)^{q_+} + A_-q_-(\partial U/\partial z)(c-U_0)^{q_-} - (\partial U/\partial z)B$$

or

$$B(1 - iR^{1/2}) = A_+q_+(c-U_0)^{q_+} + A_-q_-(c-U_0)^{q_-} \tag{9.5.15}$$

The elimination of B between (9.5.14) and (9.5.15), and the use of (9.5.10), gives

$$\frac{A_+}{A_-} = -(c-U_0)^{-2i(R-1/4)^{1/2}}\,\frac{i[R^{1/2} - (R-\frac{1}{4})^{1/2}] - \frac{1}{2}}{i[R^{1/2} + (R-\frac{1}{4})^{1/2}] - \frac{1}{2}} \tag{9.5.16}$$

A second equation for the determination of A_+ and A_- is provided by the given value of w' at the upper boundary.

Let us now discuss these amplitudes as the phase speed approaches U_0. The complex coefficient in (9.5.16) can be written as

$$(c-U_0)^{-2i(R-1/4)^{1/2}} \equiv \exp -[2i(R-\tfrac{1}{4})^{1/2}\ln(c-U_0)]$$

and we see that this has unit amplitude when $R > \frac{1}{4}$, and $c - U_0 > 0$. Thus, the magnitude of (9.5.16) and (9.5.11) are finite as $c \to U_0$, and (9.5.15) indicates that $B \to 0$. Therefore, the downward flux of energy in the lower fluid vanishes as $c \to U_0$. Note, however, that the vertical derivative of (9.5.11) behaves like $(c - U)^{-1/2}$, and therefore $\partial u'/\partial x$ will become very large at $z = 0$, as $c \to U_0$. This level, or any value of z at which

$$c - U(z) = 0 \tag{9.5.17}$$

is called the critical level. Thus, we see that at the critical level the horizontal perturbation velocities (and their vertical shear) become very large, and the downward transmission of energy is prohibited. The large shears mentioned above imply that the previously neglected viscous effects will become important in dissipating some of the energy which is incident upon the critical level. See Bretherton (1966) for further discussion.

9.6 Instability of a Stratified Shear Flow

The question of the stability of the horizontal flow in the stratified thermocline also bears on the turbulence problem. We are concerned here with relatively small scales of motion for which the Coriolis force is negligible, and shall inquire into the conditions when a given shear flow $U(z)$, having a given $s(z)$, is unstable to infinitesimal perturbations $w(z) \exp ik(x - ct)$. Suppose that the basic flow occurs between two horizontal boundaries at $z = (0, H)$, so that the boundary condition on the vertical velocity is

$$\hat{w}(0) = 0 = \hat{w}(H)$$

From the general wave equation (9.5.3), we obtain the following equation for the eigenfunction $\hat{w}(z)$:

$$(U(z)-c)^2(d^2\hat{w}/dz^2) + [gs(z)-(U-c)U_{zz}-k^2(U-c)^2]\hat{w} = 0 \qquad (9.6.1)$$

and this equation is also valid for complex phase speeds

$$c = c_r + ic_i \qquad (9.6.2)$$

We will prove the Miles–Howard theorem, which states that if

$$gs(z)/U_z^2 > \tfrac{1}{4} \qquad (9.6.3)$$

for all z, then an amplifying wave or

$$c_i > 0 \qquad (9.6.4)$$

is *impossible*.

The theorem will be proved by assuming that there exists an amplifying solution (9.6.4), and by showing that this implies a condition (9.6.3) on the Richardson number. If a complex c exists, then $U(z) - c$ does not vanish anywhere, and therefore the function $G(z)$ defined by

$$w(z) = (U(z)-c)^{1/2}G(z) \qquad (9.6.5)$$

is finite and differentiable. By dividing (9.6.1) with $(U - c)$, and by substituting (9.6.5), we get

$$(U-c)^{1/2}(d^2/dz^2)[(U-c)^{1/2}G] + G[gs/(U-c)-U_{zz}-k^2(U-c)] = 0 \qquad (9.6.6)$$

The first term is rewritten as

$$(U-c)^{1/2}\frac{d}{dz}\frac{d}{dz}[(U-c)^{1/2}G] \equiv \frac{d}{dz}\left[(U-c)^{1/2}\frac{d}{dz}(U-c)^{1/2}G\right]$$
$$-\frac{1}{2}(U-c)^{-1/2}U_z\frac{d}{dz}[(U-c)^{1/2}G]$$

$$= \frac{d}{dz}\left[(U-c)\frac{dG}{dz} + \frac{1}{2}U_zG\right] - \frac{1}{2}U_z\left[\frac{dG}{dz} + \frac{U_zG}{2(U-c)}\right]$$

$$= \frac{d}{dz}\left[(U-c)\frac{dG}{dz}\right] + G\left[\frac{1}{2}U_{zz} - \frac{1}{4}\frac{U_z^2}{U-c}\right]$$

and thus (9.6.6) simplifies to

$$\frac{d}{dz}\left[(U-c)\frac{dG}{dz}\right] + G\left[\frac{gs-\frac{1}{4}U_z^2}{U-c} - \frac{1}{2}U_{zz} - k^2(U-c)\right] = 0, \qquad G(0) = G(H) = 0 \tag{9.6.7}$$

This has a suitable form for an application of the Rayleigh integral technique (1.5.19), and thus we multiply (9.6.7) by the complex conjugate eigenfunction $G^*(z)$. By integrating the result from $z = 0$ to $z = H$ and by using the boundary conditions $G^*(0) = 0 = G^*(H)$, we get

$$-\int_0^H dz\left[\frac{dG^*}{dz}\frac{dG}{dz}(U-c)\right] + \int_0^H dz\, GG^*\left\{\frac{gs-\frac{1}{4}U_z^2}{(U-c)} - k^2(U-c) - \frac{1}{2}U_{zz}\right\} = 0$$

The imaginary part of this equation is

$$c_i\int_0^H dz\left[\frac{dG^*}{dz}\frac{dG}{dz} + k^2GG^*\right] + c_i\int_0^H dz\, GG^*\frac{gs-\frac{1}{4}U_z^2}{(U-c_r)^2+c_i^2} = 0 \tag{9.6.8}$$

Since (9.6.4) has been assumed, the first term in (9.6.8) is always positive, and therefore (9.6.8) can only be satisfied if $U_z^2 > 4gs(z)$ for some z. On the other hand, if the Richardson number (9.6.3) is everywhere greater than $\frac{1}{4}$, then (9.6.8) cannot be satisfied, and this contradiction implies that no amplifying wave is possible for large Richardson number.

The integral equation (9.6.8) implies that a necessary condition for instability is a Richardson number which falls below $\frac{1}{4}$, but the latter is by no means sufficient. The point is nicely illustrated by the homogeneous Couette flow ($s = 0$, U_z = constant) of Chapter I. In Section 1.5, we showed that this (zero Richardson number) flow is stable, and so we see that the breakdown of a stratified shear flow depends on the vorticity gradient, as well as on the Richardson number.

Figure 9.4 shows the form of a shear layer across which the velocity $U(z) = \Delta U f(z/h)$ changes by $2\,\Delta U$, so that $f(-\infty) = 1 = -f(\infty)$. The thickness h of the shear layer is defined by the slope at the inflection point, so that $f'(0) = -1$. We assume that the density, $\rho(z) = \rho_0 + \Delta\rho f(z/h)$, has the same form as the velocity profile. Thus the undisturbed static stability is $-h^{-1}(\Delta\rho/\rho_0)f'(z/h)$, and the basic shear is $(\Delta U/h)f'(z/h)$. Since the maximum

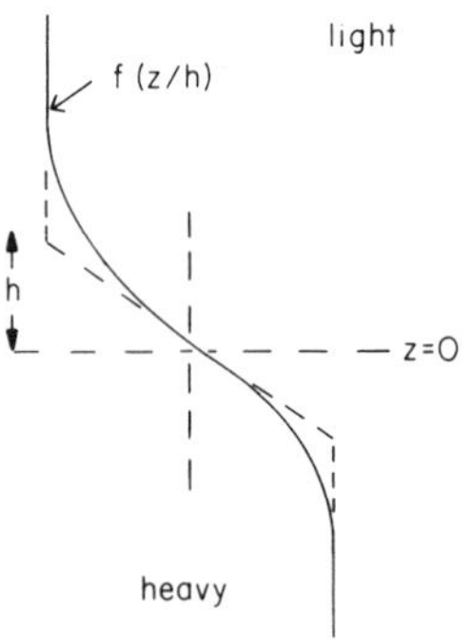

FIG. 9.4 The vertical profile of a shear layer and a similar density field. See Section 9.6.

value of $-f'(z/h)$ is unity at the inflection point ($z = 0$), then the minimum value of the Richardson number (9.6.3) is

$$\min R = g(\Delta\rho/\rho_0)h/(\Delta U)^2 \tag{9.6.9}$$

Let us note that as $h \to 0$, Fig. 9.4 approaches the Kelvin–Helmholtz model [cf. (1.6.28)] with $\Delta\rho$ corresponding to $\rho_w - \rho_a$ and ρ_0 corresponding to $(\rho_w + \rho_a)/2$. From (1.6.30), we see that instability always occurs in that model for sufficiently large k [for all values of $(\Delta U, \Delta\rho)$] and thus we conclude that the shear layer (Fig. 9.4) is unstable for sufficiently small values of (9.6.9). On the other hand, we have just shown that the shear layer is definitely stable when (9.6.9) exceeds $\frac{1}{4}$. Consequently, there exists a finite critical Richardson number below which the stratified shear layer is unstable, and the precise value of this number depends on the form of f (Thorpe, 1969).

The right hand side of (9.6.9) furnishes a useful "overall" measure of the Richardson number in a shear layer even when the velocity and density profiles are *not* exactly similar. In this case, however, instability can occur when (9.6.9) is greater than $\frac{1}{4}$, because the *local* Richardson number (9.6.3) may be smaller than $\frac{1}{4}$ in part of the fluid.

Laboratory experiments (Turner, 1973) confirm that stratified shear flows only become unstable when the Richardson number is sufficiently small. When this occurs in a shear layer (Fig. 9.6), then a regular train of amplifying Kelvin–Helmholtz waves is observed. Large amplitude waves break in the crests and troughs, and the subsequent mixing process transfers both momentum and density down the respective mean gradients. Part of the kinetic energy released from the mean flow is used to raise the center of gravity of the density field, and the remainder is dissipated in the turbulent process. Thus the layer thickness h tends to increase as a result of the instability, and (9.6.9) tends to increase. When the turbulence in the breaking wave decays, we have a new and *stable* state of laminar flow.

Amplifying Kelvin–Helmholtz waves have been photographed by introducing dye tracers in the upper thermocline (Woods, 1968). The waves form only when the thickness of the shear layer is rather small (~10 cm) and the corresponding density gradient is also much larger than normal. Although the process leading to the formation of such thin layers is not known, it is probably related to the mechanism discussed in Section 3.6.

9.7 The Effect of Shear on Unstable Stratification–Langmuir Cells

A basic state having negative s, or

$$\beta \equiv -s > 0 \tag{9.7.1}$$

is in unstable equilibrium because a parcel will accelerate in the direction in which it is displaced (cf. Section 9.1). Although this *convective* case really belongs in Chapter X, it is convenient to discuss here an important qualitative influence of a superimposed basic shear flow $U(z)$. Accordingly, we shall presently compare the growth rates of longitudinal modes (9.5.4) and transverse modes (9.5.3) in a layer which is bounded by two flat walls at $z = 0, H$. The basic flow has the uniform shear ϵ specified by (9.7.4), and the wavelength of the perturbation is again given by $2\pi/k$. As mentioned previously, the shear flow has no influence on the longitudinal modes (9.5.4) and substitution of $w = \hat{w}_0 \exp iky$ gives the eigenfunction equation

$$d^2\hat{w}_0/dz^2 - k^2\hat{w}_0 + (g\beta/\sigma_0^2)w_0(z) = 0, \qquad \hat{w}_0(0) = 0 = \hat{w}(H) \tag{9.7.2}$$

where $c = i\sigma_0$. The solutions of (9.7.2) are sine functions having integral multiples of H as their half wavelengths, and the first mode is

$$w_0(z) = \sin(\pi z/H) \qquad \text{and} \qquad g\beta/\sigma_0^2 = k^2 + (\pi^2/H^2) \qquad \text{(longitudinal mode)} \tag{9.7.3}$$

The time dependent part of w' is $\exp -ikct = \exp k\sigma_0 t$, and we see from (9.7.3) that the maximum value of the growth rate $k\sigma_0$ equals $(g\beta)^{1/2}$ at large k.

Equation (9.5.3) applies for the transverse modes $w' = \hat{w}(z) \exp ikx$, and we shall show that the growth rate $k\sigma$, for the same k, is *less* than the corresponding value obtained from (9.7.3). The basic flow under consideration is given by

$$U(z) = \epsilon\mu(z) = \epsilon(z - H/2) \tag{9.7.4}$$

and since $\partial^2 U/\partial z^2 = 0$, $s = -\beta$, and $c = i\sigma$, the eigenfunction equation obtained from (9.5.3) is

$$d^2\hat{w}/dz^2 - k^2\hat{w} + g\beta\hat{w}/(\sigma + i\epsilon\mu(z))^2 = 0, \qquad \hat{w}(0) = 0 = \hat{w}(H) \tag{9.7.5}$$

For small values of ϵ, we can obtain a power series solution of (9.7.5) having the form

$$\hat{w} = \hat{w}_0 + \epsilon\hat{w}_1 + \epsilon^2\hat{w}_2 + \cdots \qquad \text{and} \qquad \sigma = \sigma_0 + \epsilon\sigma_1 + \epsilon^2\sigma_2 + \cdots \tag{9.7.6}$$

where the leading terms $\hat{w}_0$ and σ_0 are given by (9.7.2) and (9.7.3). In this expansion, one must keep some amplitude factor fixed, and such a normalization can be achieved by using the series

$$\hat{w} - \hat{w}_0 = \sum_{m \neq 1} B_m \sin(m\pi z/H)$$

where the amplitude of the fundamental mode $\hat{w}_0$ is taken to be equal to unity for all ϵ. For a velocity profile (9.7.4) which is antisymmetric about the midplane ($z = H/2$), there will be "no difference" between positive and negative values of ϵ. Therefore σ is an even function of ϵ, or

$$\sigma_1 = 0 \tag{9.7.7}$$

and the vanishing of this term in (9.7.6) can easily be verified from the expansion procedure given below. In order to compute the next term σ_2 in (9.7.6), we first substitute $k^2 = g\beta/\sigma_0^2 - \pi^2/H^2$ in (9.7.5) and rearrange terms to get

$$d^2\hat{w}/dz^2 + (\pi^2/H^2)\hat{w} = g\beta[\sigma_0^{-2} - (\sigma + i\epsilon\mu)^{-2}]\hat{w}$$

Upon multiplying by the complex conjugate eigenfunction w^* and integrating, we have

$$\int_0^H dz\, \hat{w}^*(d^2\hat{w}/dz^2 + (\pi^2/H^2)\hat{w}) = g\beta \int_0^H dz\, \hat{w}^*\hat{w}[\sigma_0^{-2} - (\sigma + i\epsilon\mu)^{-2}] \tag{9.7.8}$$

Note that when *any* function

$$\sum_{m=1}^{\infty} D_m \sin(m\pi z/H)$$

that satisfies the boundary conditions is substituted for $\hat{w}$ on the left hand side of (9.7.8), the result is

$$\int_0^H dz\, \hat{w}^*(d^2\hat{w}/dz^2 + (\pi^2\hat{w}/H^2)) = -\sum_1^{\infty} (m^2 - 1)D_m^* D_m \int_0^H \sin^2(m\pi z/H) \leqslant 0$$

Therefore, (9.7.8) is a negative real number, or

$$\operatorname{Re} \int_0^H dz\, \hat{w}^*\hat{w}[\sigma_0^{-2} - (\sigma_0 + i\epsilon\mu + \epsilon^2\sigma_2 + \cdots)^{-2}] \leqslant 0 \tag{9.7.9}$$

where the expansion $\sigma = \sigma_0 + \epsilon^2 \sigma_2 + \epsilon^4 \sigma_4 + \cdots$ has been used in the denominator. The leading terms in the ϵ expansion of the bracketed factor in (9.7.9) are $2\epsilon i\mu\sigma_0^{-3} + \epsilon^2 (2\sigma_2\sigma_0^{-3} + 3\mu^2\sigma_0^{-4}) + \cdots$, and therefore the leading term in the expansion of (9.7.9) is

$$\epsilon^2 \int_0^H dz\, \hat{w}^* \hat{w} [2 \text{ Re } \sigma_2 \sigma_0^{-3} + 3\mu^2 \sigma_0^{-4}] \leqslant 0$$

The positive σ_0 root of (9.6.3) is relevant here, and therefore the above relation can only be satisfied if

$$\text{Re } \sigma_2 < 0 \tag{9.7.10}$$

Thus, we conclude that the real part of $\sigma = \sigma_0 + \epsilon^2 \sigma_2 + \cdots$ for a transverse wave of given length $2\pi/k$ is less than the corresponding value of σ_0 for a longitudinal mode, and the maximum possible growth rate $(g\beta)^{1/2}$ occurs for a longitudinal mode.

What is the physical significance of the disturbance having the maximum growth rate? Suppose we start with a perfect state of hydrostatic equilibrium, and apply a "white noise" initial perturbation. All the Fourier wave numbers k will then start to amplify at this point, with growth rates $k\sigma$ given by the foregoing theory. After a sufficiently long interval of time, however, the spectrum will no longer be "white" since the amplitude of the mode having the largest growth rate will "dominate" the field. As this disturbance attains finite amplitude, it will modify the mean stratification, as indicated further in Chapter X, and also modify the growth of the underdeveloped modes. According to this neo-Darwinian conception, the mode that grows the fastest will be predominant in the nonlinear convection field that ensues. Thus, we argue that when convection is initiated in a shear flow (9.7.4), the cells should consist of longitudinal rolls whose axes are parallel to the flow (see Gage and Reid, 1968).

This effect (Avsec, 1939) offers a possible explanation of the *Langmuir cells* frequently observed in the surface layer of the ocean and in lakes when a strong wind blows. In addition to the evaporative cooling of the surface (Section 10.7), the wind induces a downstream drift in the water. Accordingly, the cold and heavy surface water is influenced by the vertical shear, and thus the convection in the water prefers to take the form of longitudinal roll cells, whose axes are aligned parallel to the wind. It should be stated that completely different mechanisms have been proposed to account for the Langmuir cells.

9.8 Wave–Wave Interactions

In Section 9.4, we used the linearized equations of motion to obtain the dispersion relation (9.4.7) connecting horizontal wavelength $2\pi/k$, vertical wavelength $2\pi/l$, and frequency kc. This relation also applies to free waves in an unbounded fluid having uniform stability s. If the amplitudes are infinitesimal, this dispersion relation can be applied to each Fourier component in an arbitrary initial state, and the results can be added to compute the total perturbation at any subsequent time. In this section, we examine the effect of *finite* amplitude. For the case of a *single* plane wave [subsection (a)], we will show that the nonlinear terms in the equations of motion vanish identically, so that the frequency of oscillation is the same as that given by the linear dispersion relation. The nonlinear terms do not vanish, however, when more than one Fourier component is present in the initial state. Consequently, these modes interact to generate new wave numbers and new frequencies, as described in subsection (b). We will also show how the oscillation of each mode can be modulated over time intervals which are large compared with the Brunt–Väisälä period, so that there can be a "slow" exchange of energy between properly oriented triads of waves [subsection (d)]. Subsection (c) contains a derivation of a useful invariant which, together with the conservation of energy, allows us to expound the main idea without excessive calculations. We may also remark here that the ideas developed and the procedures used are relevant to several other problems, such as the interaction of surface gravity waves (Chapter I) and the problem of homogeneous turbulence (Chapter VI).

(a) *One Plane Wave.* The wave number vector $\mathbf{b}_m$ is constructed normal to the wave fronts (Fig. 9.5a) with a magnitude $b_m = |\mathbf{b}_m|$ determined by the given wavelength $(2\pi/b_m)$. θ_m denotes the given angle between $\mathbf{b}_m$ and the unit vertical vector $\mathbf{k}$. In such a plane wave, the vector velocity $\mathbf{V}_m(\mathbf{r}, t)$ is parallel to, and uniform within, the front passing through point $\mathbf{r}$. The magnitude V_m of the velocity only varies with the distance $\xi = \mathbf{b}_m \cdot \mathbf{r}/b_m$ measured normal to the front, and the same is true for both the pressure perturbation ϕ_m' and the density perturbation ρ_m'. Thus, we look for solutions $\mathbf{V}_m$ and ρ_m' which are only functions of the argument $\mathbf{b}_m \cdot \mathbf{r} - \omega_m t$, where the frequency ω_m of the plane wave is to be determined.

From the geometry of Fig. 9.5a, we see that the wavelength of the disturbance measured along the $\mathbf{k}$ axis is $(\cos\theta_m)^{-1}(2\pi/b_m)$, whereas $(\sin\theta_m)^{-1}(2\pi/b_m)$ is the wavelength along the horizontal axis. Thus, for *infinitesimal* amplitudes, we may substitute $l = b_m \cos\theta_m$, and $k = b_m \sin\theta_m$ in the dispersion relation (9.4.7), and the solution for the frequency kc is then

$$\omega_m = (gs)^{1/2} \sin\theta_m \tag{9.8.1}$$

It will now be shown that (9.8.1) is also valid for finite amplitude.

Since $\nabla\rho_m'$ is orthogonal to the wave front and orthogonal to $\mathbf{V}_m$, we have $\mathbf{V}_m \cdot \nabla\rho_m' \equiv 0$, and thus the corresponding nonlinear term in the density equation (9.3.11) vanishes identically for any plane wave. Both ρ_m' and the vertical velocity $(w_m = V_m \sin\theta_m)$ are sinusoidal functions of $\mathbf{b}_m \cdot \mathbf{r} - \omega_m t$,

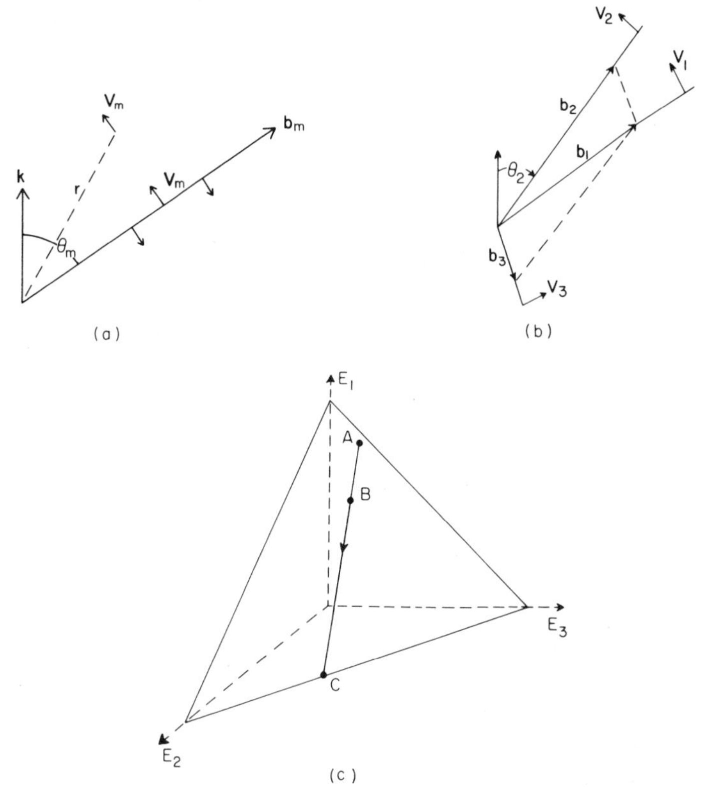

FIG. 9.5 (a) A single plane wave having fronts normal to $\mathbf{b}_m$, with wavelength $2\pi|\mathbf{b}_m|^{-1}$ in an unbounded stratified fluid. (b) Three plane waves that satisfy the resonance relations. (c) Phase space diagram showing the slow exchange of energies E_1, E_2, E_3 among the triad of waves.

and therefore the horizontal average of their product is independent of the vertical coordinate z. Consequently, the $\partial(\overline{\rho' w})/\partial z$ term in (9.3.11) vanishes, and (9.3.12) also implies that the static stability ($s = -\rho_0^{-1}\,\partial\bar{\rho}/\partial z$) is independent of time. Because of these kinematic considerations, all the nonlinear terms in (9.3.11) vanish independently of the amplitude, and we have

$$(\partial/\partial t)(\rho_m'/\rho_0) - sV_m \sin\theta_m = 0 \tag{9.8.2}$$

Likewise, the Lagrangian derivative of $\mathbf{V}_m$ is identically equal to the Eulerian derivative because $(\mathbf{V}_m \cdot \nabla)\mathbf{V}_m \equiv 0$. Since $\nabla\phi_m'$ is normal to the wave front, the only components of the momentum equation (9.3.7) lying parallel to the front are

$$\partial V_m/\partial t = -g(\rho_m'/\rho_0)\sin\theta_m \tag{9.8.3}$$

By eliminating the density perturbation in (9.8.2) and (9.8.3), we have

$$\partial^2 V_m/\partial t^2 + (gs \sin^2 \theta_m) V_m = 0$$

Equation (9.8.4) gives a solution of this equation, where the frequency ω_m is indeed identical to (9.8.1), and the amplitude a_m and phase factors ψ_m are arbitrary. The solution (9.8.5) for the density is obtained from (9.8.3), and the vorticity (9.8.6) is obtained by differentiating (9.8.4). The total wave energy [see (9.4.5)] is given by (9.8.7).

$$V_m = a_m \cos(b_m \xi - \omega_m t + \psi_m) \tag{9.8.4}$$

$$\rho_m'/\rho_0 = +a_m (g/s)^{-1/2} \sin(b_m \xi - \omega_m t + \psi_m) \tag{9.8.5}$$

$$\zeta_m = \partial V_m/\partial \xi = -a_m b_m \sin(b_m \xi - \omega_m t + \psi_m) \tag{9.8.6}$$

$$E_m = \tfrac{1}{2} V_m^2 + \tfrac{1}{2}(g/s)(\rho_m'/\rho_0)^2 = \tfrac{1}{2} a_m^2 \tag{9.8.7}$$

(b) *Exchange of Energy among Several Modes.* We now consider the case in which several plane waves are present, these being denoted by wave number vectors $\mathbf{b}_1, \ldots, \mathbf{b}_m, \mathbf{b}_p, \mathbf{b}_q, \ldots$ and for the sake of simplicity, we will assume that these vectors are coplanar (Fig. 9.5b), so that the total velocity vector $\mathbf{V}$ is confined to the vertical xz plane. The typical plane wave $\mathbf{b}_m$ is represented by the velocity

$$\mathbf{V}_m = a_m(t) \cos(\mathbf{b}_m \cdot \mathbf{r} - \omega_m t + \psi_m(t))$$

and by a density field

$$\rho_m'/\rho_0 \equiv \rho_m'' \tag{9.8.8a}$$

having the same wavelength. In this spatial Fourier representation, the amplitude a_m and the phase ψ_m may be functions of time. Two different modes p and q, with $\mathbf{b}_p \neq \mathbf{b}_q$, are orthogonal in the sense

$$\langle \mathbf{V}_p \cdot \mathbf{V}_q \rangle = 0 = \langle \rho_p' \rho_q' \rangle = \langle \mathbf{V}_p \rho_q' \rangle \tag{9.8.8b}$$

where brackets denote an average over all xz, and the total fields are given by

$$\mathbf{V} = \sum_{m=1} \mathbf{V}_m, \qquad \rho'/\rho_0 = \sum_{m=1} \rho_m'' \tag{9.8.9}$$

Since $\mathbf{b}_m \cdot \mathbf{r}$ is a linear function of x and z, it follows that the horizontal average of the products of functions like (9.8.8a) is independent of z, and we denote this symmetry property by

$$(\partial/\partial z)\overline{(\ )} = 0 \tag{9.8.10}$$

Note that the term $\rho_0^{-1}\,\overline{\rho' w}$ in (9.3.11) is equal to the sum of $\overline{\rho_m'' w_m}$ over the entire set of orthogonal modes, where $w_m = \mathbf{k} \cdot \mathbf{V}_m$ is the vertical velocity associated with each mode. The symmetry condition (9.8.10) then implies $\overline{\partial \rho' w}/\partial z = 0$, and (9.3.11) reduces to

$$(\partial/\partial t)(\rho'/\rho_0) - ws = -\mathbf{V}\cdot\nabla\rho'/\rho_0 \tag{9.8.11a}$$

where the static stability is an absolute constant.

When the right hand side of (9.8.11a) is evaluated for any one mode m it vanishes, or

$$\mathbf{V}_m \cdot \nabla\rho''_m \equiv 0 \tag{9.8.11b}$$

but for two different modes $\mathbf{b}_2$ and $\mathbf{b}_3$, the right hand side of (9.8.11a) has the value

$$(\mathbf{V}_2 + \mathbf{V}_3)\cdot\nabla(\rho''_2 + \rho''_3) = \mathbf{V}_2\cdot\nabla\rho''_3 + \mathbf{V}_3\cdot\nabla\rho''_2 \neq 0 \tag{9.8.11c}$$

Suppose, for example, that the initial ($t = 0$) state contains only these two modes 2 and 3, with amplitudes sufficiently small so that (9.8.11c) can be neglected in (9.8.11a), as a first (linear) approximation. The frequencies ω_2 and ω_3 of these waves are then determined by the linear dispersion relation (9.8.1). The effect of the quadratically small terms can now be determined by the following iterative procedure.

When the linear eigenfunctions are used to evaluate (9.8.11c), and when the result is substituted in the right hand side of (9.8.11a), we obtain the *inhomogeneous* linear equation

$$\partial(\rho'/\rho_0)/\partial t - ws \simeq -\mathbf{V}_2\cdot\nabla\rho''_3 - \mathbf{V}_3\cdot\nabla\rho''_2 \tag{9.8.12a}$$

whose solution gives the second approximation to ρ' and $\mathbf{V}$. When a similar procedure is applied to the nonlinear terms in the momentum equation (9.3.7), we have

$$\partial\mathbf{V}/\partial t + g(\rho'/\rho_0)\mathbf{k} + \nabla\phi = -\mathbf{V}\cdot\nabla\mathbf{V} \simeq -\mathbf{V}_2\cdot\nabla\mathbf{V}_3 - \mathbf{V}_3\cdot\nabla\mathbf{V}_2 \tag{9.8.12b}$$

as a simultaneous inhomogeneous linear equation for the second approximation. The inhomogeneous terms in Eqs. (9.8.12) are *products* of sinusoidal functions of $(\mathbf{b}_2\cdot\mathbf{r} - \omega_2 t)$ and $(\mathbf{b}_3\cdot\mathbf{r} - \omega_3 t)$. Those terms can therefore be expressed as the *sum* of sinusoidal functions whose arguments are

$$(\mathbf{b}_2 \pm \mathbf{b}_3)\cdot\mathbf{r} - (\omega_2 \pm \omega_3)t \tag{9.8.13}$$

By elimination of ρ'/ρ_0 and ϕ, we then obtain an inhomogeneous wave equation, with the quadratically small terms (9.8.13) on the right hand side, and the first approximation for $\mathbf{V}_2$ and $\mathbf{V}_3$ is recovered from the homogeneous solution. But the *forced* solution leads to new components having (9.8.13) as their arguments and $|\mathbf{V}_2|\,|\mathbf{V}_3|$ as their order of magnitude. This nonlinear iteration procedure fails, however, if the inhomogeneous terms "resonate" in the linear wave equation. Such resonance will occur if the wave number $\mathbf{b}_2 \pm \mathbf{b}_3$ of the right hand sides of Eqs. (9.8.12) and their frequencies $\omega_2 \pm \omega_3$ satisfy the linear dispersion relation for free waves. Thus, if

$$\mathbf{b}_2 + \mathbf{b}_3 \equiv \mathbf{b}_1 \qquad \text{and} \qquad \omega_2 + \omega_3 \equiv \omega_1 \tag{9.8.14}$$

and if ω_1 happens to equal the *free* oscillation frequency for wave number $\mathbf{b}_1$, then the formal solution of the inhomogeneous wave equation (9.8.12) will amplify with time, as in an ordinary resonant oscillator. Under these conditions the simple iterative procedure fails, even when the amplitudes are small, and we must seek another method to analyze this interesting case.

The wave numbers $\mathbf{b}_2$ and $\mathbf{b}_3$ are shown together with their sum $\mathbf{b}_1 = \mathbf{b}_2 + \mathbf{b}_3$ in Fig. 9.5b, and the frequencies are determined by the angles θ_1, θ_2, and θ_3, according to the linear dispersion relation (9.8.1). Thus, we see that the resonance condition will be satisfied if

$$\sin\theta_2 + \sin\theta_3 = \sin\theta_1 \tag{9.8.15}$$

We will return to the problem of the resonant triad (9.8.14) interaction after deriving a more general system of equations which shows how the energy is exchanged between different modes.

By multiplying (9.8.11a) with $gs^{-1}\rho''_m$, by averaging the result, and by using Eq. (9.8.8b) and (9.8.9) in the first two terms of (9.8.11a), we obtain

$$(g/2s)(\partial/\partial t)\langle(\rho''_m)^2\rangle - g\langle w_m \rho''_m\rangle = -(g/s)\langle\rho''_m \mathbf{V}\cdot\nabla\rho'/\rho_0\rangle$$

By taking the scalar product of (9.3.7) with $\mathbf{V}_m$, and by using a similar procedure, we also have

$$\tfrac{1}{2}(\partial/\partial t)\langle \mathbf{V}_m^2\rangle + g\langle\rho''_m w_m\rangle = -\langle \mathbf{V}_m \cdot \mathbf{V}\cdot\nabla\mathbf{V}\rangle$$

When these two equations are added, one obtains the energy equation

$$\partial E_m/\partial t = -\langle \mathbf{V}_m \cdot \mathbf{V}\cdot\nabla\mathbf{V}\rangle - (g/s)\langle\rho''_m \mathbf{V}\cdot\nabla\rho'/\rho_0\rangle \tag{9.8.16a}$$

$$= -\sum_p \sum_q \langle \mathbf{V}_m \cdot \mathbf{V}_p \cdot \nabla\mathbf{V}_q + (g/s)\rho''_m \mathbf{V}_p \cdot \nabla\rho''_q\rangle \tag{9.8.16b}$$

where E_m is defined by (9.8.7) and the summation extends over all values of p and q. If (9.8.16a) be summed over all m, the last term becomes

$$\left(\sum \rho''_m\right)\mathbf{V}\cdot\nabla\rho'/\rho_0 = \langle\rho'/\rho_0\, \mathbf{V}\cdot\nabla\rho'/\rho_0\rangle = \tfrac{1}{2}\langle\nabla\cdot\mathbf{V}(\rho'/\rho_0)^2\rangle = 0$$

and since

$$\left\langle\sum \mathbf{V}_m \cdot \mathbf{V}\cdot\nabla\mathbf{V}\right\rangle = \langle\mathbf{V}\cdot\mathbf{V}\cdot\nabla\mathbf{V}\rangle = 0$$

we obtain

$$(\partial/\partial t)\sum_m E_m = 0 \tag{9.8.17}$$

Thus, the total wave energy is conserved. But for any wave number, say $m = 1$, Eq. (9.8.16a) shows that the energy changes at the rate

$$\partial E_1/\partial t = -\sum \langle \mathbf{V}_1 \cdot \mathbf{V}_p \cdot \nabla \mathbf{V}_q + (g/s)\rho_1''\mathbf{V}_p \cdot \nabla \rho_q'' \rangle \tag{9.8.18a}$$

Consider the contribution of the $\rho_1''\mathbf{V}_2 \cdot \nabla \rho_3''$ in (9.8.18a). Such a product of three sinusoids equals the sum of sinusoids having $(\mathbf{b}_1 \pm \mathbf{b}_2 \pm \mathbf{b}_3) \cdot \mathbf{r}$ in their arguments. Therefore, the spatial average will vanish unless $\mathbf{b}_1 \pm \mathbf{b}_2 \pm \mathbf{b}_3 = 0$. And, more generally, only those combinations in (9.18a) that satisfy the "triad relation" $(\mathbf{b}_p \pm \mathbf{b}_q \pm \mathbf{b}_1 = 0)$ can lead to a nonvanishing integral. Suppose, for example, that the initial state consists of only three waves (as in Fig. 9.5b). Then the terms $\mathbf{V}_1 \cdot \mathbf{V}_2 \cdot \nabla \mathbf{V}_3$, etc., do *not* vanish, and the right hand side of (9.8.18a) is proportional to the product of the amplitudes of the triad. Since these amplitudes are proportional to the square roots of their energies, we may write (9.8.18a) in the abbreviated form

$$dE_1/dt = \beta_1 E_1^{1/2} E_2^{1/2} E_3^{1/2} \tag{9.8.18b}$$

where the coefficient β_1 is discussed further in subsection (d). Thus, we see how the nonlinear interaction of triads can lead to a change of energy in each mode, while leaving the total energy invariant, and we now consider a second invariant.

(c) *Vorticity and Wave "Action."* The components of a two-dimensional field $\mathbf{V}$ satisfy the continuity equation

$$\partial u/\partial x + \partial w/\partial z = 0 \tag{9.8.19}$$

and in Eq. (5.1.5b), we showed that $\nabla \times d\mathbf{V}/dt = (d/dt)(\nabla \times \mathbf{V})$. Thus, if

$$\zeta = \partial u/\partial z - \partial w/\partial x \tag{9.8.20}$$

denotes the y component of vorticity, and if the curl of the Boussinesq equation (9.3.3) be taken, then we obtain

$$d\zeta/dt = (g/\rho_0)\, \partial\rho/\partial x \tag{9.8.21}$$

where the last term is the y component of the curl of the buoyancy force. Since $d\rho/dt = 0$, the product of (9.8.21) with ρ can be written as

$$(d/dt)(\rho\zeta) = (g/2\rho_0)\, \partial\rho^2/\partial x$$

and a horizontal integration then gives

$$(\partial/\partial t)\overline{\rho\zeta} + (\partial/\partial z)\overline{w\rho\zeta} = 0$$

where (9.8.19) has been used. By substituting $\rho = \rho' + \bar{\rho}(z)$, and by noting that

$\overline{w} = 0 = \overline{\zeta}$ because of the cyclic variation of these fields in the x direction, we obtain

$$(\partial/\partial t)\overline{\rho'\zeta} = -(\partial/\partial z)[\overline{\rho}(\overline{w\zeta})] - (\partial/\partial z)\overline{\rho' w\zeta}$$

The horizontal averages of $\rho'\zeta$, $w\zeta$, and $(\rho' w\zeta)$ are all independent of z, because of the symmetry property (9.8.10), and therefore the above equation simplifies to

$$(\partial/\partial t)\overline{\rho'\zeta} = \rho_0 s\overline{w\zeta}$$

where s is the static stability. But

$$\overline{w\zeta} = \overline{w\ \partial u/\partial z} = \overline{-u\ \partial w/\partial z} = \overline{u\ \partial u/\partial x} = 0$$

according to (9.8.19) and (9.8.20), and therefore we obtain the nonlinear invariant

$$(\partial/\partial t)(\overline{\rho'\zeta}) = 0 \tag{9.8.22a}$$

Let ζ_m denote the component of vorticity which is in (time) phase with the density fluctuation ρ''_m of the mode having wave number $\mathbf{b}_m$. Then the above equation is equivalent to

$$(\partial/\partial t) \sum_m \overline{\rho''_m\zeta_m} = 0 \tag{9.8.22b}$$

The quantity $\rho''_m\zeta_m$ can be exactly evaluated using (9.8.5)–(9.8.7) and (9.8.1) for the case of a single mode. By doing this, and by then introducing the unit vector $\mathbf{i}$ in the x direction, we have

$$\begin{aligned} \overline{-\rho''_m\zeta_m} &= \tfrac{1}{2}a_m^2 b_m (g/s)^{-1/2} = E_m b_m (g/s)^{-1/2} \\ &= E_m (g/s)^{-1/2}((\mathbf{b}_m \cdot \mathbf{i})/\sin\theta_m) = s(E_m \mathbf{b}_m/\omega_m)\cdot\mathbf{i} \end{aligned} \tag{9.8.23}$$

Although (9.8.22b) is satisfied when several interacting waves are present, (9.8.23) is only an approximation to $\overline{\rho_m''\zeta_m}$, and therefore the combination of the two, or

$$\mathbf{i} \cdot (d/dt) \sum E_m \mathbf{b}_m/\omega_m = 0 \tag{9.8.24}$$

is only asymptotically valid, as indicated below. The quantity $E_m\mathbf{b}_m/\omega_m$ is sometimes called the "wave action."

(d) *Triad Interaction.* Let us now consider an initial $(t = 0)$ state consisting of the three wave numbers shown in Fig. 9.5b, the amplitudes $E_1^{1/2}$, $E_2^{1/2}$ and $E_3^{1/2}$ of which are assumed to be "small" to the same order of magnitude $E^{1/2}$. This triad satisfies the resonance relation (9.8.14), and therefore the energy in wave number $\mathbf{b}_1$ will start to change at the rate given by (9.8.18b).

Similar equations can also be obtained for dE_2/dt and dE_3/dt. The coupling coefficient β_1 does not depend on the amplitude of the fields on the right hand side of (9.8.18a), but only on their wavelengths and relative orientation θ_1, θ_2, and θ_3. The calculation of this coupling coefficient is quite tedious (see Davis and Acrivos, 1967, for example) and need not concern us here. The main point is that E_1, E_2, and E_3 change slowly, at a rate proportional to $E^{3/2}$, as a result of their mutual interaction. The triad will also give rise to other Fourier terms in $\mathbf{V} \cdot \nabla \mathbf{V}$, but these wave components do not resonate and consequently their energy is negligible to order E^2. Therefore, (9.8.18b) and the analogous equations for dE_2/dt and dE_3/dt, constitute a closed system of nonlinear equations for the determination of $E_1(t)$, $E_2(t)$, and $E_3(t)$. These asymptotic equations describe the first-order nonlinear effects on the waves. We need not write down the equations for dE_2/dt and dE_3/dt corresponding to (9.8.18b), because simpler (and equivalent) relations are provided by the invariants (9.8.17) and (9.8.22b). Since the energy in waves other than b_1, b_2, and b_3 is small to order E^2, the conservation of energy is approximated by

$$dE_1/dt + dE_2/dt + dE_3/dt = 0 \tag{9.8.25}$$

and a similar error is introduced by using (9.8.24) or

$$\mathbf{i} \cdot [(\mathbf{b}_1/\omega_1)\, dE_1/dt + (\mathbf{b}_2/\omega_2)\, dE_2/dt + (\mathbf{b}_3/\omega_3)\, dE_3/dt] = 0 \tag{9.8.26}$$

as the truncated version of (9.8.22b)

When dE_1/dT is eliminated between (9.8.25) and (9.8.26), and when $\mathbf{b}_1$ and ω_1 are eliminated using (9.8.14), we obtain

$$\mathbf{i} \cdot \left[\frac{dE_2}{dt}\left(\frac{\mathbf{b}_2}{\omega_2} - \frac{\mathbf{b}_2 + \mathbf{b}_3}{\omega_2 + \omega_3}\right) + \frac{dE_3}{dt}\left(\frac{\mathbf{b}_3}{\omega_3} - \frac{\mathbf{b}_2 + \mathbf{b}_3}{\omega_2 + \omega_3}\right)\right] = 0$$

$$\mathbf{i} \cdot [\omega_2^{-1}\, dE_2/dt - \omega_3^{-1}\, dE_3/dt]\,(\omega_3 \mathbf{b}_2 - \omega_2 \mathbf{b}_3) = 0$$

Therefore,

$$\omega_2^{-1}\, dE_2/dt = \omega_3^{-1}\, dE_3/dt \tag{9.8.27}$$

and we conclude that E_2 and E_3 must either increase or decrease together (while E_1 changes in the opposite sense). The appropriate sign depends on the initial value of the relative phases ψ_1, ψ_2, and ψ_3, and $E_1(t)$ can be made to decrease with time by appropriate choice of these phases. [The reader can verify this from (9.8.18b), (9.8.25) and (9.8.27) by calculation of β_1. This long calculation is made by truncating (9.8.18a) with $p, q = 2, 3$ and by using the linear eigenfunctions for $\mathbf{V}_1$, $\mathbf{V}_2$, $\mathbf{V}_3$, ρ_1'', ρ_2'', ρ_3'' given in (9.8.4) – (9.8.5).]

The foregoing conclusion means that a single plane wave $\mathbf{b}_1$ is *unstable* with respect to a perturbation consisting of two waves $\mathbf{b}_2$ and $\mathbf{b}_3$ which satisfy the triad relation, as shown by the phase space diagram of Fig. 9.5c. The coordinates of point A give the respective values of the energy in the triad after the original wave is disturbed by the perturbation E_2, E_3. Since $E_1 + E_2 + E_3$ is independent of time, the subsequent trajectory ABC of the representative point must lie on the plane surface shown, and this trajectory must be a straight line because of (9.8.27) (or dE_2/dE_3 = constant). For some perturbations (depending on the phases ψ_1, ψ_2, and ψ_3) the phase point moves in the direction of the arrow, and the energy E_1 in the original wave decreases.

The nonlinear growth rate (9.8.18b) is clearly small compared with the Brunt–Väisälä frequency when the amplitude is small, and therefore the phase point in Fig. 9.5c will take a long time to reach C. From (9.8.27) one can show that the phase point retraces its path when it reaches C, and thus we see how the inertial oscillations of the linear theory are modulated as energy is shifted back and forth among the triad. However, the nonlinear theory given above is not valid over arbitrarily long intervals of time, because more and more wave numbers are generated in the nonlinear cascade. These additional modes will eventually find "partners" for exchanging energy with one of the original triad, and thus the energy will eventually spread to a large number of modes.

9.9 Inertia–Gravity Waves

The equations for a *rotating* stratified fluid will be illustrated by examining the propagation of inertia–gravity waves between the two rigid level surfaces ($z = 0, H$) shown in Fig. 9.2. The effective force of gravity g points normal to the undisturbed isopycnal surfaces, these being indicated by the dashed lines, and the basic static stability $s = -\rho_0^{-1}\,\partial\bar{\rho}/\partial z$ is assumed constant. Let $\mathbf{V}' = (u', v', w')$ denote the infinitesimal velocity perturbation and its x, y, z components in the rotating coordinate system, where the x axis is directed into the plane of the page (Fig. 9.2). The horizontal wavelength of the perturbation is assumed to be sufficiently small so that the curvature of the level surfaces can be neglected in that which follows. The density perturbation is denoted by ρ', the pressure perturbation by $\rho_0\phi'$, and the x, y, z components of the Coriolis force are readily found to be

$$2\hat{\omega} \times \mathbf{V}' = (-fv' + f_H w', fu', -f_H u')$$

where f is the component of $2\hat{\omega}$ along the z axis and f_H is the horizontal component. Thus, the linearized form of the three momentum equations, the density equation, and the continuity equation, respectively, are given by

$$\begin{aligned} \partial u'/\partial t - fv' + f_H w' &= -\partial\phi'/\partial x \\ \partial v'/\partial t + fu' &= -\partial\phi'/\partial y \\ \partial w'/\partial t - f_H u' &= -\partial\phi'/\partial z - g\,\rho'/\rho_0 \end{aligned} \tag{9.9.1}$$

$$(\partial/\partial t)(\rho'/\rho_0) - w's = 0$$

$$\partial u'/\partial x + \partial v'/\partial y + \partial w'/\partial z = 0$$

If $2\pi/k$ is the wavelength of a normal mode in the x direction, and if ω is the frequency, then the pressure perturbation has the form

$$\phi' = \text{Re}\, \phi(y, z) \exp(ikx + i\omega t) \qquad (k, \omega > 0) \tag{9.9.2}$$

and the corresponding amplitudes of u', v', w' are denoted by u, v, w. From the fourth equation in (9.9.1), we see that the amplitude of ρ'/ρ_0 is given by $(i\omega)^{-1}sw$, and therefore the remaining equations become

$$\begin{aligned} i\omega u - fv + f_H w &= -ik\phi \\ fu + i\omega v &= -\phi_y \\ -if_H \omega u + (gs - \omega^2)w &= -i\omega\phi_z \\ iku + v_y + w_z &= 0 \end{aligned}$$

By solving the first three equations for u, v, w in terms of ϕ, we obtain

$$\begin{aligned} [(gs - \omega^2)(f^2 - \omega^2) - f_H^2\omega^2]\, w &= -i\omega\phi_z (f^2 - \omega^2) - if_H\omega(f\phi_y - k\omega\phi) \\ [(gs - \omega^2)(f^2 - \omega^2) - f_H^2\omega^2]\, u &= (gs - \omega^2)(k\omega\phi - f\phi_y) - f_H\omega^2\phi_z \\ [(gs - \omega^2)(f^2 - \omega^2) - f_H^2\omega^2]\, v &= -i(gs - \omega^2)(\omega\phi_y - kf\phi) - if_H\omega(f\phi_z + f_H\phi_y) \end{aligned} \tag{9.9.3}$$

Assuming that the frequency computed below is such that

$$(gs - \omega^2)(f^2 - \omega^2) - f_H^2\,\omega^2 \neq 0$$

then the substitution of (9.9.3) in the continuity equation, and the simplification of the result, gives

$$(f^2 - \omega^2)\phi_{zz} + 2ff_H\phi_{yz} + (gs - \omega^2 + f_H^2)\phi_{yy} - (gs - \omega^2)k^2\phi = 0 \tag{9.9.4}$$

Since $w(0) = w(H) = 0$, Eq. (9.9.3) gives the boundary condition

$$(f^2 - \omega^2)\phi_z + f_H(f\phi_y - k\omega\phi) = 0 \qquad (z = 0, H) \tag{9.9.5}$$

If we now neglect lateral boundary effects by assuming a periodic oscillation in the y direction with arbitrary wave number b, then the separated solution of (9.9.4) is

$$\phi = e^{iby}[A_1 e^{im_1 z} + A_2 e^{im_2 z}] \tag{9.9.6}$$

where

$$\begin{pmatrix} m_1 \\ m_2 \end{pmatrix} = \frac{-ff_H b \pm \{(ff_H b)^2 - (f^2 - \omega^2)[(gs - \omega^2)(k^2 + b^2) + b^2 f_H^2]\}^{1/2}}{f^2 - \omega^2} \tag{9.9.7}$$

and where A_1 and A_2 are constants of integration. By substituting (9.9.6) in the boundary condition (9.9.5), we find that the two homogeneous linear equations for A_1 and A_2 can only be satisfied if

$$e^{im_1 H} = e^{im_2 H} \qquad \text{or} \qquad m_1 - m_2 = 2n\pi/H \tag{9.9.8}$$

where n is an integer. The frequency equation obtained by substituting (9.9.7) in (9.9.8) is

$$(ff_H b)^2 - (f^2 - \omega^2)[(gs - \omega^2)(k^2 + b^2) + b^2 f_H^2] = (n\pi/H)^2 (f^2 - \omega^2)^2 \tag{9.9.9a}$$

and this quartic has two distinct solutions for ω^2.

In the ocean, the Brunt-Väisälä frequency is much greater than the Coriolis parameters, or

$$gs \gg f^2 \quad \text{or} \quad f_H^2 \tag{9.9.9b}$$

and for this asymptotic case, one of the roots

$$gs > \omega^2 > f^2$$

of (9.9.9a) may be isolated by expanding ω^2/f^2 in the small parameter $f_H{}^2/gs$. Thus we find that the value of the leading term can be obtained by setting $f_H = 0$ in (9.9.9a), or

$$(\omega^2 - f^2)(n\pi/H)^2 = (gs - \omega^2)(k^2 + b^2)$$

and therefore

$$\omega^2 = \frac{f^2 + [gs(k^2 + b^2)/(n\pi/H)^2]}{1 + [k^2 + b^2]/(n\pi/H)^2} \tag{9.9.10}$$

gives the frequency of "traditional" inertia–gravity waves (Eckart, 1960). By setting $f_H = 0$ in (9.9.7) and by using (9.9.10), we see that the values of m_1 and m_2 are equal to $\pm n\pi/H$, so that the first ($n = 1$) of these modes varies sinusoidally in z with a half wavelength equal to H.

The second root of (9.9.9a) is slightly less than f^2 when (9.9.9b) applies, and we can isolate it by using (9.9.9b) and

$$(gs - \omega^2)(k^2 + b^2) + b^2 f_H^2 \simeq gs(k^2 + b^2)$$

in (9.9.9a). A quadratic equation for $(f^2 - \omega^2)$ is then obtained, and the root which approaches zero with f_H^2/gs is

$$(f^2 - \omega^2) = \frac{[(gs)^2(k^2 + b^2)^2 + 4(ff_H bn\pi/H)^2]^{1/2} - gs(k^2 + b^2)}{2(n\pi/H)^2}$$

or

$$\frac{f^2 - \omega^2}{f^2} = \frac{f_H^2}{gs}\,\frac{b^2}{k^2 + b^2} + \ldots \tag{9.9.11}$$

These modes depend on the local horizontal component f_H of the earth's rotation vector, and the corresponding value of (9.9.7) is much larger than the vertical wave number of the "traditional" mode.

The terms containing f_H in the equations of motion (9.9.1) are usually neglected for the large scale and slowly varying motions in the strongly stratified thermocline because the vertical velocities are small compared to the horizontal velocities, and therefore $f_H w' \ll fv'$ in (9.9.1). We have shown, however, that oscillations near the inertia frequency ($\omega \sim f$) are an exception, and the reason for this appears to be the singular behavior of Eq. (9.9.4) when $\omega = f$.

References

Archimedes, P. D. Q. (213 B.C.). The Role of Bathtubs in Natural Philosophy. *Proc. Greek Fluid Dyn. Summer Program* (Rex Midas, ed.), pp. 1781–1782. Olympian Press, 38° 9′2″N, 30° 1′8″E.

Avsec, D. (1939). "Thermoconvective Eddies in Air with Application to Meterorology" (in French). Sci. and Tech. Pub. Air Ministry, Works Inst. Fluid Mech. Sci. Paris, No. 155. Gauthier-Villars, Paris.

Bretherton, F. D. (1966). The Propagation of Groups of Waves in a Shear Flow. *Quart. J. Roy. Soc.* **92**, 466–480.

Cox, C. S., and Sandstrom, H. (1962). Coupling of Internal and Surface Waves in Water of Variable Depth. *J. Oceanogr. Soc. Japan* 20th Anniv. Vol. 499–513.

Davis, R. E., and Acrivos, A. (1967). The Stability of Oscillatory Internal Waves. *J. Fluid Mech.* **30**, 723–736.

Eckart, C. (1960). "Hydrodynamics of Oceans and Atmospheres." Pergamon, Oxford.

Gage, K. S., and Reid, W. H. (1968). The Stability of Thermally Stratified Shear Flow. *J. Fluid Mech.* **33**, 21–23.

Pollard, R. (1970). On the Generation by Winds of Inertial Waves in the Ocean. *Deep Sea Res. Oceanogr. Abstr.* **17**, 795–812.

Rattray, M. (1960). On the Coastal Generation of Internal Tides. *Tellus* **12**, 54–62.

Thorpe, S. A. (1969). Experiments on the Stability of Stratified Shear Flows. *Radio Sci.* **4**, 1327–1331.

Webster, F. (1968). Observations of Inertial Period Motion in the Deep Sea. *Rev. Geophys.* **6**, 473–490.

Woods, J. D. (1968). Wave Induced Shear Instability in the Summer Thermocline. *J. Fluid Mech.* **32**, 791–800.

CHAPTER X

Convection

10.1 Introduction

The combined atmosphere–ocean system provides an example of "free" convection, since density differences provide the *primum mobile*. But the transformations of the incoming solar energy are so diverse and complex, that we must isolate different aspects before attempting to put them together. In this chapter, we consider the "vertical" convection which arises when a broad horizontal layer is heated *uniformly* from below and cooled from above. The term "horizontal" convection is used when the fluid is differentially heated on a level surface, and this case is considered in Chapter XII.

A vivid impression of vertical convection can be obtained by gazing at the surface of a cup of tea when it is illuminated by the bright sun and cooling by evaporation. Because the index of refraction varies with temperature one can see the shadows of thin cold plumes, as they plunge downward from the top heavy surface.

The parallel plate convection experiment (Fig. 10.1) provides a simpler version of this problem. Here we maintain a uniform temperature difference ΔT

between the lower ($z = 0$) and upper ($z = H$) boundaries, and the state can be in conductive and hydrostatic equilibrium, as shown by the dashed temperature profile in Fig. 10.1a. According to Fourier's law, the general expression for the molecular flux of heat is $-K_q \nabla T$, where K_q (cal cm^{-2} sec^{-1} /°C cm^{-1}) is the thermometric conductivity, and therefore $K_q \, \Delta T/H$ is the uniform upward flux of heat in the state of conductive equilibrium.

If there be a net convergence of heat per unit volume, or $\nabla \cdot K_q \nabla T$, then the conservation of energy requires a corresponding increase of internal energy (plus the work of compression). But if the conditions of the Boussinesq approximation (Chapter IX) apply, then the compressibility can also be neglected in the first law of thermodynamics, and thus the rate of increase of internal energy per unit volume is given by $\rho_0 c_p \, dT/dt$, where c_p (cal/gm/°C) is the specific heat. By setting this equal to the convergence of the molecular flux, we obtain the heat equation

$$dT/dt = k_T \nabla^2 T \qquad \text{and} \qquad k_T = K_q/\rho_0 c_p \ (\text{cm}^2/\text{sec}) \tag{10.1.1}$$

where k_T is called the kinematical thermal diffusivity. Besides adding the heat diffusion term to (9.3.8), we must now add the Navier–Stokes (Chapter V) viscous force $\nu \nabla^2 \mathbf{V}$ to (9.3.7), and the Boussinesq momentum equation then becomes

$$dV/dt = -\nabla\phi + g\alpha T'\mathbf{k} + \nu \nabla^2 \mathbf{V}, \qquad \nabla \cdot \mathbf{V} = 0 \tag{10.1.2}$$

where

$$\phi = p'/\rho_0 - w^2, \qquad w = \mathbf{k} \cdot \mathbf{V}, \qquad \rho' = -\rho_0 \alpha T',$$

and

$$T'(x, y, z, t) = T(x, y, z, t) - \overline{T}(z, t) \tag{10.1.3}$$

These equations are used in Section 10.2 to determine the critical condition for the instability of the conductive equilibrium state. The nonlinear effect on the convection cell is considered in Section 10.4, and the associated heat flux is computed. The similarity theory of Section 10.6 is used to determine the variation of heat flux with ΔT when the latter is moderately large. This also sets the stage for the similarity theory of fully developed thermal turbulence (Section 10.6), and the reader should be warned that the arguments here become deceptively simple and increasingly tentative.

We then return to our cup of tea, with (Section 10.8) or without (Section 10.7) sugar, and consider the important problem of evaporative convection. When a solute is present, the evaporating H_2O molecules tend to increase the concentration and therefore the density, at the free surface. Thus, the salinity S (gm solute/gm of solution) also contributes to the convection.

According to Fick's law for a dilute solution, the conservation of salt is expressed by

$$dS/dt = k_s \nabla^2 S \tag{10.1.4}$$

where k_S is the molecular diffusity (cm^2/sec).

Although the effects of thermal buoyancy $\alpha T'$ and salt buoyancy $\beta S'$ are coupled in the natural evaporation process, pure haline convection is approximately realized when salt water freezes, because α is small at low temperatures. The salt molecules which are left in solution when the surface freezes provide the buoyancy for convection (Foster, 1969).

10.2 The Critical Rayleigh Number

In the parallel plate convection experiment (Fig. 10.1), the undisturbed value of $\partial T/\partial z$ is $-\Delta T/H$, and we now examine the stability of this conductive and

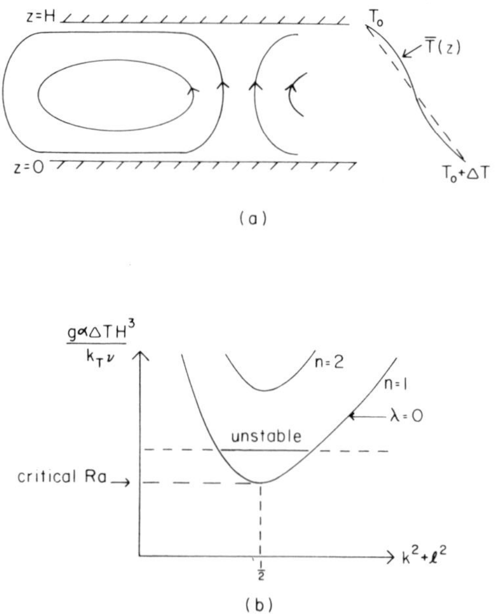

FIG. 10.1 (a) Schematic diagram of steady convection cell in the parallel plate experiment. $\bar{T}(z)$ is the horizontally averaged temperature profile as modified by convection. (b) The Rayleigh number for neutral stability plotted as a function of the horizontal wave numbers k, l of the cell. The vertical mode number $n = 1$ corresponds to the cell shown in (a).

hydrostatic equilibrium. Let $w_1(x, y, z, t)$ denote the vertical component of the perturbation velocity $\mathbf{V}_1$, so that the linearization of (10.1.1) is

$$\partial T_1'/\partial t + w_1(-\Delta T/H) = k_T \nabla^2 T_1' \tag{10.2.1a}$$

and the linearization of (10.1.2) is

$$\partial \mathbf{V}_1/\partial t = -\nabla\phi_1 + g\alpha T_1'\mathbf{k} + \nu \nabla^2 \mathbf{V}_1 \tag{10.2.1b}$$

where T_1' is the temperature perturbation, and ϕ_1 the pressure.

All of the perturbation fields can be represented by the normal modes of the form

$$e^{\lambda t} \cos(\pi kx/H) \cos(\pi ly/H) \cdot \text{function of } (z) \tag{10.2.2}$$

where k and l are arbitrarily assigned wave numbers, and the vertical eigenfunctions are to be determined subject to subsequently stated boundary conditions. The growth rate $\lambda = \lambda(k, l, \Delta T, \ldots)$ will then appear as the eigenvalue, and if λ is positive for any k or l, then this mode will amplify in time, whereas if $\lambda < 0$ for all k and l, then the basic state is stable. By setting the growth rate equal to zero, the equation $0 = \lambda(k, l, \Delta T, \ldots)$ will then determine the value of ΔT for which a given k, l mode is *neutral.* By considering all values of k and l, and by finding the smallest value of $\Delta T = \Delta T_c$ for which the relation $0 = \lambda(k, l, \Delta T)$ can be satisfied, we determine the minimum critical temperature difference, or the *Rayleigh number*

$$R_a \equiv g\alpha\, \Delta T H^3/k_T \nu \tag{10.2.3}$$

for instability. Thus, when (10.2.3) exceeds the critical value, some infinitesimal disturbance will amplify and the conductive equilibrium state will no longer be realized.

In order to determine the critical condition, we may, therefore, set $\partial/\partial t = 0$ in (10.2.1). If u_1, v_1, w_1 denotes the x, y, z velocity components of a neutral ($\lambda = 0$) mode, then we have

$$0 = -\partial\phi_1/\partial x + \nu \nabla^2 u_1 \tag{10.2.4a}$$

$$0 = -\partial\phi_1/\partial y + \nu \nabla^2 v_1 \tag{10.2.4b}$$

$$0 = -\partial\phi_1/\partial z + g\alpha T_1' + \nu \nabla^2 w_1 \tag{10.2.4c}$$

$$\partial u_1/\partial x + \partial v_1/\partial y + \partial w_1/\partial z = 0 \tag{10.2.5}$$

$$-(\Delta T/H) w_1 = k_T \nabla^2 T_1' \tag{10.2.6}$$

The elimination of T_1' between (10.2.4c) and (10.2.6) gives

$$0 = -\nabla^2\, \partial\phi_1/\partial z + \nu \nabla^4 w_1 - (g\alpha\,\Delta T/k_T H) w_1$$

The elimination of u_1 and v_1 in Eqs. (10.2.4a), (10.2.4b), and (10.2.5) yields

$$0 = -(\partial^2\phi_1/\partial x^2 + \partial^2\phi_1/\partial y^2) - \nu\,\nabla^2\,\partial w_1/\partial z \tag{10.2.7}$$

and when these two equations are combined, we have

$$(\partial^2/\partial x^2 + \partial^2/\partial y^2)(\nu\,\nabla^4 w_1 - (g\alpha\,\Delta T/k_T H)w_1) + \nu\,\nabla^4\,\partial^2 w_1/\partial z^2 = 0 \tag{10.2.8}$$

The metal boundaries at $z = 0$, H are assumed to be perfect conductors of heat, so that they remain isothermal even when the perturbation is introduced, and thus we have the four boundary conditions

$$w_1 = 0 = T_1' \qquad \text{at} \quad z = 0, H \tag{10.2.9}$$

for the solution of the sixth-order equation (10.2.8). In a realistic laboratory experiment, the two additional boundary conditions come from the no slip condition $u_1 = 0 = v_1$, and the calculations for this case can be found in the literature cited. For our purpose, however, it suffices to consider the simpler "no stress" boundary conditions, as used by Lord Rayleigh in his original paper. The x component of the stress on a horizontal boundary is $\partial u_1/\partial z$, and therefore the slippery boundary conditions require $\partial u_1/\partial z = 0 = \partial v_1//\partial z$ for all x and y on $z = 0, H$. The vertical derivative of (10.2.5) then gives

$$\partial^2 w_1/\partial z^2 = 0 \qquad (z = 0, H) \tag{10.2.10}$$

and we see that $\nabla^2 w_1 = 0$ on the boundaries. Since T_1' also vanishes on the boundaries, (10.2.4c) implies $\partial\phi_1/\partial z = 0$ on $z = 0, H$. By taking the z derivative of (10.2.7), we then obtain

$$\partial^4 w_1/\partial z^4 = 0 \qquad (z = 0, H) \tag{10.2.11}$$

For any integer n, the boundary conditions (10.2.9)–(10.2.11) are satisfied by

$$w_1 = A\cos(\pi kx/H)\cos(\pi ly/H)\sin(nz\pi/H) \tag{10.2.12}$$

where A is a constant, and the result of substituting (10.2.12) in (10.2.8) is

$$g\alpha\,\Delta T H^3/\pi^4 k_T\nu = n^2(n^2 + k^2 + l^2)^2/(k^2 + l^2) + (n^2 + k^2 + l^2)^2 \tag{10.2.13}$$

Figure 10.1b is a sketch of (10.2.13) as a function of $k^2 + l^2$ for $n = (1, 2)$. The minimum value of (10.2.13) occurs for

$$n = 1, \qquad k^2 + l^2 = \tfrac{1}{2} \tag{10.2.14}$$

and therefore the minimum critical value of the Rayleigh number is

$$(\mathrm{Ra})_c = (27/4)\pi^4 \tag{10.2.15}$$

[and we mention that the corresponding value for the experimental case of "no slip" boundaries is $(\mathrm{Ra})_c = 1708$]. When $\mathrm{Ra} > \mathrm{Ra}_c$, all of the points inside the

$n = 1$ curve (Fig. 10.1b) correspond to amplifying ($\lambda \geqslant 0$) modes, as can be verified by including the time dependent terms in (10.2.4)–(10.2.6).

A two-dimensional convection cell is obtained by setting $l = 0$ in (10.2.12), and the wave number $k = 1/\sqrt{2}$ is then given by (10.2.14). By substituting (10.2.12) (with $l = 0$) in (10.2.6), we solve for the thermal fluctuation listed in (10.2.16), and from (10.2.5) we obtain the horizontal velocity. Thus, the neutral eigenfunctions for a two-dimensional "roll" cell are

$$w_1 = A \cos\left(\frac{\pi x}{H\sqrt{2}}\right) \sin\left(\frac{\pi z}{H}\right), \qquad u_1 \quad -\sqrt{2}A \sin \frac{\pi x}{H\sqrt{2}} \cos\left(\frac{\pi z}{H}\right)$$

$$T_1' = \left[\frac{H\,\Delta T_c}{\frac{3}{2}\pi^2 k_T}\right] A \cos \frac{\pi x}{H\sqrt{2}} \sin\left(\frac{\pi z}{H}\right), \qquad g\alpha\,\Delta T_c = \tfrac{27}{4}\pi^4 \left(\frac{k_T \nu}{H^3}\right) \tag{10.2.16}$$

In Section 10.4, we shall determine the steady state amplitude A of the convection when ΔT is slightly greater than ΔT_c, and for that purpose, we find it convenient to list here the following formal properties of (10.2.16). By direct substitution it is easy to verify

$$\begin{aligned} (u_1\,\partial/\partial x + w_1\,\partial/\partial z)w_1 &= \text{independent of } x \\ (u_1\,\partial/\partial x + w_1\,\partial/\partial z)u_1 &= \text{independent of } z \end{aligned} \tag{10.2.17a}$$

$$(u_1\,\partial/\partial x + w_1\,\partial/\partial z)T_1' \equiv (\partial/\partial z)\overline{(w_1 T_1')} \tag{10.2.17b}$$

and also

$$\begin{aligned} \left[H^{-1}\int_0^H \overline{w_1 T_1'}\,dz\right]^2 &= \left(\frac{A^2 H\,\Delta T_c}{6\pi^2 k_T}\right)^2 \\ H^{-1}\int_0^H \overline{[w_1 T_1']^2}\,dz &= \frac{3}{8}\left(\frac{A^2 H\,\Delta T_c}{3\pi^2 k_T}\right)^2 \\ H^{-1}\int_0^H k_T^2 \overline{(\nabla T_1')^2}\,dz &= \frac{3}{8}\left(\frac{A\,\Delta T_c}{\frac{3}{2}\pi}\right)^2 \end{aligned} \tag{10.2.18}$$

where the bar again indicates an average value over an infinite horizontal plane.

10.3 Power Integrals

Before turning to the finite amplitude problem, we will derive some general integral properties, such as are applicable to any (statistically) steady state. The only assumption is that the horizontally averaged fields are independent of time.

In Section 9.2, we obtained separate equations for the mean temperature $\overline{T}$ and the fluctuation T' in (10.1.3). It is a simple matter to generalize those equations by inserting the diffusion term

$$k_T \nabla^2 T = k_T \nabla^2 T' + k_T \partial^2 \overline{T}/\partial z^2$$

into Eqs. (9.3.8) and (9.3.9). Thus, the equations for the fluctuating field $\partial T'/\partial t$ and the mean field $\partial \overline{T}/\partial t$ are now given by

$$\partial T'/\partial t + \mathbf{V} \cdot \nabla T' - (\partial/\partial z)\overline{wT'} + w\, \partial \overline{T}/\partial z = k_T \nabla^2 T' \tag{10.3.1}$$

$$\partial \overline{T}/\partial t + (\partial/\partial z)\overline{wT'} = k_T\, \partial^2 \overline{T}/\partial z^2 \tag{10.3.2}$$

In a statistically steady state ($\partial \overline{T}/\partial t = 0$), the vertical integral of (10.3.2) is

$$\overline{wT'} - k_T\, \partial \overline{T}/\partial z = F \tag{10.3.3}$$

where the constant $F = -k_T(\partial \overline{T}(0)/\partial z$ represents the total heat that enters and leaves the fluid at the boundaries. Although $\overline{wT'}$ must vanish at the boundaries, it represents the portion of the total flux F which is transported by convection at other values of z. The result of averaging (10.3.3) vertically is

$$\langle \overline{wT'} \rangle + k_T\, \Delta T/H = F \tag{10.3.4}$$

where brackets denote a vertical average, or

$$\langle (\ ) \rangle \equiv H^{-1} \int_0^H (\ )\, dz$$

Let us now consider the steady state mechanical energy equation obtained by taking the scalar product of $\mathbf{V}$ with the momentum equation (10.1.2), and by averaging the result over the whole fluid. The pressure-work term vanishes, or $\langle \overline{\mathbf{V} \cdot \nabla \phi} \rangle = 0$, and because of the continuity and boundary conditions, we also have

$$\langle \overline{\mathbf{V} \cdot d\mathbf{V}/dt} \rangle = \tfrac{1}{2}\langle \overline{d\mathbf{V}^2/dt} \rangle = \tfrac{1}{2}(\partial/\partial t)\langle \overline{\mathbf{V}^2} \rangle + \tfrac{1}{2}\langle \overline{\nabla \cdot \mathbf{V}(\mathbf{V}^2)} \rangle = 0$$

Thus the only nonvanishing terms in the mechanical energy equation are

$$\langle \overline{g\alpha T' \mathbf{V} \cdot \mathbf{k}} \rangle = -\nu \langle \overline{\mathbf{V} \cdot \nabla^2 \mathbf{V}} \rangle = \nu \langle \overline{\mathbf{V} \cdot \nabla \times \nabla \times \mathbf{V}} \rangle$$

or

$$\langle g\alpha \overline{wT'} \rangle = \nu \langle \overline{(\nabla \times \mathbf{V})^2} \rangle + \nu \langle \overline{\nabla \cdot (\mathbf{V} \times \nabla \times \mathbf{V})} \rangle$$

The first of these terms represents the rate at which the force of gravity does work on warm rising and cold sinking elements. The second term represents the rate of dissipation of kinetic energy by the mean square vorticity of the fluid. The last term, being the volume integral of the divergence of a vector field, represents the rate at which work is done on the fluid by stresses acting at a moving boundary. The latter term clearly vanishes if the boundaries are either "no slip" ($\mathbf{V} = 0$) or "no stress" ($\partial u/\partial z = 0 = \partial v/\partial z = w$), and thus we have

$$g\alpha\langle\overline{wT'}\rangle = \nu\langle\overline{(\nabla \times \mathbf{V})^2}\rangle \tag{10.3.5}$$

Since the right hand side of (10.3.5) is positive, the average value of the convective *heat* flux in (10.3.5) and (10.3.4) must be upward.

Equation (10.3.3) is merely an expression of the first law of thermodynamics, and a quantitative expression of the second law is now obtained by multiplying (10.3.1) with T' and by averaging the result over the whole fluid. In the statistically steady state, the term $\overline{T'\,\partial T'/\partial t}$ must vanish, and we also have $\langle \overline{T'\mathbf{V}\cdot\nabla T'}\rangle = \frac{1}{2}\langle\overline{\nabla\cdot\mathbf{V}(T')^2}\rangle = 0$. From the next term in (10.3.1), we obtain $\langle\overline{T'\,\partial(\overline{wT'})/\partial z}\rangle = \langle\overline{T'}\,\partial\overline{wT'}/\partial z\rangle$, and this vanishes because $\overline{T'} = 0$. Therefore, the terms which remain, upon averaging the product of T' with (10.3.1), are

$$-\langle\overline{wT'}\,\partial\overline{T}/\partial z\rangle = k_T\langle\overline{(\nabla T')^2}\rangle \tag{10.3.6}$$

The right hand side of (10.3.6) represents the dissipation of thermal fluctuations T' due to the conduction of heat between neighboring parcels. The first term in (10.3.6) represents a "generation of temperature fluctuations," due to the fact that a rising element tends to carry fluid to a level where its temperature is higher than the mean at the same level. Since the right hand side of (10.3.6) is positive, $\overline{wT'}$ must be positively correlated with $-\partial\overline{T}/\partial z$, in consequence of which one says that the average convective flux of heat must be down the mean temperature gradient.

When (10.3.3) is subtracted from (10.3.4), we have

$$\partial\overline{T}/\partial z = -\Delta T/H + (\overline{wT'} - \langle\overline{wT'}\rangle)/k_T \tag{10.3.7}$$

and when (10.3.7) is substituted in the *power integral* (10.3.6), the result is

$$\langle[\overline{wT'}\,]^2\rangle - \langle\overline{wT'}\rangle^2 = (k_T\,\Delta T/H)\langle\overline{wT'}\rangle - k_T^2\langle\overline{(\nabla T')^2}\rangle \tag{10.3.8}$$

This version of the power integral contains only the fluctuating temperature (and velocity) fields, and we will now use (10.3.8) to compute A in (10.2.16).

10.4 Finite Amplitude Convection

Let us now return to the problem of determining the steady state amplitude of a two-dimensional convection cell (Fig. 10.1). When the Rayleigh number is *slightly* larger than its critical value, a mode having the form given by (10.2.16) will amplify with a small positive growth rate λ. Although the power integral (10.3.6) with $\partial\overline{T}/\partial z = -\Delta T/H$ is satisfied at the critical Rayleigh number, it is obviously not satisfied for supercritical conditions because $\partial(\overline{T'})^2/\partial t \neq 0$, and the generating term on the left hand side of (10.3.6) will exceed the dissipation term. Therefore, if the growing wave of the linear theory is to reach a steady state [in which (10.3.6) must be satisfied] the nonlinear terms must modify either the generation term or the dissipation term.

Since the vertical velocity and temperature of the linear mode are perfectly correlated (10.2.16), there is an upward flux of heat ($\overline{wT'} > 0$), and therefore the upper fluid ($z > H/2$) tends to increase in mean temperature, whereas $\overline{T}$ tends to decrease for $z < H/2$ [cf. (10.3.2)]. Thus, an amplifying linear mode tends to modify the mean $\overline{T}$ profile as shown by the curve in Fig. 10.1a.

The nonlinear tendency for $-\partial\overline{T}/\partial z$ to be reduced in the center and increased at the boundaries implies a reduction in the left hand side of 10.3.6 (relative to that which would be computed from the linear $\overline{T}$) because $\overline{wT'}$ is largest at the center. These preliminary observations suggest that the $\overline{T}$ modification is the main mechanism for bringing about the equilibration of the linear mode.

Before verifying this statement, we should note that another qualitative mechanism is also possible. The amplifying fundamental mode (10.2.16) would generate harmonics by the $\mathbf{V}\cdot\nabla\mathbf{V}$, $\mathbf{V}\cdot\nabla T'$ interactions (see Section 9.8), and the increased dissipation of these harmonics might achieve the balance required by (10.3.6). But we will show that this effect is relatively small, and that the steady state amplitude of (10.2.16) can be computed by merely substituting those functions into (10.3.8) and by then solving the resulting *algebraic* equation for A as a function of $\Delta T - \Delta T_c$. This statement also requires justification, and so we turn to some order of magnitude considerations.

A steady state velocity field u_1, w_1 having the form of (10.2.16) produces a local acceleration $\mathbf{V}_1 \cdot \nabla \mathbf{V}_1$ which is irrotational, according to the calculation in (10.2.17a). Therefore, when (10.2.16) is substituted into the curl of the nonlinear momentum equation, we get

$$0 = \nu\nabla^2\, \nabla \times \mathbf{V}_1 + g\alpha\, \nabla \times \mathbf{k}T_1'$$

and this is satisfied by (10.2.16) for all A. Therefore, no harmonics of (10.2.16) are generated on account of the nonlinear momentum terms, and we turn next to the nonlinear terms in the steady state heat equation (10.3.1). By using (10.3.7) to eliminate $\partial\overline{T}/\partial z$, we can rewrite (10.3.1) as

$$k_T\,\nabla^2 T' + w(\Delta T/H) = \{\mathbf{V}\cdot\nabla T' - (\partial/\partial z)\overline{wT'}\} + k_T^{-1} w\{\overline{wT'} - \langle\overline{wT'}\rangle\} \qquad (10.4.1a)$$

The first nonlinear term on the right hand side also happens to vanish identically when (10.2.16) is substituted therein because of (10.2.17b). But the last term, or

$$k_T^{-1} w\{\overline{wT'} - \langle\overline{wT'}\rangle\} = O(A^3) \qquad (10.4.1b)$$

does not vanish identically, and we note that (10.4.1b) is cubic in the amplitude

A. From (10.2.16) we also see that the left hand side of (10.4.1a) vanishes only when $\Delta T = \Delta T_c$ and is otherwise equal to

$$(A/H)(\Delta T - \Delta T_c)\cos(\pi x/H\sqrt{2})\sin \pi z/H \tag{10.4.1c}$$

This Fourier term must be balanced by the corresponding Fourier component of (10.4.1b), and thus we see that $(\Delta T - \Delta T_c)A \propto A^3$, or $A \propto (\Delta T - \Delta T_c)^{1/2}$ for small $\Delta T - \Delta T_c > 0$.

The other Fourier components in (10.4.1b) will generate vertical harmonics of (10.2.16), the magnitude of which are obtained by the iterative solution of the inhomogeneous equation (10.4.1a). Thus, the fundamental produces harmonics of magnitude $O(A^3)$ on the right hand side of (10.4.1a), and therefore the corresponding harmonics of T' and w are $O(A^3)$. Since these harmonics are orthogonal to w_1 and T_1', it follows that the heat flux and thermal dissipation terms in (10.3.8) have the form

$$\overline{wT'} = \overline{w_1 T_1'} + O(A^6) \qquad \text{and} \qquad \langle\overline{(\nabla T')^2}\rangle = \langle\overline{(\nabla T_1')^2}\rangle + O(A^6)$$

This implies that if (10.1.6) be used for the exact values of w and T' in (10.3.8), then the resulting algebraic equation for A is only in error by $O(A^6)$. When (10.2.18) is used to evaluate the integrals, we then obtain

$$\frac{1}{24}\left(\frac{A^2 H\,\Delta T_c}{\pi^2 k_T}\right)^2 - \frac{1}{36}\left(\frac{A^2 H\,\Delta T_c}{\pi^2 k_T}\right)^2 = \frac{k_T\,\Delta T}{H}\left(\frac{A^2 H\,\Delta T_c}{6\pi^2 k_T}\right) - \frac{A^2(\Delta T_c)^2}{6\pi^2} + O(A^6)$$
$$= (\Delta T - \Delta T_c)A^2\,\Delta T_c/6\pi^2$$

Therefore, the amplitude of the cell is

$$A = \sqrt{12\pi^2}\,(k_T/H)((\Delta T - \Delta T_c)/\Delta T_c)^{1/2} \tag{10.4.2}$$

and the error is small to order $(\Delta T - \Delta T_c)/\Delta T_c$. In this case, the asymptotic value of the heat flux, or the nondimensional *Nusselt* number is

$$N = \frac{F}{k_T(\Delta T/H)} = 1 + \frac{\langle\overline{wT'}\rangle}{k_T(\Delta T/H)} = 1 + \frac{A^2 H^2\,\Delta T_c}{6\pi^2 k_T^2\,\Delta T} = 1 + 2\left[\frac{\Delta T - \Delta T_c}{\Delta T}\right] \tag{10.4.3}$$

The last term is the fraction of the total heat transport which is due to *convection,* and the increased slope of the mean field $\overline{T}(z)$ at the boundaries can be computed from (10.3.3).

This calculation shows how the convection tends to remove the original source of instability by concentrating the mean field $\overline{T}(z)$ gradients near the boundaries, but the calculation is rather special, and therefore the reader is referred to the more systematic nonlinear iteration techniques that are available in the literature. We may also remark that (10.2.16) is by no means the only possible solution in the vicinity of the critical Rayleigh number, since (10.2.14) has an infinite number of solutions each of which is distinguishable by the shape of their nodal curves ("planform"). Steady nonlinear solutions have been obtained for each permissible combination of k and l, and the question then arises as to which solution is actually realized. This question is especially interesting because the manifold of the solutions increases with Ra, thereby providing the rationale for seeking a new statistical formulation when the Rayleigh number is very large and the field is turbulent. See Turner (1973) for references.

10.5 Convection at Large Prandtl Number

The region of validity of (10.4.3), and the character of the convection depends on the *Prandtl* number

$$\mathrm{Pr} = \nu/k_T \tag{10.5.1}$$

An example of a high Prandtl number fluid is a viscous silicone oil. The solar atmosphere is a good example of small Pr, because, ν is determined by the the small molecular mean free path, whereas k_T is determined by the large radiation (photon) mean free path. For water, the Prandtl number is a modest seven.

Let us cast (10.1.1) and (10.1.2) into a nondimensional form which will reveal the dynamical significance of the high Pr limit. By taking H as the length unit, the nondimensional coordinates become

$$(x_*, y_*, z_*) = H^{-1}(x, y, z)$$

and by taking H^2/k_T as the time unit, we have the nondimensional $t_* = tk_T/H^2$. The nondimensional velocity is $\mathbf{V}_* = (H/k_T)\mathbf{V}$, the nondimensional viscous force is $\nu(k_T/H^3)\,\nabla_*{}^2\mathbf{V}_*$, and the nondimensional pressure is $\rho_0\nu(k_T/H^2)\phi_*$. Although the mean temperature field $\bar{T}$ is set equal to $\Delta T\,\bar{T}_*(z_*, t_*)$, so that $\bar{T}_*(0, t_*) = 1$ and $0 = \bar{T}_*(1, t_*)$ are the boundary values, we use $\mathrm{Ra}^{-1}\,\Delta T$ as the scale for the temperature fluctuation, so that

$$T_*' = (\Delta T/\mathrm{Ra})T'$$

is the nondimensional temperature fluctuation. By substituting the total temperature field

$$\Delta T[\bar{T}_*(z_*, t_*) + \mathrm{Ra}^{-1}T_*'(x_*, y_*, z_*, t_*)]$$

in (10.1.1), we get

$$((d/dt_*) - \nabla_*^2)(\bar{T}_* + \mathrm{Ra}^{-1}T_*') = 0 \tag{10.5.2}$$

and the nondimensional form of the momentum equation is

$$\mathrm{Pr}^{-1}\, d\mathbf{V}_*/dt_* = -\nabla_*\phi_* + \nabla_*^2\mathbf{V}_* + T_*'\mathbf{k}, \qquad \nabla\cdot\mathbf{V}_* = 0 \tag{10.5.3}$$

Suppose that Ra be fixed at any finite value, and let

$$\mathrm{Pr} \to \infty$$

so that the asymptotic form of (10.5.3) is

$$0 = -\nabla_* \phi_* + \nabla_*^2 \mathbf{V}_* + T_*' \mathbf{k} \tag{10.5.4}$$

The heat equation (10.5.2) is unaltered, however, because Pr does not appear therein. Thus, the asymptotic equations (10.5.2) and (10.5.4) involve only one nondimensional number Ra, and therefore the nondimensional heat transport will only be a function of Ra.

In Fig. 10.1 we see a tendency for thermal boundary layers (at $z = 0, H$) to form when Ra is increased above its critical value, and we now consider what happens when Ra is much larger. The asymptotic equation (10.5.4) applies if we consider the limit $\nu \to \infty$, $k_T \to 0$, with νk_T held constant at a value determined by the preassigned Ra. Such "small" values of k_T will favor the formation of the thermal boundary layer mentioned above, and these are anticipated by the dashed horizontal lines in Fig. 10.2. The slippery boundary conditions of Section 10.4 are again used, so that the horizontal velocity of the two-dimensional cell will not vanish on $z = 0, H$. The horizontal distance between two contrarotating cells is assumed to be of the same order of magnitude H as the fundamental in Fig. 10.1, but now the effect of the higher harmonics will be prominent in the vertical thermal boundary layers separating adjacent cells. No momentum boundary layers will appear, however, because the "large" viscosity would diffuse them away.

We shall first amplify this tentative picture (Fig. 10.2) of the steady state circulation, by following the motion of a parcel whose initial location is just above the cell corner *a*. When this parcel enters the lower boundary layer it is heated by conduction from the plate, but the smallness of k_T prevents heat from diffusing into the isothermal core. Thus, all of the added heat moves with the parcel to *b*, and then the thin horizontal boundary layer is converted into a thin

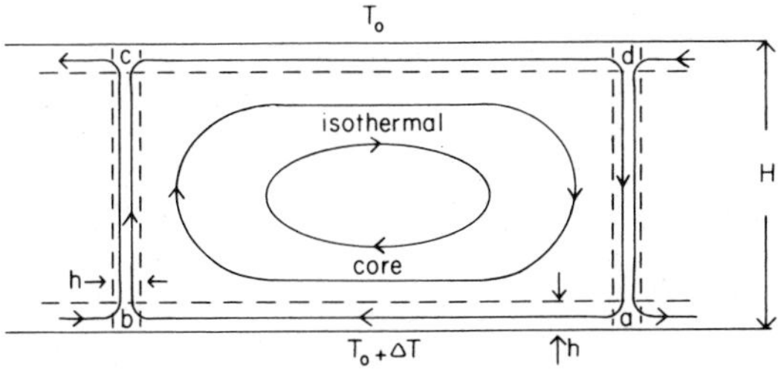

FIG. 10.2 Cellular convection at large Rayleigh and Prandtl number. The dashed lines indicate the horizontal and vertical boundary layers of width h. The curved lines are the streamlines.

vertical boundary as the parcel is carried upward to c. For reasons of continuity, we will assume that both boundary layer thicknesses have the same order of magnitude $h \ll H$. All the heat acquired by the parcel at the lower boundary is retained until it is deflected horizontally at c and cooled by the upper boundary. The parcel then arrives at d, whereupon it sinks toward a in another boundary layer of horizontal width $\sim h$, and so on. Let us now examine this picture for dynamical consistency, and also estimate the order of magnitude of the heat flux F (10.3.3).

We have stated that the isothermal core ($T' = 0$) does not participate in the heat transfer, and therefore its temperature equals the average of T_0 and $(T_0 + \Delta T)$. The velocity field $\mathbf{V}$ is not zero in the core, however, because of the strong viscous coupling to the vertical boundary layers, upon which the buoyancy torque acts. The vanishing of the buoyancy force ($T' = 0$) in the core of a high Prandtl number fluid (10.5.4) merely implies an irrotational viscous force ($\nabla \times \nabla^2 \mathbf{V} = 0$), and thus we see that $\mathbf{V} \neq 0$ is possible. If $w_n = |\mathbf{V}|$ denotes the order of magnitude of the smoothly varying velocity field, then the right hand side of (10.3.5) is of order $\nu(w_n/H)^2$, and most of this dissipation is due to the vorticity in the isothermal core. However, the buoyancy work term on the left hand side of Eq. (10.3.5) is mainly due to the thin vertical thermal boundary layers. Since these layers only occupy the small fraction h/H of the total horizontal distance over which the averages in (10.3.5) are taken, and since the temperature fluctuation T' across the layer is of the same order as the temperature difference ΔT between the plates, we conclude that the left hand side of (10.3.5) is of order $g\alpha\, \Delta T\, w_n(h/H)$. By equating the buoyancy-work and dissipation terms given above, we have

$$w_n \sim g\alpha\, \Delta T\, hH/\nu \tag{10.5.5}$$

A second relation between the unknowns w_n and h can be obtained by equating the conductive heat flux ($F = -k_T\, \partial\bar{T}(0)/\partial z$) at $z = 0$ to the convective heat flux $F \sim \overline{wT'}$ at $z = H/2$. Thus we have

$$F = -k_T\, \partial\bar{T}(0)/\partial z \sim k_T\, \Delta T/h \quad \text{and} \quad F \sim \overline{wT'} \sim w_n\, \Delta T\, (h/H) \tag{10.5.6}$$

or

$$w_n \sim k_T H/h^2 \tag{10.5.7}$$

The simultaneous solution of (10.5.7) and (10.5.5) gives

$$h \sim (k_T\nu/(g\alpha\, \Delta T))^{1/3} \qquad \text{and} \qquad w_n \sim (\nu/H)\, (\mathrm{Ra}^{2/3}/\mathrm{Pr}) \tag{10.5.8}$$

From (10.5.6), we see that the heat flux

$$g\alpha F \sim (g\alpha\, \Delta T)^{4/3}(k_T^2/\nu)^{1/3} \tag{10.5.9}$$

increases as the $\frac{4}{3}$ power of ΔT ($\gg \Delta T_c$).

The dimensional theory given above will obviously not be valid if the neglected inertial terms ($\mathbf{V} \cdot \nabla\mathbf{V} \sim w_n^2/H$) should be larger than the viscous term

$(\nu\nabla^2 \mathbf{V}) \sim \nu w_n/H^2)$. According to (10.5.8), the ratio of these forces is

$$w_n H/\nu \sim \mathrm{Ra}^{2/3}/\mathrm{Pr}$$

and therefore, the requirement for the validity of the high Prandtl number regime is

$$\mathrm{Pr} \gg \mathrm{Ra}^{2/3}$$

The reader is referred to the literature (see Chapter VII of Turner, 1973) for numerical calculations related to the theory described above. The author would also like to acknowledge helpful conversations with Dr. Herbert Huppert on this subject.

The high Pr fluid is important because of the insight it provides into one aspect of the total problem. Thus, the thermal equation (10.5.2) contains *all* the nonlinear terms, and we also have some insight into the thermal boundary layers which will appear again in Section 10.6.

10.6 Similarity Theory of Thermal Turbulence

The two-dimensional cells discussed above have been realized at moderately large Ra in high Pr fluids (Turner, 1973), but if Ra is increased further, new modes appear, and the convection becomes three-dimensional. Eventually, the cells start oscillating, and finally the convection becomes fully turbulent. In general, the nondimensional heat flux is a complicated function of $\mathrm{Ra} = g\alpha\,\Delta T H^3/k_T\nu$ and ν/k_T, but we expect a simpler form in the asymptotic regime $\mathrm{Ra} \to \infty$. This regime can be obtained by allowing H to increase, with other things held constant.

Only the lower boundary ($z = 0$) is shown in Fig. 10.3, and we suppose that the entire system has uniform temperature at some initial instant $t = 0$. The temperature of the lower (upper) plate is then raised (lowered) by ΔT, and maintained at that value for $t > 0$. We now discuss the temporal development of the field, and then use a similarity theory to estimate some integral properties of the statistically steady state.

Since the fluid is resting at $t = 0$, a horizontally uniform temperature field $\overline{T}(z, t)$ will start to diffuse upward from $z = 0$, according to the well known solution of the Fourier heat conduction equation

$$\partial\overline{T}/\partial t = k_T\, \partial^2\overline{T}/\partial z^2 \qquad (10.6.1)$$

and two successive T profiles are shown in Fig. 10.3a. The only length scale arising in the solution of (10.6.1) is $(k_T t)^{1/2}$, and therefore the thermal boundary layer thickness

$$h \equiv \frac{-\Delta T}{(\partial\overline{T}/\partial z)_{z=0}} \qquad (10.6.2)$$

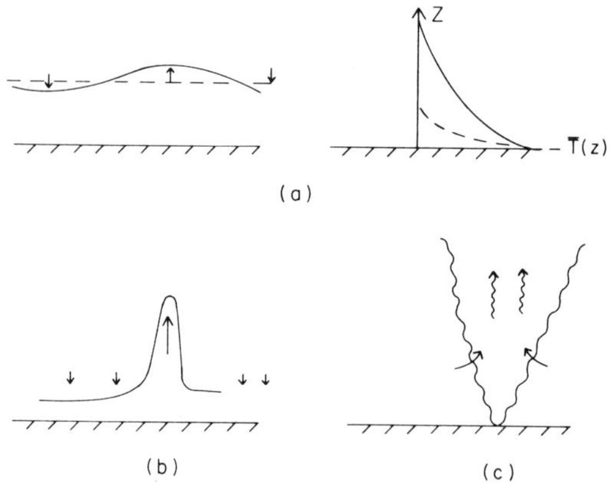

FIG. 10.3 (a) Small displacements of an isotherm at the point of initial instability of an upward diffusing temperature profile $\bar{T}$. (b) A later stage of development showing the warm fluid leaving the boundary layer and being replaced by cold descending fluid. (c) A turbulent plume.

is proportional to $(k_T t)^{1/2}$. Although the diffusive states are in hydrostatic equilibrium for all t, the density stratification becomes increasingly unstable because the *boundary layer Rayleigh number*

$$\mathrm{Ra}_* \equiv g\alpha\, \Delta T\, h^3/k_T \nu \tag{10.6.3}$$

increases as $t^{3/2}$. Infinitesimal velocity perturbations will start to amplify at a value of h (or a value of t) which can be estimated by setting

$$\mathrm{Ra}_* \sim 10^3 \tag{10.6.4}$$

this being the order of magnitude of the critical Rayleigh number calculated in Section 10.2. Although that calculation applies to a vertical temperature gradient which is constant in space-time (whereas the present $\bar{T}(z, t)$ is not), the value of h computed from (10.6.3), or

$$h = (\mathrm{Ra}_* \nu k_T/g\alpha\, \Delta T)^{1/3} \tag{10.6.5}$$

is not very sensitive to the precise value of the critical Rayleigh number, and therefore (10.6.4) can be used for order of magnitude purposes.

Figure 10.3a also shows the small perturbations in the position of an isotherm when the boundary layer first becomes unstable, and Fig. 10.3b corresponds to a later time. We see that the warm plume is *free* to accelerate upward through its surroundings, but the downward velocities of the perturbation are *limited* in amplitude by the proximity of the boundary. Thus, a relatively weak and broad sinking motion replaces the rapidly ascending warm plume. The instability acts relatively fast to remove the warm boundary layer, thereby again bringing cold fluid in close contact with $z = 0$. The upward diffusion of heat then recommences, according to (10.6.1), and we have a kind of "relaxation–oscillation" of the thermal boundary layer. The latter builds up to the value (10.6.5) by diffusion, and the subsequent instability causes a sudden decrease in the thickness of the thermal boundary layer. The order of magnitude of the average heat flux in the statistically steady state is

$$F = -k_T \, \partial \bar{T}/\partial z = k_T \, \Delta T/h$$

and by eliminating h with (10.6.5), we have

$$(g\alpha F) = (g\alpha \, \Delta T)^{4/3} (k_T^2/\nu)^{1/3} \, \mathrm{Ra}_*^{-1/3} \tag{10.6.6}$$

The content of this similarity theory appears with the assertion that Ra_* is independent of $H \to \infty$, or $\mathrm{Ra} \to \infty$. Therefore, the heat flux should be proportional to $(\Delta T)^{4/3}$. If we go further by assuming Ra_* to be a constant (10.6.4) determined by stability considerations, then the resulting value of (10.6.6) is in surprisingly good agreement with experiments (in a limited range).

The foregoing similarity argument concentrates on the stability properties of the boundary layer (see also Section 6.3), and ignores the feedback effects of the motions induced at $z \gg h$. Therefore, we will now examine the way in which the rms vertical velocity

$$\hat{w}(z) \equiv (\overline{w^2})^{1/2} \tag{10.6.7}$$

increases with z, and the way in which the rms temperature fluctuation

$$\hat{T}(z) \equiv (\overline{(T')^2})^{1/2} \tag{10.6.8}$$

decreases with z. Then we will reconsider the heat flux law (10.6.6).

In the stage of development shown in Fig. 10.3b, the accelerating plume will start to develop additional instabilities, these being similar to those which occur in the smoke jet rising from a cigarette. The amplifying wave-like disturbances in such a jet lead to an *entrainment* of surrounding fluid. The subsequent mixing of the cold surroundings then reduces the temperature variance (10.6.8) of the jet, and also increases the horizontal width. Thus we expect that (10.6.8) should decrease to zero as $z \to \infty$. But the total vertical flux of heat F must be independent of z in the statistically steady state, and since the conductive component is negligible at large z, we have

$$\overline{wT'} \equiv \gamma \hat{w} \hat{T} \simeq F \tag{10.6.9}$$

where

$$\gamma \equiv \overline{wT'}/\hat{w}\hat{T} \tag{10.6.10}$$

is the correlation coefficient between temperature and vertical velocity. Since warm fluid rises on the average, we have

$$0 < \gamma < 1 \tag{10.6.11}$$

and we conclude from the constancy of (10.6.9) that $\hat{w} \to \infty$ as $\hat{T} \to 0$.

The value of $\hat{w}(z)$ depends on the buoyant acceleration $|g\alpha T'\mathbf{k}| \sim g\alpha\hat{T}(z)$ of a parcel in the region beneath z. The order of magnitude of the time that it takes a parcel to go from h to z is $t \sim z/\hat{w}(z)$, and consequently

$$|dw/dt| \sim \hat{w}/(z/\hat{w}) = (\hat{w})^2/z \tag{10.6.12}$$

is the upward acceleration of a parcel at large z. This must be of the same order as the buoyancy term, and therefore the ratio

$$\Gamma = zg\alpha\hat{T}(z)/\hat{w}^2(z) \tag{10.6.13}$$

must be constant for $z \gg h$. By combining this with (10.6.9), we obtain

$$\hat{w}(z) = (\gamma\Gamma)^{-1/3}\,[gz\alpha F]^{1/3} \tag{10.6.14}$$

and thus we conclude that the rms vertical velocity increases as the cube root of the distance from the boundary. Kraichnan (1962) has estimated the constants γ and Γ to be such that

$$(\gamma\Gamma)^{-1/3} \simeq (2)^{1/3}$$

and this coefficient is consistent with velocity measurements made by Deardorff and Willis (1967) over a limited range in air.

Equation (10.6.14) implies that if the (finite) vertical dimension H of a convection chamber increases, then the maximum vertical velocities at $z = H/2$ increase as $H^{1/3}$. On the basis of continuity, we might then expect that the comparable values of the *horizontal* velocity appear near the boundary ($z \sim h$). This implies large shears and *dynamic* instabilities in the thermal boundary layer, and such an effect will modify F in a way that depends on H. This conclusion conflicts with the basic similarity law (10.6.6), since the latter assumes that F is independent of $H \to \infty$. Therefore, the heat flux should have a weak dependence on Ra $\to \infty$(Kraichnan, 1962).

10.7 Evaporative Convection

Evaporation from the sea is usually accompanied by short or long wave radiation, in contrast with the simpler process which will be discussed in this section. Suppose we have a semi-infinite body of pure water. Let us evaporate E (gm/cm^2/sec) by *pumping* the moist air away from the interface and by continually adding unsaturated air. In this "adiabatic" case the heat of evaporation EL (cal/cm^2/sec) is supplied by the internal energy of the liquid. We want to compute the steady state value of the temperature difference

$$\Delta T = \bar{T}(-\infty) - \bar{T}(0) \tag{10.7.1}$$

between the surface and great depths. Since the evaporation causes the free surface to move downward with speed

$$\bar{w} = E/\rho_0 \tag{10.7.2}$$

where ρ_0 is the mean density of the water, we employ a Cartesian coordinate system which moves downward with the same speed $\bar{w}$, and the origin $z = 0$ is placed on the free surface. A basic upward velocity of $\bar{w}$ then appears at all $z < 0$ in this moving system.

The heat of evaporation EL at the flat free surface is obtained by molecular conduction in the liquid and by referring to (10.1.1), we get the flux boundary condition

$$EL = -K_q\, \partial\bar{T}(0)/\partial z \qquad \text{or} \qquad -k_T\, \partial\bar{T}(0)/\partial z = EL/\rho_0 c_p \tag{10.7.3}$$

where $\bar{T}(z)$ denotes the horizontally averaged temperature. For $z < 0$, the total upward heat flux is given by

$$F = -k_T\, \partial\bar{T}/\partial z + \bar{w}\bar{T} + \overline{w'T'} \tag{10.7.4}$$

where the last term is the familiar convection term, while the preceding term is due to the "upwelling" of internal energy by the (apparent) *mean* vertical velocity.

Just for the fun of it, let us compute the steady state value of ΔT by assuming that there is no convection $[w'(x, y, z, t) = 0]$, and by assuming a balance between upwelling and conduction. From (10.7.4) and (10.7.2), we then have

$$(\partial/\partial z)[k_T\, \partial\bar{T}/\partial z - (E/\rho_0)\bar{T}] = 0$$

By referring to (10.7.1), we see that the solution of the above equation can be written as

$$\bar{T}(z) - \bar{T}(-\infty) = -\Delta T \exp(zE\rho_0^{-1}k_T^{-1})$$

and substitution in (10.7.3) gives $\Delta T = L/c_p = 600°\,\mathrm{C}$! The absurdity of this result serves to convince us that the thermal boundary layer in water is unstable, and turbulent velocities ($w' \neq 0$) will be produced.

Let us note that the vertical derivative of the diffusive term in (10.7.4) is of order $k_T \partial^2 \bar{T}/\partial z^2 \sim k_T\, \Delta T/h^2$, where h denotes the thickness of the boundary layer. The corresponding upwelling term is $\bar{w}\, \partial\bar{T}/\partial z \sim \bar{w}\, \Delta T/h$, and its ratio to the diffusive term is

$$\bar{w}h/k_T \sim (E/\rho_0)(h/k_T) \sim (E/\rho_0 k_T)(k_T\, \Delta T/(EL/\rho_0 c_p))$$

where (10.7.3) has been used to obtain $h \sim k_T \Delta T(EL/\rho_0 c_p)^{-1}$. Therefore, $\bar{w}h/k_T \sim \Delta T(L/c_p)^{-1}$ is small if the temperature variation across the thermal boundary layer is much less than $L/c_p = 600°\,\mathrm{C}$. We shall tentatively assume this to be the case, and consequently the upwelling term $\overline{w}$ will be neglected in the following considerations of the turbulent boundary layer. We will take $\overline{T}(-\infty) = 0$ as the reference temperature, and thus the uniform value of the heat flux (10.7.4) now becomes

$$F = -k_T\, \partial\overline{T}(z)/\partial z + \overline{w'T'} = EL/\rho_0 c_p \tag{10.7.5}$$

where (10.7.3) has been used. Thus the evaporation problem reduces to an upside-down version of Section 10.6, in which the heat flux $F = EL/\rho_0 c_p$ is given. The similarity law relating F and ΔT should therefore have the same form as (10.6.6), or

$$(\alpha\,\Delta T) = g^{-1}\,[\mathrm{Ra}_*(g\alpha EL/\rho_0 c_p)^3\, \nu k_T^{-2}]^{1/4} \tag{10.7.6}$$

but the numerical value of Ra_* may be different because we now have a free surface boundary condition. Although the author knows of no relevant empirical determinations of Ra_*, the value of (10.7.6) depends only on its $\frac{1}{4}$ power, and therefore (10.6.4) may be used for the purpose of an estimate of ΔT. The approximate value of the physical parameters are

$$\nu = 10^{-2}\ \mathrm{cm^2/sec}, \qquad \alpha = 2 \times 10^{-4}\ (^\circ\mathrm{C})^{-1}$$

$$k_T = 10^{-3}\ \mathrm{cm^2/sec}, \qquad c_p = 1\ \mathrm{cal/gm/^\circ C}$$

and the typical oceanic evaporation rate is

$$E/\rho_0 = 1\ \mathrm{m/yr} = \tfrac{1}{3} \times 10^{-5}\ \mathrm{cm/sec} \tag{10.7.7}$$

From (10.7.6), we then estimate $\Delta T \sim 1^\circ$C, and the corresponding value of the boundary layer thickness (10.6.5) is $h \sim 5$ mm.

Ewing and McAlister (1960) have made radiometric measurements of the evaporative and radiative boundary layers in both the laboratory and sea. When air of 55% relative humidity was blown over a salt solution, they measured a $\Delta T \sim 1^\circ$C. When field observations were taken on a clear night with 83% relative humidity, they measured $\Delta T = 0.6^\circ$C, $h < 1$ mm, and the radiative effect was probably dominant. The value of ΔT in the latter case can readily be estimated by replacing the evaporation in (10.7.6) with the long wave radiation flux.

10.8 Haline Convection

When evaporation takes place over salt water, the salt molecules are left behind, and a steady state stratification $\overline{S}(z)$ can arise in a semi-infinite liquid having uniform initial salinity $S_0 = \overline{S}(-\infty)$. We again use the coordinate system moving downward with the speed (10.7.2), so that there appears to be an upwelling of

$$\bar{\rho}\,\overline{w}\bar{S} = E\bar{S}(z) \text{ gm salt/cm}^2\text{/sec} \tag{10.8.1}$$

at any z. But only H_2O molecules flow upward, and therefore the total flux of salt across the free surface ($z = 0$) must vanish. Thus, the sum of the diffusive and convective fluxes, or

$$\rho_0 F_S \equiv (k_S\,\partial\bar{S}/\partial z - \overline{w'S'})\rho_0 \tag{10.8.2}$$

must be equal and opposite to (10.8.1). We want to estimate the steady state value of the surface salinity $\bar{S}(0)$, or the fractional variation

$$\frac{\Delta S}{S_0} \equiv \frac{\bar{S}(0) - \bar{S}(-\infty)}{\bar{S}(-\infty)} \ll 1 \tag{10.8.3}$$

when this quantity is small, or when E is sufficiently small. Since (10.8.2) must equal (10.8.1) at all z, the approximation (10.8.3) implies

$$F_S = k_S\,\partial\bar{S}/\partial z - \overline{w'S'} = ES_0/\rho_0 \tag{10.8.4}$$

This boundary condition is similar to (10.7.5), and we proceed to develop the analogy by considering the following problem of "isothermal–haline" convection. Suppose the evaporation from the salt solution is produced by the same pumping arrangement that was used in Section 10.7, with the exception that the free surface is to be kept at uniform temperature. Thus, the thermal fluctuation vanishes throughout the fluid and the buoyancy force is $-g\rho'/\rho_0 = -g\beta S'$. This problem is analogous to that of Section 10.7 when $-\alpha T$ is replaced by βS, when αF is replaced by βF_S (10.8.4), and when k_T is replaced by k_S. Therefore, the replacement of $\alpha(EL/\rho_0 c_p)$ in (10.7.5) by $\beta(ES_0/\rho_0)$ yields

$$\beta\,\Delta S = g^{-1}\,[\mathrm{Ra}_*(g\beta ES_0/\rho_0)^3 \nu k_S^{-2}]^{1/4} \tag{10.8.5}$$

as the mean salinity excess of the surface.

It is instructive to compare the value of (10.8.5) with the density difference (10.7.6) when the evaporation rate is the same in the *two different* problems. Thus, we divide (10.8.5) by (10.7.6) to obtain the formal density ratio

$$\beta\,\Delta S/\alpha\,\Delta T = (\beta S_0 c_p/\alpha L)^{3/4}(k_T/k_S)^{1/2} \tag{10.8.6}$$

These numbers will be quite significant in the *coupled* thermohaline problem which occurs when the surface of the ocean evaporates. In a salt free ocean ($S_0 = 0$), the value of (10.8.6) is zero and the thermal effect dominates. If the thermal expansion coefficient vanishes ($\alpha = 0$) the value of (10.8.6) is infinite and the salt buoyancy dominates. The interesting case occurs when (10.8.6) is of order unity, for then the buoyancy effects due to temperature and salinity will be comparable. By using $k_T/k_S = 100$ and a mean oceanic salinity $S_0 = 36 \times 10^{-3}$, we have

$$\beta S_0 c_p/\alpha L \simeq \tfrac{1}{4} \tag{10.8.7}$$

so that (10.8.6) is greater than unity in the ocean.

References

Deardorff, J. W., and Willis, G. E. (1967). Investigation of Turbulent Thermal Convection between Horizontal Plates. *J. Fluid Mech.* **28**, 675–704.

Ewing, G., and McAlister, E. D. (1960). On the Thermal Boundary Layer of the Ocean. *Science* **131**, 1374–1376.

Foster, T. D. (1969). Experiments on Haline Convection Induced by Freezing of Salt Water. *J. Geophys. Res.* **74**, 6967.

Kraichnan, R. H. (1962). Turbulent Thermal Convection at Arbitrary Prandtl Number. *Phys. Fluids* **5**, 1374–1389.

Turner, J. S. (1973). "Buoyancy Effects in Fluids." Cambridge Univ. Press, London and New York.

CHAPTER XI

Thermohaline Convection

11.1 Introduction

The warmest and saltiest surface waters are found at those latitudes having an excess absorption of solar energy and an excess evaporation, whereas the cold fresh waters are found at those latitudes having excess long wave radiation and precipitation. These source regions are rather effectively mixed in the upper fifty meters by wind and wave turbulence, but the mixing of temperature and salinity in the underlying thermocline is much different. The extremely low rate of the global mixing process can be illustrated by means of the following calculation.

Consider a basin of depth $H = 1000$ m, which is initially filled with water having uniform salinity $S_0 = 35\ ^0/_{00}$ ($\rho_0 \simeq 1.035$), and suppose that the uniform evaporation rate is 1 m/yr. At the end of one decade, the water depth will then be decreased by $\Delta H = 10$ m, and the salinity will be increased by ΔS. Since the invariant mass of salt in the basin is given by the product of salinity, density, depth, and cross-sectional area, we obtain the result

$$\Delta S/S_0 \simeq \Delta H/H = 10/1000$$

or $\Delta S = 0.4\ ^0/_{00}$. Let us now note that the salinity difference between the upper kilometer of the subtropical ocean and the underlying deep water is about $\Delta S = 1\ ^0/_{00}$. If the real ocean were suddenly mixed to a uniform salinity S_0, then the preceding calculation shows that it would take several decades before the normal evaporation rate could reestablish the present day salinity differential. Of course the salinity (and sea level) in the tropical ocean will not continue to change because of compensating effects brought about by the general circulation, but the simple calculation does show that the oceanic time constant is measured in decades (or longer), whereas the corresponding atmospheric time constant is known to be of the order of a fortnight. This comparison between the two fluids is meaningful because of their coupling, and if the ocean should fail to return to its "proper" state after a couple of decades, then the atmospheric circulation will be altered accordingly. Thus, the slow processes which mix T and S are important for climatological reasons, as well as for an understanding of the density stratification of the ocean.

A significant correlation between temperature and salinity variations is observed in certain large volumes of the ocean, and these are called *water masses* (Sverdrup *et al.*, 1942). If an isopycnal surface in a water mass is also a surface of constant (potential) temperature, then it must be an isohaline surface. In this case, there will then be some functional relation $T = T(S)$ which is characteristic of the water mass, and which is invariant under any *adiabatic* motion. The "South Atlantic central water" provides an example of a water mass having a high correlation coefficient for fluctuations in T, S. This water mass is bounded above (~50 m) by a surface layer having T, S fluctuations of a different and more irregular character. Likewise, the lower boundary (~1 km) is determined by the proximity of another water mass ("Antarctic intermediate water"), the latter having a different $T(S)$ relation because of its different mode of origin.

The "central waters" occur in the subtropical latitudes of high evaporation, and therefore the salinity decreases downward from the surface. But the corresponding decrease of temperature compensates by a factor of 2, and thus the density of the central water increases downward. In the polar water masses, on the other hand, the salinity increases downward, and this field compensates for the corresponding downward increase of temperature. Thus, the static stability of the ocean is (almost) everywhere positive.

The mean $T = T(S)$ curve for a water mass is obtained by averaging space–time measurements, and the fluctuations are a measure of the mixing process. These fluctuations are relatively small in the middle of the central water mass, and relatively large at the boundary separating it from the "slope water." Thus, one finds geostrophic eddies embedded in the North Atlantic central water which are in the initial stages of mixing, and which originated in the slope water. Such large scale processes, however, are not discussed in this chapter, but we will concentrate on the smaller scales of motion and the final stage of mixing.

When the density $\bar{\rho}(z) = \rho_0(\beta\bar{S} - \alpha\bar{T})$ increases downward, the fluid may still be unstable because of the difference between the molecular heat conductivity k_T and the salt diffusivity k_S. In Section 11.2, we discuss the "salt finger" instability, which occurs when $\partial\bar{S}/\partial z > 0$, and the inverse case, which occurs when $\partial\bar{T}/\partial z < 0$. In Section 11.3, we examine some larger scale motions which are brought about by the average effect of the small scale motions (salt fingers). The turbulent regimes for the two types of "double diffusion" are then discussed by means of similarity theories (Sections 11.4 and 11.5).

These "double diffusive" effects release the *potential energy* which is associated with either the $\bar{S}$ or the $\bar{T}$ field, and consequently these processes tend to lower the center of gravity of the fluid. On the other hand, the mixing produced by a shear flow instability in a density stratified fluid will tend to *raise* the center of gravity. Such a case of mechanically driven turbulence is considered in Section 11.7.

Very little is known about the oceanic mixing process, and therefore the thermodynamical relations derived in Section 11.6 provide useful constraints which the *microsctructure* must conform to, independent of the mechanism of origin.

With the exception of Section 11.6, this chapter is confined to vertical convection, and the important problem of lateral transfer is deferred until Chapter XII. The notation is the same as in the two preceding chapters, unless stated otherwise.

11.2 Instability of Temperature–Salinity Stratifications

(a) *Parcel Method Considerations.* Consider a resting fluid having uniform values of $\bar{T}_z > 0, \bar{S}_z > 0$, and $\beta\bar{S}_z - \alpha\bar{T}_z < 0$. Adiabatic perturbations will merely cause parcels to oscillate [Chapter IX], but for parcels having small diameter and small vertical velocities, we must take account of the heat conduction, and in the limiting case, the displaced parcel will tend to come to thermal equilibrium with the surrounding fluid. But the salt diffusivity k_S is much less than k_T, and consequently the salinity of the small parcel can still be nearly conserved. Therefore, a parcel displaced upward will be lighter than the surrounding fluid because of its salinity deficit, and the parcel is then accelerated in the direction of its initial displacement. Thus the basic state is unstable, because the effect of lateral heat exchange allows salty parcels to sink and fresh parcels to rise.

(b) *The Neutral Stability Problem.* The criterion for instability is now discussed by means of the Boussinesq field equations. The problem will be set in the same parallel plate geometry (Fig. 10.1a) that was used in the Rayleigh stability problem, except that the temperature $H\bar{T}_z$ of the upper boundary

$(z = H)$ now exceeds that at $z = 0$, and $H\bar{S}_z$ is the corresponding salinity excess at $z = H$. The boundaries are assumed to be isohaline as well as isothermal, so that the T_1', S_1' fluctuations vanish thereon.

For the neutral $(\partial/\partial t = 0)$ modes, the thermal equation is given by (10.2.6), or by $w_1 \bar{T}_z = k_T \nabla^2 T_1'$, and the corresponding equation for the salinity perturbation $S'(x, y, z, t)$ is $w_1 \bar{S}_z = k_S \nabla^2 S_1'$. By combining these two equations, we see that the buoyancy force $-g\rho'/\rho_0 = g(\alpha T_1' - \beta S_1')$ must satisfy

$$\nabla^2(-g\rho'/\rho_0) = g\,\nabla^2(\alpha T_1' - \beta S_1') = g w_1[\alpha \bar{T}_z/k_T - \beta \bar{S}_z/k_S] \qquad (11.2.1)$$

Having obtained this relation between ρ' and w_1, we need not repeat the eliminatory steps which lead to (10.2.8), because of a formal analogy that can now be made between the thermal and thermohaline problems. We only have to replace the factor $\alpha(\Delta T/H)k_T^{-1} = -\alpha \bar{T}_z k_T^{-1}$ appearing in (10.2.8), by the factor

$$\beta \bar{S}_z/k_S - \alpha \bar{T}_z/k_T \qquad (11.2.2)$$

which appears in (11.2.1). The resulting equation for w_1 will have the same boundary conditions as in Section 10.2, because S' and T' both vanish on the slippery boundaries. By using (11.2.2) for the corresponding factor in the Rayleigh number and by noting (10.2.15), we see that the critical thermohaline Rayleigh number is

$$(gH^4/\nu)(\beta \bar{S}_z/k_S - \alpha \bar{T}_z/k_T) = 27\pi^4/4 \qquad (11.2.3)$$

If H be large, then this instability criterion merely requires a positive value of (11.2.2), and therefore $\beta \bar{S}_z/\alpha \bar{T}_z$ must exceed $k_S/k_T = \frac{1}{100}$ for instability.

The lateral width of the cell which appears in thermal convection (Chapter X) is of the same order as H, but such wide cells are not observed when (11.2.2) is supercritical. A close packed array of long thin "salt fingers" is observed instead, and nine of these are shown schematically in Fig. 11.1. In order to explain the horizontal dimension, we now consider the growth rates of different modes when (11.2.2) is supercritical.

(c) *Finger Perturbations.* For simplicity, we shall consider only perturbations which are independent of z in a vertically unbounded $(H = \infty)$ fluid, so that $w'(x, y, t)$ denotes the vertical velocity. For these modes, the horizontal velocities vanish, and the continuity equation is satisfied identically. The absence of horizontal accelerations implies the absence of horizontal pressure gradients. The purely vertical pressure gradient can then be subtracted from the vertical momentum equation, along with the balancing gravity force $-g\bar{\rho}(z)/\rho_0$. The residual momentum equation then contains the acceleration

$$dw'(x, y, t)/dt = \partial w'/\partial t$$

the fluctuating buoyancy term, and the viscous term, as indicated in the first of Eqs. (11.2.4). Note that the nonlinear terms which ordinarily would appear in the acceleration (above) vanish identically because of the kinematic structure of the mode. The nonlinear terms in the thermal equation

$$(d/dt)(T' + \bar{T}) = k_T \nabla^2 (T' + \bar{T})$$

also vanish because $dT'(x, y, t)/dt = \partial T'/\partial t$, and because $d\bar{T}(z)/dt = w' \, \partial \bar{T}/\partial z$. Thus, we obtain the second equation in (11.2.4), and by analogous considerations of the salt field, we get

$$\begin{aligned} \partial w'/\partial t &= \nu(\partial^2/\partial x^2 + \partial^2/\partial y^2)w' - g(\beta S' - \alpha T') \\ \alpha \, \partial T'/\partial t + w'\alpha \bar{T}_z &= k_T(\partial^2/\partial x^2 + \partial^2/\partial y^2)T'\alpha \\ \beta \, \partial S'/\partial t + w'\beta \bar{S}_z &= k_S(\partial^2/\partial x^2 + \partial^2/\partial y^2)S'\beta \end{aligned} \tag{11.2.4}$$

The normal modes of these linear (but not linearized) equations are expressed as

$$\begin{aligned} w' &= \hat{w}e^{\lambda t} \sin(k_1 x) \sin k_2 y, \\ \alpha T' &= \hat{T}e^{\lambda t} \sin(k_1 x) \sin k_2 y, \qquad k_1^2 + k_2^2 = k^2 \\ \beta S' &= \hat{S}e^{\lambda t} \sin(k_1 x) \sin k_2 y, \end{aligned} \tag{11.2.5}$$

and substitution in (11.2.4) gives

$$(\lambda + \nu k^2)\hat{w} = g(\hat{T} - \hat{S}) \tag{11.2.6a}$$

$$(\lambda + k_T k^2)\hat{T} = -\hat{w}\alpha\bar{T}_z \tag{11.2.6b}$$

$$(\lambda + k_S k^2)\hat{S} = -\hat{w}\beta\bar{S}_z \tag{11.2.6c}$$

where k^2 is the sum of the squares of the horizontal wave numbers and λ is the growth rate of the mode. By eliminating $\hat{T}$ and $\hat{S}$ from (11.2.6a), we can obtain a cubic equation for λ as a function of k. For zero growth rate, we set $\lambda = 0$ in Eqs. (11.2.6), and the solution for the neutral wave number is

$$k_*^{-1} = [(g/\nu)(\beta\bar{S}_z/k_S - \alpha\bar{T}_z/k_T)]^{-1/4} \tag{11.2.7}$$

We shall now compute the value of k which maximizes λ.

When $\hat{w}$ is eliminated in (11.2.6b–c), and when the nondimensional numbers

$$R_S = g\beta\bar{S}_z/\nu k_T k^4, \qquad N = \alpha\bar{T}_z/\beta\bar{S}_z, \qquad \tau = k_S/k_T, \qquad P = \nu/k_T, \qquad \lambda_0 = \lambda/k_T k^2 \tag{11.2.8}$$

are introduced, we obtain

$$\begin{aligned} [\lambda_0 + 1 + NR_S(1 + \lambda_0/P)^{-1}]\hat{T} &= NR_S(1 + \lambda_0/P)^{-1}\hat{S} \\ [\lambda_0 + \tau - R_S(1 + \lambda_0/P)^{-1}]\hat{S} &= -R_S(1 + \lambda_0/P)^{-1}\hat{T} \end{aligned} \tag{11.2.9a}$$

Multiplication of these then gives

$$0 = [\lambda_0 + 1 + NR_S(1 + \lambda_0/P)^{-1}]\,[\lambda_0 + \tau - R_S(1 + \lambda_0/P)^{-1}] + NR_S^2(1 + \lambda_0/P)^{-2} \tag{11.2.9b}$$

and the roots of this equation for λ_0 will be discussed for the interesting asymptotic case in which

$$\nu \gg k_T \gg k_S \tag{11.2.10}$$

When (11.2.10) applies, $\tau \ll 1$, the term containing P^{-1} is also negligible, and therefore (11.2.9b) reduces to a quadratic equation

$$\lambda_0{}^2 + \lambda_0(1 + NR_S - R_S) - R_S = 0 \tag{11.2.11a}$$

We must restrict N and R_S to be of order unity (as $P \to \infty$ and $\tau \to 0$) to ensure that the root of (11.2.11a) is the proper asymptotic approximation of (11.2.9b).

Having established this asymptotic relation for the growth rate, we now find it convenient to revert back to physical units by substituting (11.2.8) in (11.2.11a), and the latter equation then becomes

$$\lambda^2 + \lambda[k_T k^2 + g(\alpha\bar{T}_z - \beta\bar{S}_z)/k^2\nu] - g\beta\bar{S}_z(k_T/\nu) = 0 \tag{11.2.11b}$$

The growth rate relation can be readily obtained by solving this quadratic equation, but we are mainly interested in the value of k^2 which maximizes $\lambda = \lambda(k^2)$. Therefore, if (11.2.11b) be differentiated with respect to k^2, and if $d\lambda/dk^2$ be set equal to zero in the result, then the wave number corresponding to maximum growth rate is given by

$$(d/dk^2)[k_T k^2 + ((g\alpha\bar{T}_z - g\beta\bar{S}_z)/\nu k^2)] = 0$$

or

$$k^{-1} = ((g\alpha\bar{T}_z - g\beta\bar{S}_z)/\nu k_T)^{-1/4} \tag{11.2.12}$$

In Section 9.7, we mentioned the general idea that the scale of the motion which appears in an unstable state will be determined by the (linear) disturbance having maximum growth rate, and this idea finds some support in the particular case of thermal convection (Chapter X) for which the nonlinear calculations have been made and compared with experiment. We expect, therefore, that (11.2.12) determines the order of magnitude of the width of "salt fingers" in the supercritical thermohaline instability. This conclusion agrees with experiment Linden (1973). The linear results obtained above can also be used to compute the ratio of the horizontally averaged heat flux $(-\overline{w'T'}\alpha > 0)$ and the salt flux $(-\overline{w'S'}\beta > 0)$. This ratio equals $\hat{T}/\hat{S}$, according to (11.2.5), and by using (11.2.9a) together with the approximations ($\tau \ll 1, P \gg 1$), we get

$$\alpha\overline{w'T'}/\beta\overline{w'S'} = (R_S - \lambda_0)/R_S \tag{11.2.13}$$

From (11.2.8) and (11.2.12), we see that $R_S = (N-1)^{-1}$ for the wave of maximum growth rate, and the value of λ_0 is determined from (11.2.11a), or

$$\lambda_0^2 + 2\lambda_0 - [1/(N-1)] = 0, \qquad \text{is} \quad \lambda_0 = -1 + [N/(N-1)]^{1/2} > 0$$

By substituting these values of R_S and λ_0 in (11.2.13), the flux ratio becomes

$$\frac{\alpha\overline{w'T'}}{\beta\overline{w'S'}} = 1 - \frac{[N/(N-1)]^{1/2} - 1}{(N-1)^{-1}} = N - [N(N-1)]^{1/2} \tag{11.2.14}$$

and thus we see that the flux ratio decreases from unity when $N = 1$ to a value of

$$\alpha\overline{w'T'}/\beta\overline{w'S'} = \tfrac{1}{2} \tag{11.2.15}$$

when N is much larger than unity. The above theory gives no information, however, about the absolute amplitude or the individual fluxes, and we defer this question until Section 11.4.

(d) *Heating a Stable Salt Gradient from Below.* Let us now consider the stability of a basic state in which $\bar{S}_z < 0$ and $\bar{T}_z < 0$, this being the inverse of the problem considered above. We will again restrict the analysis to modes having the form (11.2.5) and refer the reader to Baines and Gill (1969) for a more complete calculation. In the present problem, the results are more conveniently expressed in terms of

$$R_T = -NR_S = -g\alpha\bar{T}_z/k^4 k_T \nu \tag{11.2.16}$$

than in terms of R_S.

Let us note that if $k_S \ll k_T$, then the *neutral* stability condition (11.2.3) cannot be satisfied for positive values of $-\alpha\bar{T}_z > -\beta\bar{S}_z$. However, this does not mean that a top heavy thermal stratification is stable, because *oscillatory* modes could lead to instability, and therefore the existence of complex λ must be investigated. Under supercritical conditions, moreover, we can have real positive roots ($\lambda > 0$), and these modes are called "convective" to distinguish them from the amplifying oscillations associated with complex λ. We will show that such convective modes occur in a deep fluid when $-\alpha\bar{T}_z > -\beta\bar{S}_z$.

When (11.2.10) is satisfied, (11.2.11a) can be used to compute growth rates, and the substitution of (11.2.16) then gives

$$\lambda_0^2 + \lambda_0(1 - R_T + R_T/N) + R_T/N = 0 \tag{11.2.17}$$

The necessary and sufficient condition for a real positive value of λ_0 is

$$1 - R_T + R_T/N < -2(R_T/N)^{1/2} \quad \text{or} \quad N^{1/2} > R_T^{1/2}/(R_T^{1/2} - 1) \tag{11.2.18}$$

Since the last term $R_T^{1/2}/(R_T^{1/2-1})$ approaches unity for small k, we see that instability can occur for $N > 1$.

Under oceanic conditions, the direct manifestations of these double-diffusive processes are restricted to extremely small scales (1 cm), and the typical variation of density in a salt finger is measured in parts per million. The evidence for their existence has, therefore, been based upon such secondary effects as are discussed in Section 11.4. But Williams (1974) has photographed the refraction pattern of groups of salt fingers in the oceanic pycnocline.

11.3 Interaction of Salt Fingers with Larger Scales of Motion

We have mentioned that the fields (11.2.5) are exact solutions of the nonlinear equations of motion for an unbounded stratification. Note that the nondivergent vertical flux of heat and salt (11.2.13) cannot modify linear $\overline{T}(z)$, $\overline{S}(z)$ fields. The question arises as to the mechanism which actually limits and determines the amplitudes of these modes. In this section, it will be shown that a *second* instability can occur because of the horizontal variations of density within the laminar fingers, and this consideration will be used for the determination of amplitudes in Section 11.4.

When the new perturbation is introduced, it will interact with the basic finger field as well as with the mean vertical gradients $\overline{T}_z$, $\overline{S}_z$, and we restrict the horizontal wavelength of this perturbation to be much larger than the finger width. Let $\mathbf{V}_m = (u_m, v_m, w_m)$ denote the infinitesimal velocity of the new perturbation, and S_m, T_m, ρ_m and $\rho_0\phi_m$ denote the corresponding salinity, temperature, density, and pressure perturbations. The values of $\mathbf{V}_m$ (etc.) are obtained by averaging the total fields over a distance which is large compared to a finger and small compared to the wavelength of $\mathbf{V}_m$. The disturbed finger velocity $\mathbf{V}'(x, y, z, t)$ (and the corresponding S', T', ρ', and $\rho_0\phi'$) is then obtained by subtracting the $\mathbf{V}_m$ (etc.) field from the total velocity.

Figure 11.1 shows a group of nine salt fingers which have been tilted away from the vertical unit vector $\mathbf{k}$ by a small angle, the latter depending upon the

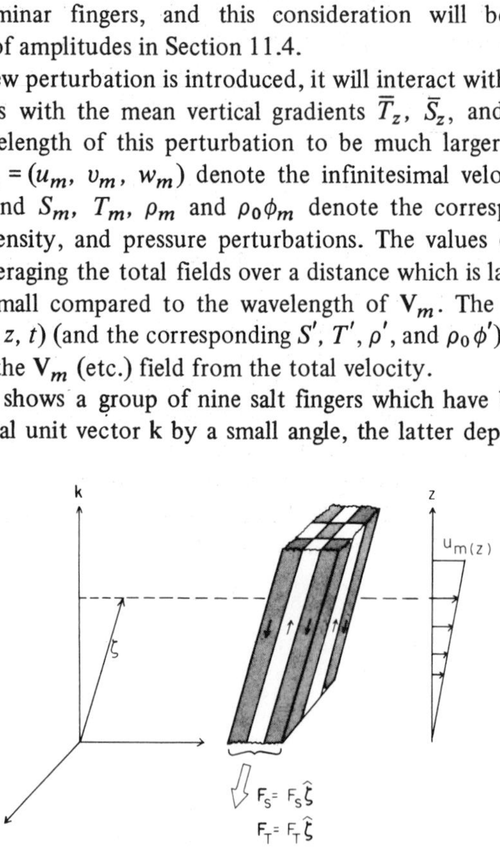

FIG. 11.1 A group of close packed salt fingers being tilted by the shear $\partial u_m/\partial z$ of a large scale perturbation. $\hat{\xi}$ is a unit vector drawn parallel to the instantaneous axis of the group of fingers, and $\mathbf{F}_S$, $\mathbf{F}_T$ are the average flux vectors of the group.

shear $\partial u_m/\partial z$ of the large scale perturbation, as indicated below. When averages (av) are taken over such groups of fingers, the vector fluxes

$$\mathbf{F}_S = \mathrm{av}(\mathbf{V}'S') \qquad \text{and} \qquad \mathbf{F}_T = \mathrm{av}(\mathbf{V}'T') \tag{11.3.1}$$

will vary slowly in x, y, z, and t, in accord with the spatial variations in $\mathbf{V}_m$.

In the absence of salt fingers ($F_S = 0 = F_T$), the large scale perturbation $\mathbf{V}_m, \rho_m$ would behave like the internal wave in Fig. 9.5a. If we include the viscous force in Section 9.8a, then the decay rate of the perturbation could be computed as the imaginary part of the frequency.

When the salt fingers are present, they will tend to be advected with the local velocity $\mathbf{V}_m(x, y, z, t)$ of the large scale wave, and the shear in the wave will therefore cause the fingers to tilt slightly. The direction of the finger fluxes $\mathbf{F}_S$ and $\mathbf{F}_T$ is thereby modified, and flux *divergences* are produced by $\mathbf{V}_m$. The conservation laws for heat and salt will, therefore, require an additional contribution to the large scale density perturbation ρ_m, over and above that which is produced by the action of $\mathbf{V}_m$ on $\bar{\rho}(z)$ in the absence of fingers. The buoyancy force on $\mathbf{V}_m$ will also be modified, and thus we see that the two different scales of motion are coupled. We will show now that energy can be transferred to $\mathbf{V}_m$ at a rate exceeding its dissipation, thereby leading to the instability of the basic finger field.

When the fundamental Fourier–Fick diffusion laws are averaged over a group of fingers, we obtain equations for the large scale salinity $S_m(x, y, z, t) + \bar{S}(z)$ and temperature $(T_m + \bar{T})$ fields. The averaging of the nonlinear terms gives rise to the "eddy flux" terms (11.3.1), and if the scale of $\mathbf{V}_m$ be sufficiently large, then the direct effect of the molecular diffusivities k_T and k_S can be neglected. Accordingly, the averaged conservation laws become

$$(\partial/\partial t + \mathbf{V}_m \cdot \nabla)(S_m + \bar{S}(z)) = -\nabla \cdot \mathbf{F}_S$$

and

$$(\partial/\partial t + \mathbf{V}_m \cdot \nabla)(T_m + \bar{T}(z)) = -\nabla \cdot \mathbf{F}_T$$

By combining these, we find that the large scale density perturbation $\rho_m/\rho_0 = \beta S_m - \alpha T_m$ satisfies

$$\rho_0^{-1}(\partial/\partial t + \mathbf{V}_m \cdot \nabla)(\rho_m + \bar{\rho}(z)) = -\nabla \cdot (\beta \mathbf{F}_S - \alpha \mathbf{F}_T) \tag{11.3.2a}$$

Although the diffusivities have been neglected here, the viscosity will be retained as the main dissipative mechanism because

$$\nu \gg k_T \gg k_S \tag{11.3.2b}$$

Before setting down the relevant momentum equations, we will relate the right hand side of (11.3.2a) to the infinitesimal value of $\partial \mathbf{V}_m/\partial z$.

Let ξ be a vector of unit length which is drawn parallel to the instantaneous "mean" axis of a "group" of salt fingers, as shown in Fig. 11.1. If F_S and F_T denote the flux magnitudes, then (11.3.1) can be written as

$$\beta \mathbf{F}_S - \alpha \mathbf{F}_T = -(\beta F_S - \alpha F_T)\hat{\zeta}(t) \tag{11.3.3}$$

We now introduce the basic assumption that the infinitesimal $\mathbf{V}_m$ will only change the direction of the fluxes in the salt finger "tubes", and not the magnitudes† F_S and F_T. Let us, therefore, compute $\partial\hat{\zeta}/\partial t$ from the rotation of the mean axis of a group of salt fingers. Each point on this axis moves with the local value of $\mathbf{V}_m$, and at any time, the angle between $\mathbf{k}$ and the axial vector $\hat{\zeta}$ (Fig. 11.1) is infinitesimal to the first order in $\mathbf{V}_m$. Therefore, the endpoint of $\hat{\zeta}$ moves a distance $\delta t(\partial u_m/\partial z)$ along the x axis in the time interval δt. We see that the x component of $\partial\hat{\zeta}/\partial t$ is $\partial u_m/\partial z|\hat{\zeta}|$, the y component is $\partial v_m/\partial z|\hat{\zeta}|$, but the vertical component of $\partial\hat{\zeta}/\partial t$ is quadratically small in the perturbation amplitude. Therefore, the divergence of $\partial\hat{\zeta}/\partial t$ is

$$\nabla \cdot \frac{\partial\hat{\zeta}}{\partial t} = \frac{\partial}{\partial x}\frac{\partial u_m}{\partial z} + \frac{\partial}{\partial y}\frac{\partial v_m}{\partial z} = -\frac{\partial}{\partial z}\frac{\partial w_m}{\partial z} \tag{11.3.4}$$

where the continuity relation $\partial u_m/x + \partial v_m/\partial y = -\partial w_m/\partial z$ has been used.

By taking the time derivative of (11.3.2a), and by using (11.3.3) and (11.3.4), we then obtain

$$\rho_0^{-1}(\partial/\partial t)(\partial/\partial t + \mathbf{V}_m \cdot \nabla)(\rho_m + \bar{\rho}(z)) = (\beta F_S - \alpha F_T)\,\nabla \cdot \partial\hat{\zeta}/\partial t$$
$$= -(\beta F_S - \alpha F_T)\,\partial^2 w_m/\partial z^2$$

and the linearization of the left hand side then gives

$$(\partial/\partial t)[(\partial/\partial t)(\rho_m/\rho_0) - (\alpha\bar{T}_z - \beta\bar{S}_z)w_m] + (\beta F_S - \alpha F_T)\,\partial^2 w_m/\partial z^2 = 0 \tag{11.3.5}$$

The momentum equation for $\mathbf{V}_m$ is obtained by a similar averaging procedure applied to the Boussinesq equations, and therefore the average of the $\mathbf{V}\cdot\nabla\mathbf{V}$ term will lead to Reynolds stresses associated with the fingers. But those stresses are negligible compared with the viscous stress when the Prandtl number is large, or when (11.3.2b) applies (see Section 10.5 or Stern, 1969). Therefore, the linearized momentum equation for the larger scale motion is given by

$$\partial\mathbf{V}_m/\partial t = -\nabla\phi_m - g(\rho_m/\rho_0)\mathbf{k} + \nu\,\nabla^2\mathbf{V}_m \tag{11.3.6}$$

$$\nabla \cdot \mathbf{V}_m = 0 \tag{11.3.7}$$

† In that which follows, we will also neglect the time dependence of F_S and F_T in the undisturbed state, an assumption which can be rigorously justified if one considers a model in which the finger wavelength is given by (11.2.7) rather than by (11.2.12). This assumption is also justified in the case (Section 11.4) where $\bar{S}_z = 0 = \kappa_S$.

If we regard the small scale fluxes F_S and F_T as given, then Eqs. (11.3.5)–(11.3.7) constitute a complete set†

For two-dimensional perturbations u_m, w_m in the x, z plane, Eqs. (11.3.6) and (11.3.7) become

$$\partial u_m/\partial t - \nu \nabla^2 u_m = -\partial \phi_m/\partial x$$

$$\partial w_m/\partial t - \nu \nabla^2 w_m = -\partial \phi_m/\partial z - g\rho_m/\rho_0$$

$$\partial u_m/\partial x + \partial w_m/\partial z = 0$$

and by elimination of ϕ_m and u_m, we obtain

$$(\partial/\partial t - \nu \nabla^2)\nabla^2 w_m + (g/\rho_0)(\partial^2/\partial x^2)\rho_m = 0 \tag{11.3.8}$$

When (11.3.5) is used to eliminate ρ_m, the result is

$$0 = \frac{\partial^2}{\partial t^2}\left(\frac{\partial}{\partial t} - \nu \nabla^2\right)\nabla^2 w_m + g(\alpha\overline{T}_z - \beta\overline{S}_z)\frac{\partial^3 w_m}{\partial t\, \partial x^2} - g(\beta F_S - \alpha F_T)\frac{\partial^4 w_m}{\partial x^2\, \partial z^2} \tag{11.3.9}$$

Equation (11.3.9) has the plane wave solution

$$w_m = \exp ibt + i\mu(x \sin\theta + z\cos\theta)$$

where $2\pi/\mu > 0$ is the arbitrary wavelength, θ is the arbitrary acute angle which the wave fronts make with the $+x$ axis, and the equation for the growth rate ib obtained by substitution in (11.3.9) is

$$0 = b^2(\nu\mu^2 + ib)\mu^2 - ibg(\alpha\overline{T}_z - \beta\overline{S}_z)\mu^2 \sin^2\theta - g(\beta F_S - \alpha F_T)\mu^4 \sin^2\theta \cos^2\theta \tag{11.3.10}$$

Since viscous dissipation has been included, we may turn immediately to the marginal stability problem by setting b = real. By equating to zero both the real and imaginary parts of (11.3.10), we get

$$\nu b^2 - g(\beta F_S - \alpha F_T)\sin^2\theta\cos^2\theta = 0 \quad \text{and} \quad b^3 - bg(\alpha\overline{T}_z - \beta\overline{S}_z)\sin^2\theta = 0 \tag{11.3.11}$$

† By parameterizing the effect of the small scale motions on the large scale, we have increased the (temporal) order of the differential equation (11.3.5) for the density field, and this is not unreasonable because of the additional degree of freedom which is implicit in the small scale field. Reference is also made to Section 8.2 wherein we parameterized the small scale fluxes of momentum and thereby increased the (spatial) order of the differential equation for the Ekman transport. These examples should be considered as indicative of the novel dynamical systems which result when one averages the primitive hydrodynamic equations in a nonarbitrary manner.

The elimination of b from these two equations gives the neutral stability condition

$$(\beta F_S - \alpha F_T)/(\nu(\alpha \overline{T}_z - \beta \overline{S}_z)) = (\cos^2 \theta)^{-1} \tag{11.3.12}$$

and from (11.3.11), we also see that b is equal to the Brunt–Väisälä frequency. The smallest value of the buoyancy flux which satisfies this is obtained by letting $\theta \to 0$, and thus we have the critical value

$$(\beta F_S - \alpha F_T)/(\nu(\alpha \overline{T}_z - \beta \overline{S}_z)) = 1 \tag{11.3.13}$$

for the instability of the vertical field of salt fingers.

It is apparent (and readily proven) that if the buoyancy flux of the fingers exceeds the neutral value (11.3.12), then there will be an amplification of those inertia oscillations whose fronts are inclined at an angle θ to the horizontal.

11.4 Transition to Turbulence

These ideas will now be related to laboratory experiments. The quasi-laminar field of salt fingers shown in Fig. 11.2a can be realized by the following procedure. One starts with a semi-infinite layer of fresh water having uniform temperature gradient $\overline{T}_z > 0$ and surface temperature $\overline{T}(0)$. Another semi-infinite layer, having uniform values of temperature $\Delta T + \overline{T}(0)$ and salinity ΔS, is then placed above the first layer, where $\alpha \Delta T - \beta\, \Delta S > 0$. Salt fingers form at $z = 0$ as soon as the two layers are placed in contact, and the downward moving fingers have been shaded in Fig. 11.2a. These experiments (Turner, 1973) can be most conveniently performed by using sugar/salt solutions as the doubly diffusing substances, rather than salt/heat. (In the former case, the diffusivity of sugar is $\frac{1}{3}$ that of the salt, and therefore the system is not "dynamically similar" to the heat/salt system.)

If a second experiment (Fig. 11.2b) be performed using a smaller value of $\overline{T}_z$ but the same values of ΔS and ΔT, then the initial phase of the development of the salt fingers will be qualitatively similar to that described above. But after a while the fingers between $z = 0, -H$ become unstable and eventually give way to a well stirred convective layer. This is maintained by the flux through the overlying salt finger layer, the latter now having a relatively small vertical thickness.

The finger region beneath $z = -H$ also exhibits instability in a third experiment, in which $\overline{T}_z$ is further reduced. By altering the parameters $\overline{T}_z$, ΔS, and ΔT, is it possible to obtain several convecting layers of thickness H, each of which is bounded above and below by relatively thin salt finger layers. Such regularly spaced layers have been documented in the subtropical ocean and also in the polar ocean where the T–S fields are inverted.

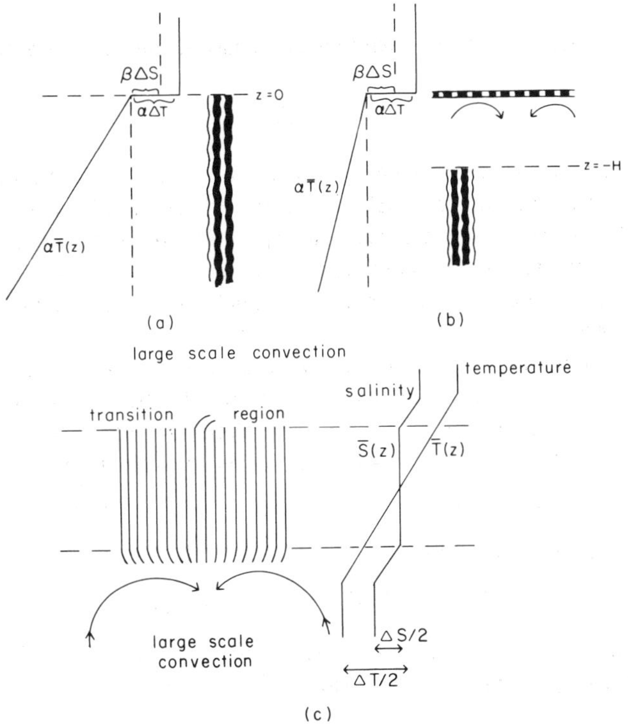

FIG. 11.2 (a) Salt fingers formed by placing a warm, salty layer above a uniform temperature gradient $\partial\bar{T}/\partial z > 0$. (b) Same as in (a) except $\partial\bar{T}/\partial z$ is reduced and the salt fingers between $z = (0, -H)$ become unstable. (c) Schematic diagram of salt fingers in an interface separating warm, salty water from the underlying cold, fresh water.

The relatively thin layer containing salt fingers can be conveniently isolated by performing an experiment in which a very deep layer of temperature $T_0 + \Delta T/2$ and salinity $S_0 + \Delta S/2$ is placed above another deep layer of temperature $T_0 - \Delta T/2$ and salinity $S_0 - \Delta S/2$, this case being schematically indicated in Fig. 11.2c. The warm salty water is drawn into the top of a salt finger through the transition region, and cold fresh water is drawn into the bottom of an adjacent finger. If

$$\nu \gg k_T \gg k_S \qquad \text{and} \qquad 1 < \alpha\, \Delta T/\beta\, \Delta S \sim 1 \tag{11.4.1}$$

then the up and downgoing fingers conserve their salinities, so that $\partial\bar{S}/\partial z = 0$ in the finger region (Linden, 1973). But $\bar{T}_z > 0$ in the finger region (Fig. 11.2c) because of the lateral heat exchange between fingers. In the statistically steady state, the order of magnitude of the finger width is determined by (11.2.12) with $\bar{S}_z = 0$, or

$$k^{-1} \sim (g\alpha\bar{T}_z/\nu k_T)^{-1/4} \tag{11.4.2}$$

and the difference in salinity S' between two adjacent fingers is a fraction of ΔS, or

$$S' \sim \Delta S \tag{11.4.3}$$

The buoyancy flux of the fingers is injected into the deep layers through the "transition" regions, thereby maintaining the "large scale" velocities. These relatively strong horizontal velocities also serve to limit the vertical dimension h of the salt finger region, and consequently the equilibrium h is determined by $g\alpha\ \Delta T, g\beta\ \Delta S,\ k_T,\ k_S$, and ν and not by the thickness or initial condition of the two deep layers.

In equilibrium, the balance between buoyancy and viscous force in the salt finger region can be obtained by setting $\lambda = 0$ in (11.2.6a), and the equilibrium heat balance is obtained from (11.2.6b). The relation between the amplitudes of the vertical velocity, temperature, and salinity fields is then given by

$$0 = -\nu w' k^2 - g\beta S' + g\alpha T' \qquad \text{and} \qquad w'\bar{T}_z = -k_T k^2 T' \tag{11.4.4}$$

and the elimination of T' yields

$$w' = -(g\beta S'/\nu k^2)(1 + (g\alpha \bar{T}_z/k^4 k_T \nu))^{-1} \tag{11.4.5}$$

The order of magnitude of the salt flux ($F_S = -\overline{wS}$) is equal to $-w'S'$, and when (11.4.5), (11.4.3), and (11.4.2) are used, we have

$$g\beta F_S \sim ((g\beta\ \Delta S)^2/\nu)(\nu k_T/g\alpha \bar{T}_z)^{1/2} \tag{11.4.6}$$

The corresponding heat flux can then be estimated by using (11.2.14) or (11.2.15).

These order of magnitude relations can be applied to the salt finger region in Fig. 11.2a because the horizontal salinity fluctuation is determined by ΔS and $\bar{T}_z$ is also "given". Since the buoyancy fluxes (11.4.6) are inversely proportional to $\bar{T}_z^{1/2}$, we see that the stability number on the left hand side of (11.3.13) is inversely proportional to the $\frac{3}{2}$ power of $\partial T/\partial z$, and therefore a reduction of $\partial \bar{T}/\partial z$ will destabilize the salt fingers. Thus, we have a qualitative explanation of the convection layer which forms in Fig. 11.2b.

In Fig. 11.2.c, however, the value of $\partial \bar{T}/\partial z$ is not "given" because the vertical thickness of the finger region must be determined by additional considerations. In this connection, the criterion (11.3.13) provides a "bounding" relation, since the salt finger layer would be unstable if the criterion were exceeded. A similar argument was used for the thermal convection problem of Section 10.6, wherein we assumed that the boundary layer was marginally stable in the Rayleigh number sense. Accordingly, we shall now assume that the salt finger layer is marginally stable in the sense of (11.3.13). Since $\bar{S}_z = 0$ in the finger region, and since $F_S - F_T \sim F_S$ [cf. (11.2.15)], the stability criterion becomes

$$\alpha \bar{T}_z \sim \beta F_S/\nu \tag{11.4.7}$$

By substituting this in (11.4.6) and by solving the result for F_S, we obtain

$$g\beta F_S \sim (g\beta\, \Delta S)^{4/3} k_T^{1/3} \tag{11.4.8a}$$

The meaning of this order of magnitude relation may now be sharpened by introducing a nondimensional constant B and by writing (11.4.8a) as

$$g\beta F_S = B(g\beta\, \Delta S)^{4/3} k_T^{1/3} \tag{11.4.8b}$$

On general dimensional grounds, B can only depend upon $\beta\, \Delta S/\alpha\, \Delta T$, ν/k_T, and ν/k_S, and the similarity theory given above states that B should be independent of changes in the molecular parameters when (11.4.1) applies. B may, however, be a function of $\beta\, \Delta S/\alpha\, \Delta T \sim 1$, and empirical values are available in the literature cited.

11.5 Cold Fresh Layers over Warm Salty Layers

The inverse of the problem considered in Section 11.4 is obtained when a deep layer of cold (temperature T_0) fresh (salinity S_0) water is placed above a warm salty (and heavier) layer (Fig. 11.3). A motionless state is then possible in which the temperature–salinity profiles merely change in time, according to the

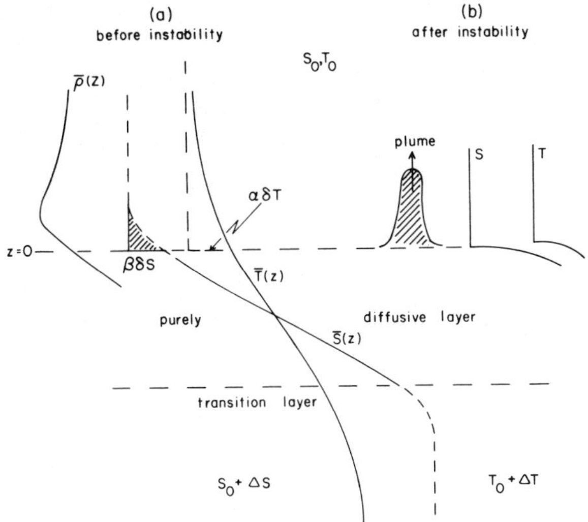

FIG. 11.3 Schematic diagram of the diffusive interface which forms when a deep layer of cold fresh water is placed above a deep layer warm light water. (a) Profiles of density, salinity and temperature just before convective instability occurs. (b) The formation of a plume in the boundary layer and the subsequent local modification of the profiles.

solutions of the respective diffusion equations. Suppose that $\bar{T}(z)$ and $\bar{S}(z)$ are geometrically similar at the initial time, so that $\rho(z)$ decreases monotonically upward to the value of $\rho(\infty)$. Since $\bar{T}$ diffuses more rapidly than $\bar{S}$ ($k_T \gg k_S$), there will be a time (Fig. 11.3a) when the density profile $\bar{\rho}$ increases upward in the "tails" (large z) because of the predominant effect of $\bar{T}$ in those regions. Convective instability can then occur in the tails when an appropriately defined Rayleigh number [cf. (11.2.18)] becomes supercritical, and the finite amplitude form of the instability is indicated by the rising plume in Fig. 11.3b. This will tend to remove the fluid having density less than $\bar{\rho}(\infty)$, and thereby transfer both heat and salt to great heights. The removal of the fluid from the transition layer also tends to produce instantaneous T, S profiles like those shown on the right hand side of Fig. 11.3. At this stage of the "relaxation oscillation" (cf. Section 10.6), we again have an upward *diffusion* of heat and salt from the stable layer beneath $z = 0$. Thus the temperature and salinity gradients in the transition layer will again increase until the point of convective instability is once more reached. We want to determine the average fluxes from the deep lower layer to the deep upper layer, and also the equilibrium thickness of the "diffusive layer" as a function of the given ($\beta\,\Delta S$, $\alpha\,\Delta T$, k_T, k_S, and ν).

We assume that $\beta\,\Delta S/\alpha\,\Delta T$ is sufficiently large, and the diffusive layer sufficiently stable, so that the "roots" of the detached plume do not penetrate into the diffusive layer. Consequently, the vertical fluxes therein are entirely due to conduction, and the thickness must then adjust so that these fluxes equal the turbulent transports in the convection layers. Although experimental values of the fluxes are available, we only have a similarity theory for their *ratio* (Shirtcliffe, 1973), but this theory is instructive and will be discussed below.

The main idea is that, when instability occurs, the plume "scoops up" all the fluid and only that fluid which is capable of rising to $z = \infty$ under its own buoyancy. The base $z = 0$ of the transition layer in which convection occurs is determined by that point in the density profile where

$$\bar{\rho}(0) = \bar{\rho}(\infty) \tag{11.5.1}$$

The shaded area in Fig. 11.3a indicates the portion of the $\bar{S}$ profile which is removed by the plume, and if $\delta S = \bar{S}(0) - \bar{S}(\infty)$ and $\delta T = \bar{T}(0) - \bar{T}(\infty)$, then (11.5.1) implies

$$\beta\,\delta S = \alpha\,\delta T \tag{11.5.2}$$

After the plume departs; the temperature–salinity profiles are vertically uniform in the transition layer, and we then obtain that phase of the relaxation–oscillation cycle in which molecular diffusion dominates. Thus, the thickness of the thermal boundary layer then increases as $(k_T t)^{1/2}$, and the corresponding salinity boundary layer will increase as $(k_S t)^{1/2}$. Therefore, the ratio of the boundary layers is $(k_S/k_T)^{1/2}$ at any time, and at the instant of convective instability, in particular, we have

$$h_S/h_T = (k_S/k_T)^{1/2} \tag{11.5.3}$$

At this time, the areas lying above $z = 0$ and bounded by the $\bar{S}$, $\bar{T}$ curves are in the ratio

$$h_S(\beta\, \delta S)/h_T(\alpha\, \delta T) \tag{11.5.4}$$

and these areas are also proportional to the convective fluxes βF_S and αF_T, which are produced by the subsequent instability. Thus, (11.5.4) gives the flux ratio

$$\beta F_S/\alpha F_T = h_S/h_T$$

when (11.5.2) is used, and (11.5.3) then implies that

$$\beta F_S/\alpha F_T = (k_S/k_T)^{1/2} \tag{11.5.5}$$

is independent of $\beta\ \Delta S/\alpha\ \Delta T$.

The key assumption (11.5.1) used in obtaining this result is not tenable when $\beta\ \Delta S - \alpha\ \Delta T$ is small because the roots of the plume may then extend into the weakly stratified "diffusive layer" and thereby "scoop up" and mix a portion of the relatively heavy fluid. But for a moderately large density difference between the two deep layers, experiments show that there is no motion in the "diffusive layer," and (11.5.5) is quantitatively confirmed.

11.6 General Thermodynamic Relations

Since little is known about oceanic microstructure, it is desirable to obtain some integral constraints which are independent of the detailed mechanism. The "power integrals" given below satisfy this requirement, and the only assumption pertains to the known climatological scale properties of the T, S distribution in a water mass. The main result (11.6.13) applies if the micromixing process is controlled by shear flow instability (see Section 11.7) or by convection.

(a) *T–S Variances in a Water Mass.* The subscript m will be used here to denote a long time average at any x, y, z point in the ocean, and primes will be used to denote the fluctuation of a field from its time average. Thus, the total salinity at a point is given by $S = S_m(x, y, z) + S'(x, y, z, t)$, the (potential) temperature is given by $T = T_m + T'$, and we have the properties

$$(T')_m = 0 = (S')_m \tag{11.6.1}$$

The conservation laws $dT/dt = k_T \nabla^2 T$ and $dS/dt = k_S \nabla^2 S$, then become

$$dT'/dt + \mathbf{V} \cdot \nabla T_m = k_T \nabla^2 T' + k_T \nabla^2 T_m \tag{11.6.2}$$

$$dS'/dt + \mathbf{V} \cdot \nabla S_m = k_S \nabla^2 S' + k_S \nabla^2 S_m \tag{11.6.3}$$

Power integrals, related to those in Section 10.3, are now obtained by multiplying (11.6.2) with T', by multiplying (11.6.3) with S', and by averaging the separate equations with respect to *time*. The last terms in each of the two averages will vanish because of (11.6.1), and thus we have

$$\tfrac{1}{2}[(d/dt)(T')^2]_m + [\mathbf{V}T']_m \cdot \nabla T_m = -k_T[(\nabla T')^2]_m + \nabla \cdot [k_T T' \nabla T']_m \tag{11.6.4}$$

$$\tfrac{1}{2}[(d/dt)(S')^2]_m + [\mathbf{V}S']_m \cdot \nabla S_m = -k_S[(\nabla S')^2]_m + \nabla \cdot [k_S S' \nabla S']_m \tag{11.6.5}$$

These equations for the production of thermal variance $(T')^2$ and salinity variance $(S')^2$, can be simplified by using the continuity equation ($\nabla \cdot \mathbf{V} = 0$) and the steady state requirement. Thus, the first term in (11.6.4) equals

$$\tfrac{1}{2}[(\partial/\partial t)(T')^2]_m + \tfrac{1}{2}[\mathbf{V} \cdot \nabla (T')^2]_m = \tfrac{1}{2}\nabla \cdot [\mathbf{V}(T')^2]_m$$

and $\mathbf{V}(T')^2$ may be called a "flux of thermal variance". This term is quite different from the heat flux, or the $[\mathbf{V}T']_m \cdot \nabla T_m$ term in (11.6.4).

The first term on the right hand side of (11.6.4) is always negative, and it represents the ultimate dissipation of the temperature fluctuations which are generated by the terms mentioned previously. The dominant scales for generating fluctuations in the ocean are the large ones (geostrophic eddies), whereas the major contribution to $[(\nabla T')^2]_m$ is due to very small (~1 cm) wavelengths. Note that the dissipation term is much larger than the last term in (11.6.4), because the latter is proportional to $\tfrac{1}{2}\nabla^2 \, [(T')^2]_m$ and because $[(T')^2]_m$ is a slowly varying function of *geographical* position. To all intents and purposes then, the variance balance (11.6.4) and (11.6.5) is given by

$$\tfrac{1}{2}\nabla \cdot [\mathbf{V}(T')^2]_m + [\mathbf{V}T']_m \cdot \nabla T_m = -k_T[(\nabla T')^2]_m \tag{11.6.6}$$

$$\tfrac{1}{2}\nabla \cdot [\mathbf{V}(S')^2]_m + [\mathbf{V}S']_m \cdot \nabla S_m = -k_S[(\nabla S')^2]_m \tag{11.6.7}$$

In addition to these equations for the thermal and salinity variances, we have another useful integral for the "production of *T–S* correlations," which is obtained as follows. By multiplying (11.6.2) with S', by multiplying (11.6.3) with T', by averaging both equations over time, and by adding the results, we get

$$[(d/dt)(T'S')]_m + [\mathbf{V}S']_m \cdot \nabla T_m + [\mathbf{V}T']_m \cdot \nabla S_m$$
$$= -(k_T + k_S)[\nabla T' \cdot \nabla S']_m + \nabla \cdot [k_T S' \nabla T' + k_S T' \nabla S']_m$$

The time average field $[k_T S' \nabla T' + k_S T' \nabla S']_m$ can only vary on a geographical scale, and therefore the last term in this equation is negligible compared to the preceding dissipation term. [The approximation is the same as that used in obtaining (11.6.6) and (11.6.7)]. Thus, the T–S correlation equation becomes

$$\nabla \cdot [\mathbf{V}T'S']_m + [\mathbf{V}S']_m \cdot \nabla T_m + [\mathbf{V}T']_m \cdot \nabla S_m = -(k_T + k_S)[\nabla T' \cdot \nabla S']_m \tag{11.6.8}$$

In certain large regions of central water masses, a plot of T_m vs S_m yields a straight line, as mentioned previously. If $\sigma = \partial T_m / \partial S_m$ denotes the slope of this mean curve, then $\nabla T_m = \sigma \nabla S_m$, and Eqs. (11.6.9) follow. The main contribution to the fluctuations T', S' are associated with the geostrophic eddies in the water mass, and the individual values of T', S' are also highly correlated by the water mass curve T_m vs S_m. Therefore, the fluxes of "thermal variance, salt variance, and $T'S'$" are proportional, as expressed in (11.6.10), and we have

$$[\mathbf{V}T']_m \cdot \nabla T_m = \sigma [\mathbf{V}T']_m \cdot \nabla S_m, \qquad [\mathbf{V}S']_m \cdot \nabla S_m = \sigma^{-1} [\mathbf{V}S']_m \cdot \nabla T_m \tag{11.6.9}$$

$$[\mathbf{V}(T')^2]_m = \sigma^2 [\mathbf{V}(S')^2]_m, \qquad [\mathbf{V}T'S']_m = \sigma [\mathbf{V}(S')^2]_m \tag{11.6.10}$$

A replacement of T' by $\sigma S'$ is *not* justified in the "dissipation" terms, because the correlation between the temperature–salinity gradients will break down on the small scales at which molecular diffusion becomes of direct importance.

When Eqs. (11.6.9) and (11.6.10) are used in (11.6.6), (11.6.7), and (11.6.8), the results are

$$\nabla \cdot (\sigma^2/2)[\mathbf{V}(S')^2]_m + \sigma [\mathbf{V}T']_m \cdot \nabla S_m = -k_T [(\nabla T')^2]_m$$

$$\nabla \cdot \tfrac{1}{2} [\mathbf{V}(S')^2]_m + \sigma^{-1} [\mathbf{V}S']_m \cdot \nabla T_m = -k_S [(\nabla S')^2]_m$$

$$\nabla \cdot \sigma [\mathbf{V}(S')^2]_m + [\mathbf{V}S']_m \cdot \nabla T_m + [\mathbf{V}T']_m \cdot \nabla S_m = -(k_T + k_S)[\nabla T' \cdot \nabla S']_m$$

If the first equation be multiplied by $-1/\sigma$, the second by $-\sigma$, and if the sum be added to the third equation, then we get

$$0 = \sigma^{-1} k_T [(\nabla T')^2]_m + \sigma k_S [(\nabla S')^2]_m - (k_T + k_S)[\nabla T' \cdot \nabla S']_m$$

When this equation is simplified by introducing the notation

$$r \equiv [\nabla T' \cdot \nabla S']_m / [(\nabla T')^2]_m^{1/2} [(\nabla S')^2]_m^{1/2} \tag{11.6.11}$$

$$q \equiv \sigma [(\nabla S')^2]_m^{1/2} / [(\nabla T')^2]_m^{1/2} \tag{11.6.12}$$

the result becomes

$$(k_T + k_S) r = (k_T / q) + k_S q \tag{11.6.13}$$

This "universal" relation between the correlation coefficient r (11.6.11) and the quotient q of the rms gradients, is independent of viscosity, etc. The major contributions to r, q come from the smallest scales of motion, as mentioned previously. But these scales are difficult to measure (Gregg and Cox, 1972), and therefore we now consider some inferences pertaining to slightly larger scales.

By choosing some point in the thermocline as the origin and by using $\mathbf{D} = (D_x, D_y, D_z)$ to denote the distance from a second fixed point, we can define a *finite difference* temperature gradient by the vector whose Cartesian components are

$$\left\{\frac{T'(D_x, 0, 0) - T'(0, 0, 0)}{D_x}, \quad \text{etc.}\right\} \tag{11.6.14a}$$

Likewise, the finite difference salinity gradient is

$$\left\{\frac{S'(D_x, 0, 0) - S'(0, 0, 0)}{D_x}, \quad \text{etc.}\right\} \tag{11.6.14b}$$

Let us define a finite difference function $R(\mathbf{D})$ by formally replacing $\nabla T', \nabla S'$ with (11.6.14a) and (11.6.14b) in (11.6.11), and $Q(\mathbf{D})$ is obtained from (11.6.12) in a similar manner. These finite difference functions reduce to r, q in the limit

$$\lim_{\mathbf{D}\to 0} R(\mathbf{D}) = r \qquad \text{and} \qquad \lim_{\mathbf{D}\to 0} Q(\mathbf{D}) = q \tag{11.6.15}$$

From (11.6.13), we then see that the function

$$(k_T + k_S)R(\mathbf{D}) - (k_T/Q(\mathbf{D})) - k_S Q(\mathbf{D}) \tag{11.6.16}$$

will vanish as $\mathbf{D} \to 0$. For large $\mathbf{D}$, on the other hand, we have pointed out that temperature–salinity fluctuations are highly correlated in the case of the central waters, and consequently (11.6.14a) equals (11.6.14b) times $\sigma = \partial T_m/\partial S_m$. From the finite difference versions of (11.6.11) and (11.6.12), we then see that

$$R(\mathbf{D}) = 1 = Q(\mathbf{D}) \qquad \text{for large } \mathbf{D} \tag{11.6.17}$$

(see Pingree, 1972), and by direct substitution of the above relations, we observe that (11.6.16) also vanishes for large $\mathbf{D}$. We have shown, therefore, that

$$(k_T + k_S)R(\mathbf{D}) - (k_T/Q(\mathbf{D})) + k_S Q(\mathbf{D}) = 0 \tag{11.6.18}$$

for values of $\mathbf{D}$ which are either very small or large.

What happens at the intervening values of $\mathbf{D}$? From measurements of vertical profiles of temperature and salinity, Evans (1974) suggests that (11.6.18) may be valid for all values of $\mathbf{D}$, when salinity increases upward. This appears plausible on similarity grounds, since there is no obvious length scale with which one can normalize $\mathbf{D}$.

(b) *Global Power Integrals.* If we consider the salinity variance budget for the whole ocean, then we must recognize that the production term is due only to the surface processes of evaporation, precipitation, freezing, and melting, whereas the dissipation of salinity variance is due only to molecular conduction. The corresponding thermal power integral will not be discussed here because more complex radiative, evaporative, and conductive processes are involved.

It is permissible to regard precipitation as the negative of evaporation, for the purpose of the salt budget, and consequently we define $E(\pm)$ as the number of grams of water per second leaving a square centimeter at any longitude and latitude. The hydrological cycle will be simplified somewhat by including runoff and the ice phase in E. Thus, if $\{\ \}$ denotes the time average of the surface integral of a quantity, then we have the water budget

$$\{E\} = 0 \tag{11.6.19a}$$

The conservation of mass is given by

$$\partial\rho/\partial t + \nabla \cdot (\rho \mathbf{V}) = 0 \tag{11.6.19b}$$

and therefore the conservation of salt can be written as

$$-\nabla \cdot \mathbf{F}_S{}^* = (\partial/\partial t)(\rho S) + \nabla \cdot (\rho \mathbf{V} S) = \rho\, dS/dt \tag{11.6.20}$$

where

$$\mathbf{F}_S^* = -\rho k_S \nabla S \tag{11.6.21}$$

is the molecular salt flux. In the following consideration of the global salt balance, the free surface will be considered steady in time, $\mathbf{k}$ will denote an upward pointing unit vector at the free surface, $S(0) = S(x, y, 0, t)$ will denote the surface ($z = 0$) salinity at any geographical position, and $\langle\ \rangle$ will denote the volume integral of a (time averaged) quantity.

By multiplying (11.6.20) with S, by integrating the result over the entire ocean, and by using (11.6.19b), we obtain

$$-\langle S\, \nabla \cdot \mathbf{F}_S{}^* \rangle = \tfrac{1}{2}\langle \rho\, dS^2/dt \rangle$$

$$\langle \mathbf{F}_S{}^* \cdot \nabla S \rangle - \langle \nabla \cdot (S\mathbf{F}_S{}^*) \rangle = \tfrac{1}{2}\langle (\partial/\partial t)(\rho S^2) \rangle + \tfrac{1}{2}\langle \nabla \cdot (\rho \mathbf{V} S^2) \rangle$$

By integrating this over the entire ocean and by using (11.6.21) in the steady state, we have

$$\tfrac{1}{2}\{\mathbf{k} \cdot \mathbf{V}\rho S^2(0)\} = -\langle \rho k_S (\nabla S)^2 \rangle - \{\mathbf{k} \cdot \mathbf{F}_S{}^* S(0)\} \tag{11.6.22}$$

The free surface value of $\mathbf{k} \cdot \mathbf{V}\rho$ equals the upward flux of salt water, of which the fraction $1 - S(0)$ consists of *pure* water. Therefore, the local evaporation rate equals

$$E = \mathbf{k} \cdot \mathbf{V}\rho(1 - S(0)) \tag{11.6.23}$$

Since no salt crosses the free surface, the upward flux of salt $\mathbf{k}\cdot\mathbf{V}S(0)\rho$ must be compensated by the downward component of (11.6.21), and therefore the total salt flux at the free surface is

$$\mathbf{k}\cdot(\rho\mathbf{V}S(0)+\mathbf{F}_S^*(0))=0$$

When this is combined with (11.6.23), we have

$$\mathbf{k}\cdot\mathbf{F}_S^*(0)=-ES(0)/(1-S(0)) \tag{11.6.24}$$

The substitution of (11.6.23) and (11.6.24) in (11.6.22) then gives

$$\tfrac{1}{2}\{ES^2(0)/(1-S(0))\}=\langle\rho k_S(\nabla S)^2\rangle \tag{11.6.25}$$

This fundamental relation will now be simplified by utilizing the basic fact that the variations in S are small compared to the mean $\langle S\rangle \ll 1$. Let an average salinity S_0 be defined by

$$S_0=A^{-1}\{S^2(0)/(1-S(0))\}^{1/2} \tag{11.6.26}$$

where A is the surface area of the ocean, and let S'' denote the fluctuation in surface salinity, so that

$$S(0)=S_0+S''$$

By using (11.6.19a) in the expansion of (11.6.25) in the small number

$$S''/S_0 \ll 1$$

we get

$$\tfrac{1}{2}\{ES''\}(\partial/\partial S(S^2/1-S))_{S=S_0}+\ldots=\langle k_S(\nabla S)^2\rho\rangle$$

Since $S_0 \ll 1$, the above equation simplifies to

$$S_0\{ES''\}=\rho_0\langle k_S(\nabla S)^2\rangle \tag{11.6.27}$$

To all intents and purposes, the value (11.6.26) of S_0 can also be taken equal to the arithmetic mean of the surface salinity (i.e., $S_0 \simeq 35^0/_{00}$).

Let us define scale values of evaporation E_* (cm/sec), halocline depth H_*, and surface salinity variation ΔS_*, by

$$E_*=\max E/\rho_0,\qquad \Delta S_*=\{ES''\}/AE_*\rho_0,\qquad H_*^{-1}=A|(\nabla S)^2|/\langle(\nabla S)^2\rangle \tag{11.6.28}$$

where $|\nabla S|$ is the rms salinity gradient in the entire ocean. By substituting (11.6.28) in (11.6.27), we then obtain a relation

$$\Delta S_*/H_*=k_S|\nabla S|^2/E_*S_0 \tag{11.6.29}$$

connecting the mean vertical gradient $\Delta S_*/H_*$ with the rms salinity gradient, provided the dissipation occurs rather uniformly throughout the depth of the main halocline. We will return to this relation in the Appendix.

11.7 Forced Convection

Having considered some very general aspects (Section 11.6) of mixing in a T–S water mass, and having considered some mechanistic examples (Sections 11.4–5) of free convection, we will now give an example of forced convection. Suppose we have an unbounded fluid which is uniformly stratified ($-S_z < 0$) with a single substance (salt, say), and suppose we generate velocities **V** by the application of external (body) forces $\hat{\mathbf{F}}(x, y, z, t)$ which are cyclic in space–time (Fig. 11.4a). These forces (of unspecified origin) will be considered as "random"

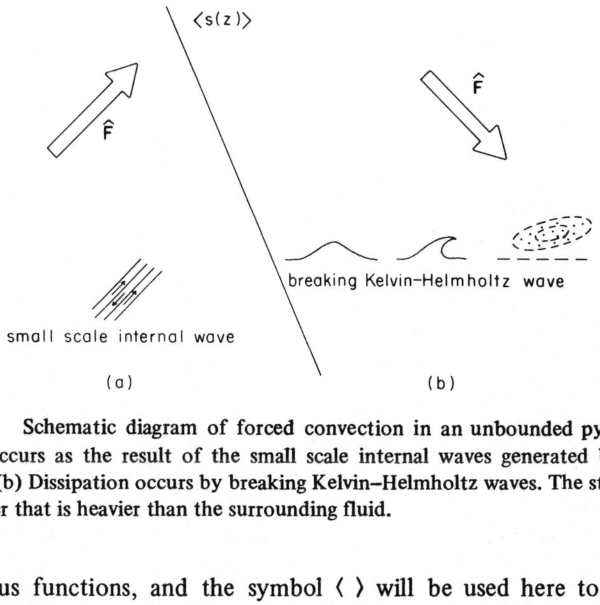

FIG. 11.4 Schematic diagram of forced convection in an unbounded pycnocline. (a) Dissipation occurs as the result of the small scale internal waves generated by nonlinear interactions. (b) Dissipation occurs by breaking Kelvin–Helmholtz waves. The stippled patch contains water that is heavier than the surrounding fluid.

homogeneous functions, and the symbol ⟨ ⟩ will be used here to denote an ensemble average at a point in the statistically steady state. Thus, the salinity fluctuation produced by $\hat{\mathbf{F}}$ is

$$S'(x, y, z, t) = S - \langle S \rangle$$

and the mean vertical gradient $\langle S_z \rangle$ is uniform by symmetry. The vertical flux of salt (11.7.3), the molecular dissipation [Eq. (11.7.2)], and the average rate at which $\hat{\mathbf{F}}$ does work (11.7.1), or

$$W \equiv \langle \mathbf{V} \cdot \hat{\mathbf{F}} \rangle \tag{11.7.1}$$

$$\epsilon \equiv \nu \langle (\nabla \times \mathbf{V})^2 \rangle \equiv \nu \zeta^2 \tag{11.7.2}$$

$$F_S \equiv \langle w'S' \rangle \tag{11.7.3}$$

are all constants.

When the external forces are added to the Boussinesq equations, we have

$$d\mathbf{V}/dt = -\nabla\phi + \nu\,\nabla^2\mathbf{V} - g\beta S'\mathbf{k} + \hat{\mathbf{F}}, \qquad \nabla\cdot\mathbf{V} = 0 \tag{11.7.4}$$

The mechanical energy equation formed by taking the scalar product of (11.7.4) with $\mathbf{V}$ and by averaging the result can be written as

$$W = \epsilon + g\beta F_S \qquad \text{since} \qquad \nabla\cdot\langle\mathbf{V}\phi\rangle = 0 = \nabla\cdot\langle\mathbf{V}(\mathbf{V}^2/2)\rangle \tag{11.7.5}$$

Equation (11.7.5) states that the mechanical energy input W is partially dissipated by the rms vorticity ζ and partially used to increase the potential energy of the salt field.

The power integral (11.6.7) can be used to obtain another relation for F_S when the time averages are replaced by the ensemble averages of this section. Since $\langle\mathbf{V}(S')^2\rangle$ is now independent of position, and since $\langle\mathbf{V}S'\rangle$ and $\nabla\langle S\rangle$ are now vertical, the power integral reduces to

$$F_S\langle -S_z\rangle = k_S\langle(\nabla S')^2\rangle \tag{11.7.6}$$

We want to find the value of F_S which is produced by the application of a given force field to $\langle S_z\rangle$, and the result can always be obtained from (11.7.6) once the rms salinity gradient is determined.

The typical amplitude $|\hat{\mathbf{F}}|$, wavelength L, and frequency ω must be stated as part of the specification of the external field. We will suppose that these are such that internal waves are generated on the large scale L, that energy is "transferred" to smaller scale internal waves by the nonlinear interactions (Section 9.8), and that the energy is dissipated on still smaller scales by the action of molecular viscosity. Within this framework, we need to distinguish between two qualitatively different regimes.

(a) *No Breaking Waves.* For sufficiently small $|\hat{\mathbf{F}}|$ or sufficiently large $\langle -S_z\rangle$, we can make the local Richardson number large (Section 9.6) and thereby neglect the contingency (Bretherton, 1969) that some of the internal waves will "break." But the wave–wave interactions (Section 9.8) can still transfer the energy W to such small scale internal waves as will produce the dissipation ϵ. The input of energy by the random force field is similar to the mechanism discussed in Chapter I (Section 1.4).

The value of $\langle(\nabla S')^2\rangle$ will be discussed for the asymptotic case $k_S \to 0$ (or $k_S/\nu \ll 1$), wherein the nondiffusive equation

$$\partial S'/\partial t + \mathbf{V}\cdot\nabla S' + w\langle S_z\rangle = 0$$

and (11.7.4) govern the wave dynamics. The solution of these equations for $\langle(\nabla S')^2\rangle$ will not depend on k_S, and therefore the asymptotic value of F_S obtained from the general equation (11.7.6) is proportional to k_S and small to that order. The coefficient of proportionality is now obtained by rewriting (11.7.6) in the nondimensional form

$$\frac{g\beta F_S}{\epsilon} = \frac{k_S}{\nu} \frac{\langle (\nabla g\beta S')^2 \rangle}{\langle -g\beta S_z \rangle \zeta^2} \tag{11.7.7}$$

and by the following determination of the asymptotic value of

$$\frac{\langle (\nabla g\beta S')^2 \rangle}{\langle -g\beta S_z \rangle \zeta^2} \tag{11.7.8}$$

The main contribution to the rms vorticity ζ and the rms salinity gradient will come from wavelengths much less than L, as mentioned previously. The small viscous damping of these waves is balanced by the energy transfer from the large scale L waves. Consequently, we assume that over any interval of the order of the Brunt–Väisälä period, the smaller scale internal waves are nearly inviscid linear oscillations. Therefore, the ratio of the velocity amplitude and the density amplitude in any wave number $\mathbf{b}_m$ can be obtained from the linear theory [Eqs. (9.8.4)–(9.8.7)]. From (9.8.5), we see that the magnitude of $\nabla \beta S'$ equals the magnitude of the vorticity of the wave times $g^{-1/2} \langle -\beta S_z \rangle^{1/2}$. Since this proportionality is independent of $\mathbf{b}_m$, the sum over all wavenumbers gives

$$\langle (\nabla g\beta S')^2 \rangle = \zeta^2 \langle -g\beta \bar{S}_z \rangle \tag{11.7.9}$$

Therefore, (11.7.8) equals unity, (11.7.7) becomes $g\beta F_S/\epsilon = k_S/\nu$, and from (11.7.5), we conclude that

$$g\beta F_S / W = k_S/\nu \ll 1 \tag{11.7.10}$$

is the asymptotic relation for the salt flux. Note that the convective flux $\overline{w'S'}$ is independent of $\langle S_z \rangle$, and a fixed fraction of the energy input is used to raise the salt field by means of the convective–diffusive effect of the field of internal waves.

(b) *Breaking Waves.* Let us consider now a regime

$$1 > g\beta F_S / W \sim 1 \tag{11.7.11}$$

with a substantial fraction (~ 1) of the energy input being used to increase the potential energy of the stratification. The existence of such a regime can be assured by specifying both F_S and W, and by then solving for the requisite $\langle -S_z \rangle$. In particular, we would like to know how this $\langle -S_z \rangle$ will change if we only change $k_S \to 0$. The only way to avoid the previous regime (11.7.10) [and the contradiction with (11.7.11)] is for the value of $\langle -S_z \rangle$ to be so small that Kelvin–Helmholtz instabilities and breaking waves occur in the random field.

The intermittent wave breaking process that forms when the local Richardson number falls below $\frac{1}{4}$, is shown schematically in Fig. 11.4b. The detachment of the crests give rise to an upward transport of relatively heavy water, as shown by the stippled eddy. The vorticity in the eddy, however, must be sufficiently large to mix this density anomaly before the high salinity parcels have a chance to "settle" back to their level of origin. The point made here may be illustrated by the fact that no net vertical transport of water occurs when a *surface* gravity wave breaks, since the droplets sprayed into the air settle back onto the surface because of their excess weight. To the extent that this effect occurs in the breaking *internal* wave, there will be a reduction of the total vertical flux of salt. To prevent this from happening and to meet the requirements of (11.7.11), the rms vorticity must be sufficiently large compared to the (curl of the) buoyancy force acting on the microstructure. On the other hand, ζ cannot be "too large" because this will produce excessive dissipation, which again leads to trouble in satisfying (11.7.11). Thus, the inertial, buoyancy, and viscous terms in the vorticity equation must all have the same order of magnitude (on the dissipation scale). The curl of the acceleration is of order ζ^2, the curl of the corresponding buoyancy force is of the same order as the rms $g\beta\,\nabla S'$, and their ratio gives the "micro-Richardson number"

$$R_\mu = \langle (g\beta\,\nabla S')^2\rangle^{1/2}/\zeta^2 \qquad (11.7.12)$$

The argument given above states that R_μ cannot become infinite or zero as $k_S \to 0$, and therefore R_μ must approach a constant. An additional argument against large R_μ is that Kelvin–Helmholtz instabilities could not then form. An additional argument for asserting that (11.7.12) cannot become very small when $k_S \to 0$, is due to the self-limiting effect of the Kelvin–Helmholtz instabilities on the local Richardson number when it starts to drop below $\frac{1}{4}$. Thus, the upshot of this similarity theory is that (11.7.12) must approach a constant as $k_S \to 0$.

Therefore, when (11.7.12) is used to eliminate $\langle(\nabla S')^2\rangle$ in (11.7.6), we obtain an expression for the vertical salinity gradient

$$-\langle g\beta S_z\rangle = \frac{k_S(R_\mu\zeta^2)^2}{g\beta F_s} = \frac{k_S}{\nu^2}\frac{\epsilon^2 R_\mu{}^2}{g\beta F_S}$$

$$= \frac{k_S}{\nu^2}\frac{(W-g\beta F_S)^2}{g\beta F_S} R_\mu^2 = \left(\frac{k_S}{\nu^2}\right) g\beta F_S\left[R_\mu\left(\frac{W}{g\beta F_S}-1\right)\right]^2 \qquad (11.7.13)$$

which is directly proportional to $k_S \to 0$. Thus, a smaller density gradient is required to produce a given buoyancy flux $g\beta F_S$ for the case of a salt stratified fluid than for the case of a fluid stratified with temperature. If (11.7.11) holds, and if $R_\mu \sim 1$, then (11.7.13) implies that the ratio of F_S to $\langle\beta S_z\rangle$ is of order ν^2/k_S.

References

Baines, P. G., and Gill, A. E. (1969). On Thermohaline Convection with Linear Gradients. *J. Fluid Mech.* **37**, 289–306.

Bretherton, F. P. (1969). Waves and Turbulence in Stably Stratified Fluids. *Radio Sci.* **4**, 1279–1287.

Evans, D. (1974). Private communication.

Gregg, M. C., and Cox, C. S. (1972). The Vertical Microstructure of Temperature and Salinity. *Deep Sea Res. Oceanogr. Abstr.* **19**, 335–376.

Linden, P. E. (1973). On the Structure of Salt Fingers. *Deep Sea Res. Oceanogr. Abstr.* **20**, 325–340.

Pingree, R. D. (1972). Mixing in the Deep Stratified Ocean. *Deep Sea Res. Oceanogr. Abstr.* **19**, 549–561.

Shirtcliffe, T. G. L. (1973). Transport and Profile Measurements of the Diffusive Interface in Double Diffusive Convection with Similar Diffusivities. *J. Fluid Mech.* **57**, 27–43.

Stern, M. E. (1969). Collective Instability of Salt Fingers. *J. Fluid Mech.* **35**, 209–218.

Sverdrup, H. U., Johnson, M. W., and Fleming, R. H. (1942). "The Oceans, Their Physics, Chemistry, and General Biology." Prentice-Hall, Englewood Cliffs, New Jersey.

Turner, J. S. (1973). "Buoyancy Effects in Fluids." Cambridge Univ. Press, London and New York.

Williams, A. J. (1974). Salt Fingers Observed in the Mediterranean Outflow. *Science* **185**, 941–943.

CHAPTER XII

Horizontal Convection and Thermoclines

12.1 Nonrotating Thermocline

A relatively intense convection develops in a fluid which is heated uniformly from below and cooled from above, whereas no motion occurs in the stable stratification which arises when a fluid is heated uniformly from above. Another fundamental case occurs when the heat source and sink lie on the same level surface, as discussed in this chapter. The ocean comes under this heading because the solar radiation is absorbed near the surface of the tropical waters, and the polar waters are cooled at the surface. In such a differentially heated fluid, the cold water tends to sink beneath the warm water and thereby establishes a density stratification (thermocline). The laminar thermocline problem in a nonrotating liquid is discussed below, and the rotating problems in Sections 12.2 and 12.3. The oceanic thermocline, on the other hand, is most dependent on the wind stress, and we are only able to touch on this coupled problem in the Appendix.

Consider a liquid contained in an infinitely long and nonrotating channel (Fig. 12.1) of width L and height H. A uniform horizontal temperature gradient

$\Delta T/L$ is maintained on the top ($z = 0$) surface, and the induced motion is assumed to be independent of distance measured normal to the plane of the page. The bottom surface ($z = -H$) and the sidewalls ($x = 0, x = L$) are *insulated,* so that the total vertical flux of heat across any horizontal surface will vanish in the steady state.

It is apparent that the coldest and heaviest water will occur at $x = 0, z = 0$. This water, of temperature $T(0, 0) = T_0$, will therefore sink along the $x = 0$ boundary and be replaced by warmer water advected along the $z = 0$ boundary. Thus we see that a closed vertical circulation cell will be completed as the cold water rises through the steady thermocline, and is heated from above.

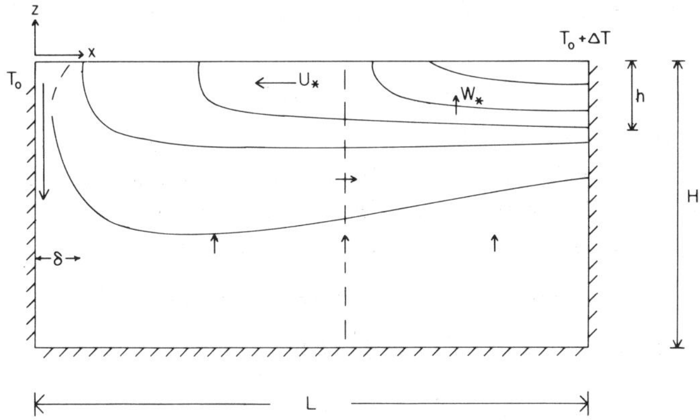

FIG. 12.1 Vertical section through a nonrotating thermocline produced by differential heating on the top ($z = 0$) of a semi-insulated channel. The typical thermocline depth is h, the typical horizontal velocity is u_*, and the typical upwelling velocity is w_*. The sinking motions are confined to the cold wall ($x = 0$) boundary layer.

The circulation cell will concentrate the isotherms in a relatively shallow depth h, so that the upwelling parcel can be heated by conduction by the time it rises to the surface. The scale depth h is to be determined as a function of the external parameters $g\alpha\,\Delta T$, H, L, k_T, and ν, where a Boussinesq fluid is assumed and standard notation is used. Only three independent nondimensional numbers can be formed from the five parameters listed above, so that the mean value of the nondimensional thermocline depth h/L can only depend on the aspect ratio H/L, the Prandtl number ν/k_T, and the Rayleigh number

$$g\alpha\,\Delta T L^3/k_T\nu \qquad (12.1.1)$$

Since we are interested in the case of relatively shallow ($h \ll H$) thermoclines in relatively wide ($H \ll L$) basins, attention is now directed toward the parametric regime which satisfies the inequality

$$h \ll H \ll L \qquad (12.1.2)$$

By referring to Fig. 12.1, we see that the first part of the inequality implies that, in the region $z < -h$, the temperature is essentially uniform and nearly equal to that ($T(0, 0) = T_0$) of the cold sinking water. Such a laminar regime has been experimentally realized and explained by means of a scale analysis (Rossby, 1965), but no analytic theory is available. Before presenting the similarity analysis, we discuss some important general integral constraints.

The horizontal average of the heat *conducted* downward across any z surface is $k_T\, \partial\overline{T}/\partial z$, and the corresponding upward *convection* of heat is $\overline{wT'}$. Since the bottom is insulated, the total heat flux must vanish in any statistically steady state, and therefore

$$\overline{wT'} - k_T\, \partial\overline{T}/\partial z = 0 \qquad (12.1.3)$$

By using brackets to denote a vertical average, we can see that the total buoyancy work $\langle g\alpha\overline{wT'}\rangle$ is strongly limited by the molecular conductivity. A corresponding limitation also occurs for the total dissipation of kinetic energy, this being given by the mechanical energy equation (10.3.5), or

$$\langle g\alpha wT'\rangle = \nu\langle\zeta^2\rangle \qquad \text{where} \qquad \zeta^2 = (\nabla \times \mathbf{V})^2 \qquad (12.1.4)$$

When (12.1.3) is substituted in (12.1.4), the result is

$$\frac{g\alpha \int_{-H}^{0} \partial\overline{T}/\partial z\, dz}{\int_{-H}^{0} \overline{\zeta^2}\, dz} = \frac{\nu}{k_T} \qquad (12.1.5a)$$

The left hand side has the form of a Richardson number, and it may be simplified by recalling that the bottom temperature is nearly equal to T_0. Since $T(x, 0)$ is a linear function of x, the numerator in (12.1.5a) equals $g\alpha\overline{T}(x, 0) - g\alpha T_0 = g\alpha\, \Delta T/2$, and therefore

$$\frac{g\alpha\, \Delta T}{\int_{-H}^{0} \overline{\zeta^2}\, dz} = 2\left(\frac{\nu}{k_T}\right) \qquad (12.1.5b)$$

Let us now define the scale depth of the thermocline by

$$h \equiv \Delta T/\max \partial\overline{T}(z)/\partial z \qquad (12.1.6)$$

We can also define a vertical velocity scale by

$$w_* = (\max \overline{wT'})/\Delta T \qquad (12.1.7)$$

and a horizontal velocity scale by

$$u_* = w_* L/h \tag{12.1.8}$$

where the maximization (max) is performed with respect to the z coordinate. By substituting the maximum of (12.1.3) in (12.1.7) and by then using (12.1.6), we obtain

$$w_* h/k_T = 1 \tag{12.1.9}$$

and therefore (12.1.8) can be written as

$$u_* h^2/Lk_T = 1 \tag{12.1.10}$$

It will be shown that (12.1.7) gives the order of magnitude of the upwelling velocity in the thermocline, and (12.1.8) the horizontal velocity near $z = 0$. But first we write (12.1.5b) in the form

$$g(\alpha\,\Delta T)h/u_*^2 = 2(\nu/k_T)I \tag{12.1.11}$$

where

$$I \equiv \frac{h^{-1} \int_{-H}^{0} \overline{\zeta^2}\, dz}{(u_*/h)^2} \tag{12.1.12}$$

When u_* in (12.1.11) is eliminated by using (12.1.10), the solution for h is

$$h = (2\nu/k_T)^{1/5} (L^2 k_T^2/g\alpha\,\Delta T)^{1/5} I^{1/5} \tag{12.1.13}$$

and the physical significance of this rather formal expression for the thermocline depth is now discussed.

We have mentioned that a large vertical temperature gradient is required (when k_T is small) to heat the parcels rising through the thermocline, and thus the latter region has large static stability. Rapid lateral variations in $w(x, z)$ are thereby precluded by the strong buoyant restoring forces which would be brought into play, and therefore the order of magnitude of the upward velocities may be denoted by $w_u(z)$. In the region near $x = 0$, on the other hand, a rapid lateral variation in $w(x, z)$ can occur because the static stability is small (or negative). Therefore, a different term $w_d(z)$ denotes the order of magnitude of the velocities in the sinking region, the latter being confined to the relatively narrow region δ, where

$$0 < \delta \ll L$$

At any z, the downward flux $w_d\delta$ must equal the upward flux $w_u(L - \delta)$, and the approximate continuity equation

$$w_d \delta = w_u L \tag{12.1.14a}$$

implies $w_d \gg w_u$. Since the temperature fluctuation on a level surface is given by $T' = T(x, z) - \bar{T}(z)$, with $\overline{T'} = 0$, the convective flux of heat can be approximated by

$$\overline{wT'} = L^{-1} \int_0^L wT'\,dx \simeq L^{-1}(-w_d) \int_0^\delta T'\,dx + L^{-1} w_u \int_\delta^L T'\,dx$$

$$\simeq (\delta/L)(-w_d)T'(0, z) = -(\delta/L)w_d(T_0 - \bar{T}(z))$$

$$\max \overline{wT'} \sim +(\delta/L)w_d\,\Delta T = w_u\,\Delta T$$

because $\bar{T}(z) - T_0 \sim +\Delta T$ in the thermocline region. From (12.1.7), we then obtain

$$w_u \sim \overline{wT'}/\Delta T = w_* \qquad (12.1.14b)$$

and this shows that the typical upwelling speed equals (12.1.7). Since the total amount of water flowing *upward*, or $w_u L$, must equal the total amount of water flowing laterally toward $x = 0$ in the upper part of the circulation cell (Fig. 12.1), we also see that the typical horizontal speeds in the thermocline are

$$w_u L/h \sim w_* L/h = u_*$$

This shows that (12.1.8) gives the order of magnitude of the horizontal velocity in the upper thermocline.

If the flow is laminar, then the rms vorticity is approximated by the mean vorticity u_*/h, or

$$\int_{-H}^{0} \overline{\zeta^2}\,dz \sim h(u_*/h)^2$$

and therefore the order of magnitude of (12.1.12) is

$$I \sim 1 \qquad (12.1.15)$$

We then conclude, from (12.1.13), that the order of magnitude of the vertical thickness of the thermocline is

$$h \sim (\nu/k_T)^{1/5}(L^2 k_T{}^2/g\alpha\,\Delta T)^{1/5} \qquad (12.1.16)$$

Just for the fun of it, let us evaluate (12.1.16) for a basin having dimensions $L = 10^9$ cm and $H = 5 \times 10^5$ cm. By also using $\Delta T = 20°$ C, $\alpha = 2 \times 10^{-4}$ (°C)$^{-1}$, $\nu = 10^{-2}$ cm^2/sec, and $k_T = 10^{-3}$ cm^2/sec, we find that the order of magnitude of the thermocline depth is 2 m. Although Rossby's experiment verifies these scaling laws for high Prandtl number fluids, the experimental range is somewhat less than 10^9 cm. The next step is to determine the conditions under which this laminar flow becomes unstable. The most important consideration appears to be the value of the overall Richardson number (12.1.11), because, if this should be very small compared to unity, then the laminar flow will not be realized and (12.1.16) will not be applicable. When the Richardson number requirement and (12.1.15) are used in (12.1.11), we see that the Prandtl number must be large. If we also assume a free upper surface in Fig. 12.1, thereby eliminating shear flow turbulence, then we establish the consistency of using the square mean vorticity instead of the mean square vorticity in evaluating (12.1.15). Under these conditions the scale depth (12.1.16) decreases as the inverse fifth power of the horizontal temperature difference.

The remarks made above do not necessarily imply a complete absence of turbulent motions in the thermocline, because there may be other mechanisms leading to instability. Apart from the possibility of eddies having variations in the y direction, we may note that the region beneath $x = 0$, $z = 0$ can become unstable in the sense of Chapter X. An inversion of the vertical temperature gradient must occur in this region because $T(0, 0)$ is the minimum temperature, as mentioned previously. Rossby noted such inversions in his experiment, but the associated Rayleigh number was too small to produce instability. The situation may be quite different, however, in the hypothetical example (above) of a basin having an overall dimension of $L = 10^9$ cm. In that case, we may have turbulent thermal convection forming immediately beneath the polar surface, and the convective plumes would then proceed to sink until they reached their own density level in the underlying (weak) thermocline. At this level, the plumes would now be spreading laterally along a constant density surface (Wu, 1969) and in a manner similar to the collapsing wedge discussed in connection with Fig. 3.2. As the thin density wedge spreads to $x \gg \delta$, it might become dynamically unstable and a weak field of small scale turbulence may appear in the main thermocline. This effect might increase the mean square vorticity and thereby modify I, but the net effect on the thermocline depth will probably be small because (12.1.13) only depends on the fifth root of I.

12.2 Thermocline in a Rotating Annulus

Figure 12.2 is a section passing through the vertical axis of symmetry of a small gap annulus, whose mean radius is much larger than the radial gap width L. This *semi-insulated* annulus rotates with angular velocity $f/2$ about the axis, and the imposed temperature on the conducting surface ($z = 0$) is again given by

$$T(x, 0) = \Delta T(x/L) \tag{12.2.1}$$

Since this case reduces to that of Section 12.1 when $f \to 0$, we will only consider the opposite limit of small Rossby number: $R_0 = v/fL \ll 1$, where $v(x, z)$ denotes the azimuthal component of velocity in the interior of Fig. 12.2.

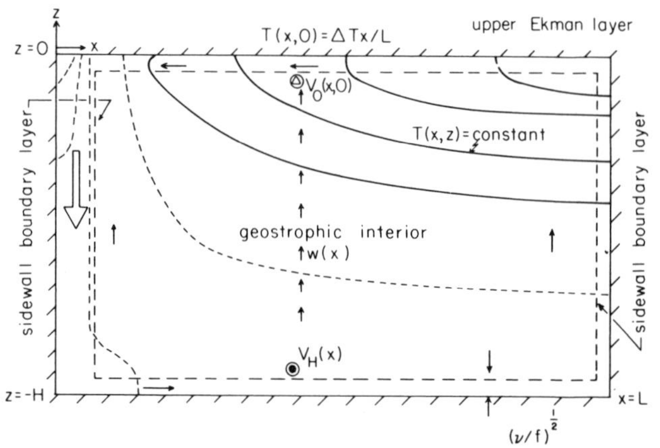

FIG. 12.2 Vertical section through a rotating annulus of radial width L. The motion is axisymmetric and the different dynamical regimes are explained in the text. The temperature difference between the solid curves is greater than the difference between dashed isotherms.

Let us recall that large restoring forces are brought into play when a rapidly rotating fluid is subject to axisymmetric displacements, these forces being responsible for the free inertia oscillations (2.2.8) in a homogeneous liquid. Such forces will also inhibit steady radial velocities u in the interior of the differentially heated fluid because there is no y-directed pressure gradient to balance the Coriolis force fu. If we start the discussion by neglecting the corresponding inertial and viscous forces, then $u = 0$ in the geostrophic interior and v satisfies the thermal wind equation

$$f\,\partial v/\partial z = g\alpha\,\partial T/\partial x \tag{12.2.2}$$

The slope of the parabolic level surfaces has also been neglected here.

Although the viscous forces have been neglected in the geostrophic interior, these forces are very important in the relatively thin horizontal boundary layers shown in Fig. 12.2, because the $v(x, z)$ obtained by integrating (12.2.2) will not satisfy the no slip boundary conditions at $z = (0, -H)$. The Ekman layer also establishes the *radial* flow component $u = u_b(x, z)$, whose convergence gives the upwelling velocities in the interior. On the basis of these remarks we will require the scale separation

$$(\nu/f)^{1/2} \ll h \ll H \ll L \tag{12.2.3}$$

This approximation, and others mentioned below, implies limits on the parametric range of validity, as may be determined a posteriori.

Because the radial velocities vanish in the geostrophic interior, and because v is independent of ν we have $\partial w/\partial z = 0$, so that the upwelling velocities $w = w(x)$ are independent of z in the interior. The kinematic constraints here also imply

$dT/dt = w\, \partial T/\partial z$, and therefore the heat conduction equation simplifies to

$$w(x)\, \partial T/\partial z = k_T\, \partial^2 T/\partial z^2 \tag{12.2.4}$$

because (12.2.3) implies

$$\nabla^2 T \simeq \partial^2 T/\partial z^2$$

A solution of (12.2.4) satisfying (12.2.1) is

$$T(x, z) = (\Delta T x/L) \exp z/h(x) \tag{12.2.5}$$

where

$$h(x) = k_T/w(x) \tag{12.2.6}$$

The exponential function in (12.2.5) is negligible near $z = -H$ and therefore the insulating boundary condition at $z = -H$ is satisfied.

Let us note here that (12.2.4) is not valid in the upper viscous boundary layer, since the radial advection of heat is important for the local heat balance of that region. But the viscous layer is so thin that even a "large" vertical temperature gradient at $z = 0$ may lead to a total temperature *difference* across the Ekman layer which is negligibly† small compared with ΔT. This is the justification for using (12.2.1) as the upper boundary condition for the interior solution (12.2.5). A similar point is involved in a subsequent discussion, wherein we replace the horizontal pressure gradient *inside* the viscous boundary layer by the pressure gradient just *outside* the layer. The resulting error is determined by the vertical integral of the hydrostatic equation, and the error is negligible because of the thinness of the boundary layer.

Before considering the dynamics of the viscous layers, we will integrate (12.2.2) vertically across the geostrophic region, making use of (12.2.5). If $v(x, z) = v_H(x)$ at the "top" of the lower Ekman layer, and $v(x, z) = v_0(x)$ at the "bottom" of the upper Ekman layer, the integration gives

$$v_0(x) - v_H(x) = (g\alpha\, \Delta T/fL)(\partial/\partial x)[x \int dz \exp z w(x) k_T^{-1}] \tag{12.2.7}$$

Because of (12.2.3), we will make a negligible error by extending the interval of integration downward to $-\infty$ and upward to $z = 0$, so that we have

$$v_0(x) - v_H = (k_T(\alpha\, \Delta T) g/fL)(\partial/\partial x)(x/w(x)) \tag{12.2.8}$$

Two additional equations connecting the geostrophic velocities v_0 and v_H

† The relevant small number is obtained from the scale values given in the text as follows. The heat conducted into the fluid, or $k_T\, \partial T(x, 0)/\partial z$ equals the convergence of the lateral heat flux (12.2.15). We then set $\partial T(x, 0)/\partial z = \delta T\, (\nu/f)^{-1/2}$ and compute

$$\delta T/\Delta T \sim f^{-1} L^{-1} (g\alpha\, \Delta T)^{1/2} (\nu/f)^{1/4} (\nu/k_T)^{1/2}$$

as the small fractional change in temperature across the upper Ekman layer, where (12.2.17), (12.2.13), and (12.2.12) have also been used.

and the "suction" velocity w are now obtained from the momentum and continuity relations for the viscous boundary layers (see Sections 5.2 and 5.3). The Coriolis force $fv_H(x)$ at the top of the lower viscous layer is equal to the horizontal pressure gradient at the same level and, for reasons mentioned above, the horizontal pressure gradient is uniform throughout the thin viscous layers. The horizontal velocities u_b and v_b in the boundary layer are, however, functions of z, and the viscous forces are approximately given by $\nu\, \partial^2 u_b/\partial z^2$ and $\nu\, \partial^2 v_b/\partial z^2$. Thus, the local balance between Coriolis, viscous, and pressure gradient force fv_H is given by

$$fu_b = \nu\, \partial^2 v_b/\partial z^2 \qquad \text{and} \qquad -fv_b = \nu\, \partial^2 u_b/\partial z^2 - fv_H(x) \qquad (12.2.9)$$

The boundary conditions are $u_b(x, -H) = 0 = v_b(x, -H)$, and for large distances above the layer, we have $u_b \to 0$ and $v_b \to v_H(x)$. This system becomes identical to Eqs. (5.2.1a–b) when G is replaced by $fv_H(x)$, and consequently the Ekman transport obtained from (5.2.6) is now given by

$$\int u_b\, dz = -v_H(x)(\nu/2f)^{1/2} \qquad (12.2.10)$$

The corresponding equations for u_b and v_b in the *upper* viscous layer can be obtained by replacing the pressure gradient $fv_H(x)$ in (12.2.9) by $fv_0(x)$. The boundary conditions are also analogous, and therefore the Ekman transport of the upper boundary layer is

$$\int u_b\, dz = -v_0(x)(\nu/2f)^{1/2} \qquad (12.2.11)$$

Since the total radial mass transport must vanish at all x, the sum of Eqs. (12.2.10) and (12.2.11) must vanish, and therefore

$$-v_0(x) = v_H(x) \qquad (12.2.12)$$

The continuity of mass also requires that the negative value of the x derivative of (12.2.10) be equal to the upward vertical velocity at the top of the lower boundary layer [see (5.3.8)], or

$$w(x) = (\nu/2f)^{1/2}\, \partial v_H/\partial x \qquad (12.2.13)$$

By substituting (12.2.12) in (12.2.8), by differentiating the result, and by using (12.2.13) to eliminate $v_H(x)$, we obtain

$$(g\alpha\, \Delta T\, k_T/fL)(\nu/8f)^{1/2}(\partial^2/\partial x^2)(x/w(x)) + w(x) = 0 \qquad (12.2.14)$$

In order to obtain the solution of this second-order nonlinear equation, boundary conditions are required at the x limits of the "geostrophic interior." Entirely different equations govern the "sidewall boundary layers," and thus we must now infer the proper conditions to apply to (12.2.14).

At any point x in the interior, the total heat transport F in the negative x direction is obtained by integrating the product of $-T(x, z)$ with the radial velocity. Since $T(0, 0) = 0 = T(x, -H)$, we can therefore see that F equals the vertical integral of $-T(x, z)u(x, z)$ in the upper Ekman layer. But $T(x, z)$ does not differ much from $x\,\Delta T$ across the viscous layer (as mentioned above), and the vertical integral of u across the layer is given by (12.2.11). Therefore, the horizontal heat flux is

$$F = (\Delta T/L)x(\nu/2f)^{1/2}v_0(x) \tag{12.2.15}$$

Since the $x = L$ sidewall is insulated, the first law of thermodynamics requires that (12.2.15) must equal the total heat which is conducted through the $z = 0$ boundary between $x = x$ and $x = L$, and the latter flux obviously tends to zero as $x \to L$. And so, therefore, we now require that (12.2.15) tend to zero, or $v_0(x) \to 0$ as $x \to L$. From (12.2.12), we then have $v_H(L) = 0$, and (12.2.8) gives

$$[(\partial/\partial x)(x/w(x))]_{x=L} = 0 \tag{12.2.16}$$

as a boundary condition for the solution of (12.2.14). For the second condition, we will require that the local depth of the thermocline (12.2.6) be finite as we approach the edge ($x \to 0$) of the geostrophic region, or the region of strong sinking motions.

By introducing the transformations

$$h(x) = k_T/w(x) = k_T L(8f/\nu)^{1/4}\,[f/(g\alpha\,\Delta T\,k_T)]^{1/2}\psi(x) \qquad (x = \xi L) \tag{12.2.17}$$

into (12.2.14), the equation for the nondimensional thermocline depth ψ becomes

$$\psi(d^2/d\xi^2)(\xi\psi) + 1 = 0 \tag{12.2.18a}$$

and the boundary condition (12.2.16) becomes

$$[(d/d\xi)(\xi\psi)]_{\xi=1} = 0 \tag{12.2.18b}$$

The second boundary condition requires ψ to be finite at $\xi = 0$, and therefore we seek a power series solution having the form

$$\psi(\xi) = \psi(0)[1 + a_1\xi + a_2\xi^2 + \cdots] \tag{12.2.19}$$

The relation between $\psi(0), a_1, a_2, \cdots$ is obtained by substituting (12.2.19) into (12.2.18a) to get

$$(1 + a_1\xi + a_2\xi^2 + \cdots)(2a_1 + 6a_2\xi + \cdots) + (\psi^2(0))^{-1} = 0$$

and by equating to zero the coefficients of the successive powers of ξ. Thus, we have

$$2a_1 + (\psi^2(0))^{-1} = 0, \qquad \text{and} \qquad 2a_1^2 + 6a_2 = 0, \ldots \tag{12.2.20}$$

and a final relation between the coefficients is provided by (12.2.18b). But an alternate and more suitable boundary condition can be obtained by dividing

(12.2.18a) with ψ, and by integrating the result from $\xi = 0$ to $\xi = 1$. Then, when the boundary condition (12.2.18b) is used, and when (12.2.19) is substituted into the result, we get

$$-\psi(0) + \frac{1}{\psi(0)} \int_0^1 \frac{d\xi}{1 + a_1\xi + a_2\xi^2 + \cdots} = 0$$

or, equivalently,

$$\int_0^{1/\psi^2(0)} \frac{d\eta}{1 + a_1\psi^2(0)\eta + a_2\psi^4(0)\eta^2 + \cdots} = 1$$

The equation for the constant ψ^2 (0) obtained by using (12.2.20) to eliminate $a_1, a_2 \cdots$ is then

$$1 = \int_0^{1/\psi^2(0)} \frac{d\eta}{1 - \eta/2 - \eta^2/12 + \cdots} \tag{12.2.21}$$

and the numerical solution of (12.2.21) gives $1/\psi^2(0) \simeq \frac{3}{4}$ to an accuracy of about 10%. Substitution of this in (12.2.17) then determines the *e*-folding thermocline depth just outside the sidewall boundary layer, and the variation of h with x can also be determined from the above theory.

The main conclusion (12.2.17) is that the thermocline depth increases as the $\frac{3}{4}$ power of f and is independent of H, provided (12.2.3) is satisfied. We also note that the increase of h with f implies that the axisymmetric flow will eventually become unstable with respect to geostrophic eddies, according to Section 4.4 and the relevant references therein. Thus, it would be most interesting to repeat the annulus experiments which have been done in the context of the atmospheric *Hadley* cell. The thermocline problem discussed above has somewhat different boundary conditions, and therefore it would be of interest to determine the effect of baroclinic instability on the scale depth h.

We are also now in a position to estimate the effect of some of the forces which have been neglected in the relatively simple theory given above. The most interesting of these is the y-directed acceleration $w\,\partial v/\partial z$ in the geostrophic interior. In order to balance this small force, in the next approximation to the foregoing theory, we require a y-directed Coriolis force or a small *radial* velocity $u \sim f^{-1} w\,\partial v/\partial z$. The reader can estimate the radial heat flux due to this ageostrophic flow and find the conditions under which $u\,\partial T/\partial x$ can be neglected in (12.2.4). Another unsolved consistency problem arises in matching the above solution to the sidewall ($x = 0$) boundary layer containing the nearly homogeneous sinking motion.

12.3 Thermocline in a Rotating Spherical Sector

In Section 12.1, we saw that the thermocline depth in a nonrotating fluid is limited by the vertical circulation cell which is established by the differential surface heating. The circulation is inhibited by rotation (Section 12.2), however, and the thermocline depth is consequently increased. The latter effect occurs because the rotational constraint does not allow geostrophic radial velocities in an axisymmetric flow. We shall now inquire into the effect of a departure from axial symmetry, as caused by the presence of meridional walls. In this case there will be an azimuthal component of pressure gradient, a meridional component of geostrophic velocity, and a corresponding lateral heat transport. To find out whether this will change the scale depth of the thermocline, we consider a geometry (Fig. 12.3b) consisting of a spherical sector bounded by meridional walls, and also by a rigid no slip boundary ($z = 0$) upon which a uniform longitudinal temperature gradient $\partial T_S/\partial\theta$ is imposed.

From our previous work, we can anticipate that the region of sinking fluid will again be confined to a relatively small area near the North Pole (Fig. 12.3b), wherein the thermocline is weakly or negatively stratified. Although the cold fluid must then proceed southward, this flow will not be axisymmetric, and therefore it need not be confined to a lower Ekman layer (as was the case in Section 12.2). On the contrary, it will be shown that the polar fluid flows southward as a thin western boundary current. The southward transport of this jet will decrease with decreasing latitude, and thus fluid will leave the jet to enter the "interior" of the basin. At this stage, the ascending fluid parcel will move northeastward, until it reaches the "base" of the thermocline. The rising parcel is heated and subjected to much larger eastward velocities, associated with the meridional temperature gradient. Eventually the rising parcel will arrive at the upper rigid boundary ($z = 0$), whereupon it will be deflected northward by the Ekman transport and cooled by conduction. Thus, the parcel returns to the pole and the cycle repeats, as indicated by the trajectory shown in Fig. 12.3b. The scheme outlined above will be shown to be dependent on the no slip boundary condition at $z = 0$, and the no stress case is deferred until subsection (c). Because of the intricacy of the kinematics of the circulation, it is necessary to discuss the different regions in Fig. 12.3b separately, and thus the abyssal dynamics is considered first in subsection (a). The dynamics of the thermocline region and the Ekman layer are then considered in subsection (b).

(a) We have mentioned that the abyss contains water of uniform temperature, and that this water rises toward the overlying thermocline. This situation is related to the circulation produced in a basin which is being continuously filled by homogeneous water entering at the pole. Reference is made here to Section 2.7, in which we examined the circulation in a pie-shaped tank as it was

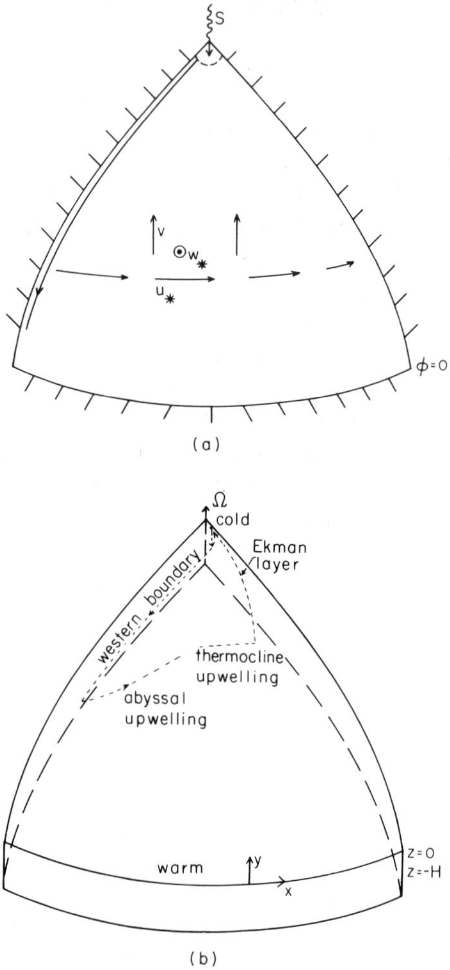

FIG. 12.3 (a) The rising motion w_* produced by filling a rotating spherical sector with *homogeneous* water entering at the pole. (b) Schematic diagram of thermocline circulation in a spherical sector when a north–south temperature gradient is imposed on a *rigid* upper surface. The dashed line is a perspective sketch of the trajectory of a parcel. The free surface case is also discussed in the text.

being slowly filled with homogeneous water. Figure 12.3a corresponds to the same problem, except that this basin is a sector of a rotating sphere, and consequently the undisturbed free surface is now separated by a *uniform* vertical distance from the bottom of the basin. Thus, a small source flow S will now only produce small variations in the height of the free surface $\hat{H}(t)$, as it rises with speed $w_* = d\hat{H}/dt$.

The inflowing water at the apex of the tank (Fig. 12.3a) is immediately deflected to the western boundary, as was the case in Section 2.7, and the flow also proceeds southward in a jet whose longitudinal width is small compared to the basin width. But the manner in which the fluid leaves the jet and fills the basin is quite different from that of Fig. 2.4. In the latter case, the local rise in the height of the free surface is accounted for by the radial variation of the parabolic free surface, whereas now we have no basic variation in sea level. Consequently, a local increase in $\hat{H}(t)$ can only be produced by convergent horizontal velocities, or vertical stretching of material columns of fluid. Conservation of mass then requires that the cross-sectional area of the material column decreases with time, and thus the product of this area with the normal component of the earth's vorticity f decreases. A compensating increase of vorticity is then required by the circulation theorem, and this is accomplished by a northward motion of the column to a latitude having a larger Coriolis parameter, as amplified below.

Since the inflow rate S is assumed small, the induced northward velocity v and eastward velocity u can be considered as infinitesimal in the vorticity equations (3.7.2) and (3.7.3), and thus we have the linearized version given by (3.7.4). When we set $\partial w/\partial z = w_*/\hat{H}$ in the latter equation, the northward velocity becomes

$$v_* = (f/\beta)(w_*/\hat{H}) = ((R \sin\theta)/(\hat{H}\cos\theta))w_* \tag{12.3.1}$$

where

$$f = 2\Omega \sin\theta \qquad \text{and} \qquad \beta = (2\Omega/R)\cos\theta \tag{12.3.2}$$

and where R is the radius of the sphere rotating at the rate Ω. Since the layer depth $\hat{H}$ has infinitesimal variations in θ, and ϕ, the continuity equation (3.7.1) becomes

$$\frac{\partial u_*}{R\,\partial\phi} + \frac{\partial(v_*\cos\theta)}{R\,\partial\theta} + \frac{w_*}{\hat{H}}\cos\theta = 0 \tag{12.3.3}$$

and the substitution of (12.3.1) then gives

$$\frac{\partial u_*}{\partial\phi} = -\frac{2w_*R\cos\theta}{\hat{H}}$$

The solution of this equation which satisfies the condition $u = 0$ on the eastern boundary ($\phi = 0$) is

$$u_* = \frac{(-2\phi)Rw_*\cos\theta}{\hat{H}} > 0 \tag{12.3.4}$$

since $\phi \leqslant 0$. Because of the necessity to connect the interior field with the incoming western boundary current, and because of the different dynamics applicable to the latter region, it is not necessary that the interior solution (12.3.4) vanish at the western boundary. The interpretation (Fig. 12.3a) is that fluid gradually leaves the southbound western boundary current, and then rises as it flows northeastward. If the basin width is $|\phi| \sim 1$, then (12.3.1) and (12.3.5) imply that v_* is of the same order as u_*.

(b) Let us now return to the thermocline problem (Fig. 12.3b) and follow an upwelling parcel as it enters the region ($z \sim -h$) containing the relatively strong meridional temperature gradients. As in Section 12.2, we assume the thermocline to be sufficiently deep so that the viscous forces can be neglected (except near $z = 0$), and thus we have the thermal wind equation

$$f\,\partial u/\partial z = -g\alpha\,\partial T/R\,\partial\theta \tag{12.3.5}$$

for the eastward velocity component. By integrating (12.3.5) from the bottom of the thermocline to the bottom of the Ekman layer ($z \simeq 0$), and by using (12.3.4) as the lower boundary condition on u, we obtain the "surface" velocity

$$u_0 = +\frac{g\alpha\,\Delta T\,h}{fR(\pi/2)} + \frac{2\phi R w_*\cos\theta}{\hat{H}} \tag{12.3.6a}$$

where $\partial T_S/\partial\theta = -\Delta T(\pi/2)^{-1}$ has been used as the surface gradient. If the depth of the ocean is relatively large, or

$$\hat{H} \gg w_* R^2 f/g\alpha\,\Delta T\,h \tag{12.3.6b}$$

Then (12.3.6a) may be approximated by

$$u_0 \simeq \frac{g\alpha\,\Delta Th}{fR(\pi/2)} \tag{12.3.6c}$$

The order of magnitude of w_*, h will be obtained on the basis of this approximation, and the consistency may then be verified by the fact that the expressions for w_*, h are independent of $\hat{H}$, so that (12.3.6b) can always be satisfied for a sufficiently deep ocean.

The heat budget of the thermocline region requires that $dT/dt \sim w_*\,\partial T/\partial z$ be of the same order of magnitude as $k_T\,\partial^2 T/\partial z^2 \sim k_T\,\Delta T/h^2$, and therefore we again obtain the scaling relation

$$w_* h/k_T \sim 1 \tag{12.3.7}$$

By referring to the kinematics used to estimate the vertical heat flux in equation (12.1.14b), it is easy to see that the total work done by gravity on the warm rising thermocline motions and the cold sinking polar motions is

$$\int_{-H}^{0} dz\, g\alpha\overline{wT'} \sim g\alpha\,\Delta T\,w_* h \tag{12.3.8}$$

per unit surface area. This rate of release of energy must be balanced by the dissipation, which we now ascribe to the upper Ekman boundary layer.

The eastward velocity (12.3.6c) (near $z = 0$) is accompanied by a northward decreasing pressure, and the velocity is therefore reduced to zero in a vertical distance $(\nu/f)^{1/2}$. Consequently, a northward transport occurs in the boundary layer, as indicated in Fig. 12.3b, and the vector vorticity has a magnitude $u_0(\nu/f)^{-1/2}$ in the Ekman layer. Therefore, the vertically integrated dissipation per unit surface area is

$$\nu[u_0(\nu/f)^{-1/2}]^2(\nu/f)^{1/2} \sim (g\alpha\, \Delta T)^2 h^2 \nu^{1/2}/R^2 f^{3/2} \tag{12.3.9}$$

where (12.3.6c) has been used. A consistent energy balance requires the equality of (12.3.8) and (12.3.9), or

$$w_* h \sim (g\alpha\, \Delta T) h^2 \nu^{1/2}/R^2 f^{3/2} \tag{12.3.10}$$

The combination of this with (12.3.7) yields the thermocline depth scale

$$h \sim [k_T R^2 f^{3/2}/g\alpha \Delta T\, \nu^{1/2}]^{1/2} \tag{12.3.11}$$

and we note that this relation is essentially the same as (12.2.17), if L be identified with R in (12.3.11). It then appears that neither the meridional boundaries, nor the variation of Coriolis parameter with latitude, changes the order of magnitude of the scale depth or its dependence on the external parameters.

Just for the fun of it, let us insert the numbers, $R = 10^9$ cm and $g\alpha\, \Delta T = 1$ cm/sec^2, and the *molecular* constants, $k_T = 10^{-3}$ cm^2/sec and $\nu = 10^{-2}$ cm^2/sec, in (12.3.11). We then obtain $h \sim 1$ km as the estimate of the scale depth of a thermocline having a *rigid* upper surface. The relevancy of this result to the ocean is, unfortunately, limited by the fact that the boundary condition on $z = 0$ is not appropriate to the real ocean, and the latter is also not laminar.

(c) In the thermocline theory of Robinson–Stommel (1959) and its later developments (Veronis, 1973), one considers a model having a *free* upper surface and no dissipation. Because of the latter restriction, a "global" solution is precluded, and only "local" balance relations are considered. The momentum balance is assumed to be strictly geostrophic as well as hydrostatic, and in the spherical coordinates we have

$$(2\Omega \sin\theta)v = (R\cos\theta)^{-1}\, \partial P/\partial\phi \tag{12.3.12}$$

$$(2\Omega \sin\theta)u = -\partial P/R\, \partial\theta \tag{12.3.13}$$

$$\partial P/\partial z = g\alpha T \tag{12.3.14}$$

$$\frac{1}{\cos\theta}\left[\frac{\partial u}{R\,\partial\phi} + \frac{\partial v\cos\theta}{R\,\partial\theta}\right] + \frac{\partial w}{\partial z} = 0 \tag{12.3.15}$$

where P is the pressure and $\mathbf{V}_2 = (u, v)$ is the horizontal velocity. When the lateral heat convection is included in the thermal equation, we have

$$w\,\partial T/\partial z + \mathbf{V}_2 \cdot \nabla T = k_T\,\partial^2 T/\partial z^2 \tag{12.3.16}$$

These equations are now cast into nondimensional form by using ΔT as the temperature scale, w_* as the scale for w, h as the scale for z

$$(R/h)w_* \sim (u, v) \tag{12.3.17}$$

as the horizontal velocity scale, and

$$\Omega R^2 w_*/h \sim P \tag{12.3.18}$$

as the pressure scale. Then Eqs. (12.3.12), (12.3.13), (12.3.15) take a nondimensional form in which all the coefficients are of order unity. The same will be true of the transformed version of (12.3.14), provided its coefficients are set equal to each other, or

$$g\alpha\,\Delta T \sim \Omega R^2 w_*/h^2 \tag{12.3.19}$$

In the nondimensional form of (12.3.16), the coefficients on the left hand side are all unity, but the coefficient of the right hand side is

$$k_T/w_* h \tag{12.3.20}$$

This conduction term is important in the Robinson–Stommel theory, and thus we set (12.3.20) equal to unity. By combining this relation with (12.3.19) we obtain the scale depth

$$h \sim (\Omega R^2 k_T/g\alpha\,\Delta T)^{1/3} \tag{12.3.21}$$

and this result may be compared with (12.3.11).

The reader is referred to the literature cited for analytical solutions of (12.3.12)–(12.3.16). Although these satisfy the given thermal boundary condition on $z = 0$, the boundary conditions on w and $\mathbf{V}_2$ cannot be satisfied because the high derivative terms which would ordinarily appear in the momentum equations have been suppressed in (12.3.12) and (12.3.13). The similarity solutions are then fitted to the observed T by choosing an effective value of k_T, the best value of which is three orders of magnitude larger than the molecular value. This procedure represents an attempt to parameterize the effects of the small scale turbulence, and implicitly assumes that the vertical flux of heat due to such scales depends mainly on the *mean* vertical temperature gradient.

In Welander's theory [see reference in the Appendix, or Veronis (1973)], the conduction term (12.3.20) is neglected, and (12.3.19) then determines the scale depth of the thermocline in terms of a "given" w_*, whose significance is discussed in the Appendix.

References

Robinson, A. R., and Stommel, H. M. (1959). The Oceanic Thermocline. *Tellus* **11**, 295–308.

Rossby, H. T. (1965). On Thermal Convection by Nonuniform Heating, etc. *Deep Sea Res. Oceanogr. Abstr.* **12**, 9–16.

Veronis, G. (1973). Large Scale Ocean Circulation. *Adv. Appl. Mech.* **13**, 1–92.

Wu, J. (1969). Mixed Region Collapse with Internal Wave Generation in a Density Stratified Medium. *J. Fluid Mech.* **35**, 531–544.

Appendix

The Ocean Pycnocline

The fundamental boundary condition for the determination of the time averaged circulation in Part I is the flux of momentum from the atmosphere to the ocean, and the order of magnitude of this stress is denoted by τ_* (cm^2/sec^2). The two layer model of Part I also contained a density deficit $\Delta\rho_*$ and a mean thickness H_* for the upper layer, but we recognize that these parameters must eventually be related to the thermodynamical boundary conditions in a theory of the thermocline and halocline.

We recall that the theory of the wind driven circulation is based upon the convergence of the Ekman drift in consequence of which we obtain a downward directed mean vertical velocity $w_m(x, y, z)$ (7.3.1) at mid latitudes, whose order of magnitude is

$$w_m \sim \tau_*/\Omega R \tag{A1}$$

where Ω is the rotation rate and R is the radius of the planet.

These downward directed velocities in the subtropics are not easily reconciled with the results of the thermocline models in Chapter XII, because the latter required *upward* vertical velocities (to balance the downward diffusion of heat in warm latitudes). But Welander (1971) has proposed an alternate thermocline theory, in which the diffusion term is neglected in (12.3.16) and the Ekman suction velocities are used to transport the warm surface waters downward. The vertical velocity scale in (12.3.19) is then given by (A1), and thus the solution for the scale depth is

$$h \sim \left(\frac{\tau_* R}{g \, \Delta\rho_* / \rho_0} \right)^{1/2} \tag{A2}$$

where $\alpha \, \Delta T$ in (12.3.17) has been replaced by $\Delta\rho_*/\rho_0$.

We now take note of the important constraint which was emphasized in Chapter XII, viz., that the total flux of heat and salt across any level surface must vanish for the *semi-insulated* ocean. The contribution of molecular conduction is justifiably neglected in all theories of the *ocean* thermocline, and consequently the total *convective* flux of heat or salt must vanish. Likewise, the flux of density must vanish, or

$$\iint \rho_m w_m \, dx \, dy + \iint (\rho' w')_m \, dx \, dy = 0 \tag{A3}$$

where $w'(x, y, z, t)$ is the instantaneous departure of the vertical velocity from $w_m(x, y, z)$, $\rho'(x, y, z, t)$ is the fluctuation from $\rho_m(x, y, z)$, and $(\rho' w')_m$ is the time average value of $\rho' w'$ at a given point. The expression for the local buoyancy flux in terms of the fluctuating salinity and temperature is

$$(-g/\rho_0)(\rho' w')_m = -g\beta(w' S')_m + g\alpha(w' T')_m \tag{A4}$$

In the Robinson–Stommel theory (Section 12.3), it is assumed that these fluctuations transport light water downward in the thermocline, the rate being proportional to $\partial\rho_m/\partial z$, or $(\rho' w')_m = -k_T' \, \partial\rho_m/\partial z > 0$, where k_T' is called the eddy diffusivity. This assumption closes the problem for the mean motion and makes it *formally* equivalent to the molecular diffusion problems considered in Chapter XII. The large scale mean motion in this theory is such as to transport heat *upward.*

Welander, on the other hand, relies on the mean vertical velocity to transport heat *downward.* The first term in (A3) is therefore positive, with the order of magnitude being given by

$$g/\rho_0 \iint \rho_m w_m \, dx \, dy \sim g(\Delta\rho_*/\rho_0)(\tau_*/\Omega R) \iint dx \, dy \tag{A5}$$

where (A1) has been used. In order to satisfy (A3) we require the second term to be negative with the order of magnitude of the associated buoyancy flux being given by

$$(g/\rho_0)(\rho' w')_m \sim -g(\Delta\rho_*/\rho_0)(\tau_*/\Omega R) \tag{A6}$$

Thus we see that this theory requires a fluctuating field whose effect is to transport cold fresh water downward, and to *release* potential energy at an average rate given by (A6). The *total* work done by gravity on the density field must vanish according to (A3), but the fluctuating component (A6) releases and dissipates the potential energy which has been generated by (A5). The ultimate source of the energy so transformed and so dissipated is to be found in the work done by the wind, as discussed in connection with (7.5.3). The budget for the mechanical energy equation is summarized in Fig. A.1. Neither the mechanics nor the scales of motion (Fig. A.1) which contribute to the release of energy has been implied, and therefore we must elaborate upon the various steps which have been assumed in the energy cascade (Fig. A.1). But first we discuss some of the other thermodynamical requirements of the global circulation.

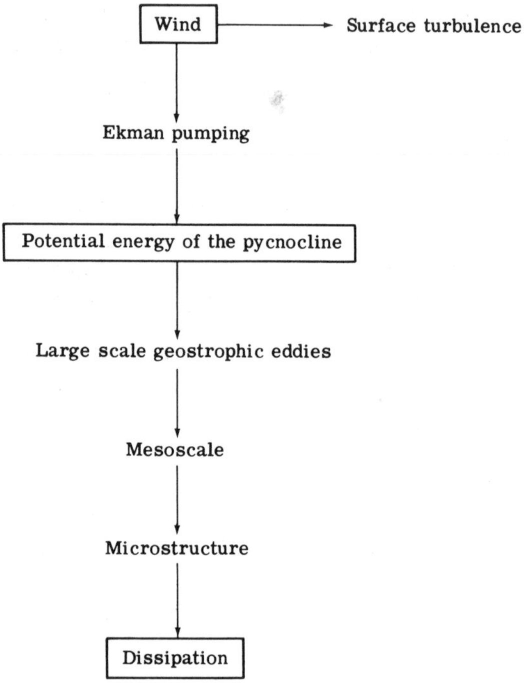

FIG. A.1 Budget for the mechanical energy equation.

We know (Von Arx, 1962) that about half of the incident solar energy is reflected, and the remainder Σ is absorbed at the surface of the earth. A major part of Σ is used for the heat of evaporation $\rho_0 E_* L$ (cal/cm^2/sec) at the ocean surface. Although the long wave radiation eventually returns all of the absorbed energy to space, there is a deficit in high latitudes which is supplied by the atmospheric circulation. However, recent determinations (Vonder Haar and Oort, 1973) show that a significant fraction of the absorbed solar energy (or of $\rho_0 E_* L$) is transferred poleward by the *ocean*. The poleward heat transport by the oceans ($\simeq 2 \times 10^{22}$ cal/yr) is roughly equal to $\frac{1}{3}$ of the heat required to evaporate the subtropical oceans.

It thus appears that the evaporation rate ($E_* \sim 1$ m/yr) is a fundamental "metric" for the oceanic heat flux. Since evaporation is obviously fundamental for the salt flux $\rho_0 E_* S_0$ (see Sections 10.8 and 11.6), we propose to use E_* as well as τ_* as the "given" boundary values for the determination of $\Delta\rho_*/\rho_0 = \beta\,\Delta S_* - \alpha\,\Delta T$.

Since $\rho_0 E_* S_0$ grams of salt are released each second by the evaporation, the near surface value of the buoyancy flux $-g\beta(w'S')_m$ is

$$g\beta E_* S_0 \tag{A7}$$

and consequently (A7) will be used as the scale for the salt buoyancy flux in the Boussinesq dynamics. We must emphasize, however, that the mechanism for distributing the flux in the vertical is unknown.

Although we have no comparable physical argument to justify the use of the evaporation for the *downward* flux of heat in low latitudes; such a procedure finds support in the measurements cited above, and therefore the metric for the oceanic thermal buoyancy flux (see Section 10.7) is taken to be

$$g\alpha E_* L/c_p \tag{A8}$$

or

$$g\beta E_* S_0(\alpha L/\beta S_0 c_p) \tag{A9}$$

Since the grand average (A6) of all the fluctuating scales of motion is such as to transport cold, fresh (and heavy) water downward, and since $\alpha L/\beta S_0 c_p$ is of order unity (10.8.7), we shall consider (A7) minus (A8) to be of order $g\beta E_* S_0$ so that

$$0 < -g(\rho' w')_m/\rho_0 \sim g\beta E_* S_0 \tag{A10}$$

is the order of magnitude of the total *release* of potential energy. The constant of proportionality which is necessary to convert (A10) into an equality will depend on other things besides $\alpha L/\beta S_0 c_p$, but the sign of (A10) is determined by (A6). When these two equations are combined, we obtain

$$\Delta\rho_*/\rho_0 R \sim \beta S_0 E_* \Omega/\tau_* \tag{A11}$$

as the scale value for the pole–equator density gradient. By using the typical values

$$E_* = 1 \text{ m/yr}, \qquad \tau_* = 1 \text{ (cm/sec)}^2$$

$$S_0 = 36‰, \qquad 2\pi/\Omega = 10^5 \text{ sec}, \qquad R = 6.5 \times 10^8 \text{ cm} \tag{A12}$$

we obtain a value of $\Delta\rho_*/\rho_0 = 3.6‰$ which is in reasonable agreement with the maximum lateral density variation observed in the pycnocline.

But this relation also implies a scale value for the north–south temperature difference on the *atmospheric* side! One might rationalize this fortuitous result by observing that the "given" values of τ_* and E_* imply much about the state of the atmosphere. Some commentary is also necessary on the fundamental climatological position which is here ascribed to the mean oceanic salinity S_0, and we make the following remarks.

(1) The heat of evaporation E_*L is about $\frac{1}{4}$ of the average long wave radiation from the earth's surface.

(2) It is reasonable to invert the ultimate climatological question which is posed in Fig. A.2 by asking for that value of Σ, or that value of the solar constant, which will yield a given (earth-like) E_*.

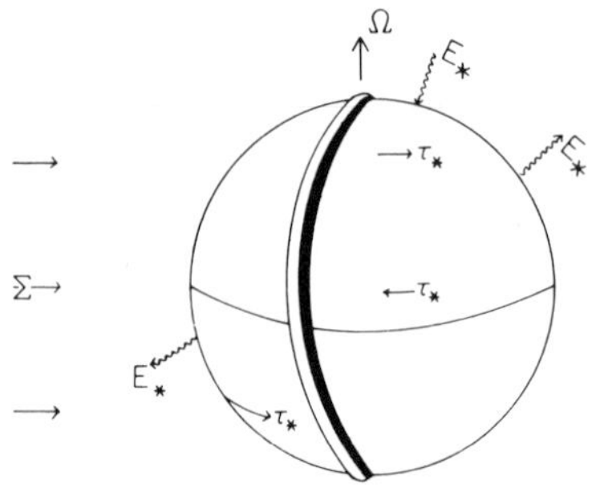

FIG. A.2 The companion of earth in perpetual spring. Σ is the solar constant, E_* the amplitude of the evaporation–precipitation, and τ_* the wind stress on the surface of the water. A thin meridional barrier provides the support for the western boundary currents.

(3) It is a fact that the global variations in $\beta\,\Delta S_*$ are comparable with those of $\alpha\,\Delta T_*$; and one also must eventually explain why $\Delta S_* \ll S_0$, considering the fact that fresh water ($S = 0$) is continually "pouring" into much more saline ocean water.

(4) The dynamical importance of the salt phase is reflected in the constant $\alpha L/c_p\beta S_0$.

(5) If the oceans were salt free ($S_0 = 0$) the slight modification of the freezing point, and also the modification of the equation of state, would undoubtably have a drastic effect on the mean ice coverage of the polar oceans and also on the mechanism by which bottom water is actually formed (Anati and Stommel, 1970).

(6) In the whimsical experiment of Fig. A.2, the (chemical) oceanographer is "permitted" to vary S_0, whereas the other external parameters belong to the meteorologists, geologists, astrophysicists, etc! Thus, we would like to know what happens to the climate of the planet in Fig. A.2, if we reduce S_0 while keeping Σ constant.

We now return to the problem of the mechanical energy budget. Several recent observations support the idea that the large scale oceanic fluctuations release the potential energy from the mean thermocline (Fig. A.1). Fuglister

(1972) has observed the detachment of large amplitude meanders from the Gulf Stream and the consequent transfer of large volumes of relatively cold fresh and heavy water to the southern side of the mean stream. Saunders (1971) has documented the reciprocal process, in which a warm fresh eddy is deposited on the northern side of the Gulf Stream. Thus, we have a visible manifestation of the transformation of the potential energy which otherwise would be part of the *semipermanent* Gulf Stream front. The previously mentioned equator–pole heat flux is also manifest in these large scale (diameter 10^7 cm, vertical dimension 10^5 cm) eddies which take a couple of years to decay (Richardson *et al.*, 1973). Although the formative stage may be mechanistically similar to atmospheric cyclones and anticyclones (Saunders, 1971), the slow dissipation of the oceanic eddy seems to be unique, in view of the absence of such strong effects (ground friction, clouds) as are present in the atmospheric counterparts.

The large decay time of the detached eddy seems to be connected with the rotational constraint on those motions which conserve potential vorticity. We pointed out in connection with Fig. 3.2a that this constraint prevents the spreading of a density wedge (or a baroclinic geostrophic vortex) on an isopycnal surface. If the potential energy of the detached Gulf Stream eddy is to decay, we must invoke "mesoscale" motions which can either transfer energy away from the larger scale (Fig. A.1) or otherwise allow it to spread laterally. There are many observations of fluctuations in temperature and velocity having vertical dimensions of 10^2–10^3 cm, and this we call the "mesoscale." We do not know to what extent such fluctuations represent inertia–gravity *waves* or small scale density *currents* of the type discussed in connection with (3.6.9) and of the type discussed by Wu (1969). Reference is also made to observations [Lambert *et al.* (1973)] which show a "mesostructure" embedded in a dissipating Gulf Stream eddy.

Thus, we have the attractive proposition that the mesoscales derive their energy from the larger geostrophic scales and pass the energy on to the microscales. While no mechanism has been proposed for such a process, there are certain useful constraints which may permit future progress. For example, the shear flow instabilities (Sections 9.6 and 11.7), may be important in the dissipative stage. But we must remember that this process leads to an increase of potential energy, whereas (A6) requires that the net effect of *all* the fluctuations be such as to *lower* the center of gravity of the pycnocline. The large scale eddies act in this sense and so, too, do the double-diffusive processes (Chapter XI). In this connection, we note that if the intermediate density lens in Fig. 3.2a is composed of T, S values lying between those of the surrounding water, then the double-diffusive effects acting across the two interfaces of the lens will alter its density and thereby cause the wedge to spread laterally, despite the rotational constraint. But such molecular processes can only be significant in the ocean when the "interfaces" are truly sharp (~10 cm) (cf. Figs. 11.2c and 11.3), and such structures can only be associated with the mesoscales (rather than with the geostrophic scale of motion). Thus, a fundamental problem is to account for the 10 cm interfaces and the concomitant mesoscale of motion. The same problem arises in connection with the Kelvin–Helmholtz instability (Section 9.6), since the Richardson number of the large scale oceanic motion is too large for instability to occur. It seems attractive to assume some kind of coupling between geostrophic meso- and microscales of motion, such that the large scale motion supplies the energy while the smaller scales provide the mechanism for release and dissipation of that energy. If this view be correct, then the *vertical* transports of heat/salt by the small scales may be more dependent on the *lateral* variations within the geostrophic eddies than upon the time averaged mean vertical density gradient $(\partial \rho_m/\partial z)$. The point at stake here involves nothing less than the effective parameterization of the small scale fluxes for the purpose of describing the geostrophic scale over long times.

Whatever be the mechanisms involved, we know that the dissipation of kinetic energy equals the volume integral of the product of the mean square vorticity ζ^2 with the molecular viscosity ν, whereas the energy supplied to the thermocline equals the volume integral of

$$g(\rho_m/\rho_0)w_m \sim g(\Delta\rho_*/\rho_0)\tau_*/\Omega R. \tag{A13}$$

Therefore, a consistent global mechanical energy budget requires the scale relation

$$\zeta^2 \sim g\beta E_* S_0/\nu \tag{A14}$$

The dissipation of T–S variances (Section 11.6) must also be accomplished by the microstructure and by considering the global hydrological cycle we obtained (11.6.29), or

$$|\nabla S| = (E_* S_0\, \Delta S_*/k_S H_*)^{1/2} \tag{A15}$$

for the rms salinity gradient $|\nabla S|$ as a function of the mean vertical salinity gradient $\Delta S_*/H_*$ in the latitudes of maximum evaporation.

The order of magnitude of the rms density gradient can be set equal to $|\beta\ \nabla S|$ because $k_S \ll k_T$ (see Section 11.6), and therefore the rms Richardson number [cf. (11.7.12)] is

$$R_\mu = g\beta|\nabla S|/\zeta^2 \tag{A16}$$

When (A.14) and (A.15) are used, this becomes

$$R_\mu \sim (S_0 E_*\ \Delta S_*/k_S H_*)^{1/2}(\nu/E_* S_0)$$

and by solving for $\Delta S_*/H_*$, we then have

$$\Delta S_*/H_* \sim (k_S/\nu^2)E_* S_0 R_\mu^{\ 2} \tag{A17}$$

In Section 11.7, we argued that a necessary condition for a significant† turbulent transport in a fluid having a large "overall" Richardson number (i.e., large static stability) is that R_μ be neither large nor small compared to unity. R_μ must adjust itself to a constant ($R_\mu \sim 1$), the precise value of which is assumed to be independent of both k_S and the large scale motion, with which the microstructure is coupled. The basis for this assumption can be related to the equality of the buoyancy force (or its "curl")

$$|\nabla \times g\mathbf{k}(\alpha T - \beta S)| \sim g\beta|\nabla S|$$

and the inertial force (or its "curl")

$$|\nabla \times d\mathbf{V}/dt| \sim \zeta^2$$

The *smallest* scale of motion gives the dominant contribution to these rms gradients, and the Coriolis term is negligible on this scale. The order of magnitude equality between the last two expressions implies $R_\mu \sim 1$, and therefore (A17) becomes

$$\Delta S_*/H_* \sim (k_S/\nu^2)E_* S_0 \tag{A18}$$

This equation states that the mean vertical salinity gradient in the subtropical halocline depends on $E_* S_0$ and on the molecular parameter $\nu^2/k_S \sim 10\ \mathrm{cm}^2/\mathrm{sec}$. By using the values given in (A13), we then find that $\Delta S_*/H_* \sim 1‰/\mathrm{km}$, a result which also agrees with observations.

It remains to comment on the difference between the scale depth obtained from (A18) [with (A11)] and the different scale depth given by (A2). The latter derives from a dimensional consideration of the importance of the lateral heat transport of the *mean* motion, and h is almost an order of magnitude smaller than the H_* computed from (A18). But the very nature of the dimensional analysis used above does not allow us to attribute much significance to this numerical difference. Alternatively, one could say that the order of magnitude *agreement* between H_* and h is not an "accident," but reflects the importance of *both* the mean horizontal velocity and the eddies in the global thermodynamics.

† We define this term to imply that the rms gradients (in temperature, salinity, and velocity) are much larger than the corresponding mean values. This condition, according to Gregg *et al.* (1973), is apparently *not* fulfilled on a local basis and the central ocean is "not often" turbulent. The significant small scale turbulence may be isolated in space–time, so that it only occurs in connection with the baroclinic eddies mentioned above.

References

Anati, D., and Stommel, H. M. (1970). The Initial Phase of Deep Water Formation in the Northwest Mediterranean etc. *Cah. Oceanogr.* **22,** 343–351.

Fuglister, F. C. (1972). Cyclonic Rings Formed by the Gulf Stream 1965–1966. *In* "Studies in Physical Oceanography" (A. Gordon, ed.), Vol. I (Wust Birthday Volume). Gordon and Breach, New York.

Gregg, M. C., Cox, C. S., and Hacker, P. W. (1973). Vertical Microstructure Measurements in the Central North Pacific. *J. Phys. Oceanogr.* **3,** 458–469.

Lambert, R. B., Kester, D. R., Pilson, M. E., and Kenyon, K. E. (1973). In Situ Dissolved Oxygen Measurements in the North and West Atlantic Ocean. *J. Geophys Res.* **78,** 1479–1483.

Richardson, P. L., Strong, A. E., and Knauss, J. A. (1973). Gulf Stream Eddies. *J. Phys. Oceanogr.* **3,** 3.

Saunders, P. M. (1971). Anticyclonic Eddies Formed from Shoreward Meanders of the Gulf Stream. *Deep Sea Res. Oceanogr. Abstr.* **18,** 1207–1219.

Von Arx, W. (1962). "An Introduction to Physical Oceanography." Addison-Wesley, Reading, Massachusetts.

Vonder Haar, T. H., and Oort, A. H. (1973). New Estimate of Annual Poleward Energy Transport by Northern Hemisphere Oceans. *J. Phys. Oceanogr.* **3,** 2.

Welander, P. (1971). The Thermocline Problem. *Phil. Trans. Roy. Soc. London,* Ser. A270, 415–421.

Wu, J. (1969). Mixed Region Collapse with Internal Wave Generation in a Density Stratified Medium. *J. Fluid Mech.* **35,** 531–544.

Index